Construction Science and Materials

Construction Science and Materials

Edited by Armando Ruiz

CLANRYE
INTERNATIONAL
www.clanryeinternational.com

Clanrye International,
750 Third Avenue, 9th Floor,
New York, NY 10017, USA

ISBN: 978-1-63240-713-9

Cataloging-in-Publication Data

Construction science and materials / edited by Armando Ruiz.
    p. cm.
Includes bibliographical references and index.
ISBN 978-1-63240-713-9
1. Building. 2. Building materials. 3. Construction equipment. 4. Construction industry--
Technological innovations. I. Ruiz, Armando.
TH145 .C66 2018
690--dc23

For information on all Clanrye International publications
visit our website at www.clanryeinternational.com

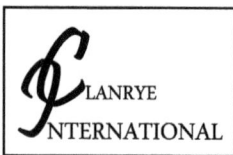

𝒞LANRYE
INTERNATIONAL

Contents

Preface

Construction science and materials deals with the designing, planning and construction of buildings, roads, canals, tunnels etc. It is important to select the right materials to be used in specific constructions as the strength and endurance of the built structure depends on it. Advancements in the field of technology have resulted in scientifically engineered materials that are used primarily for construction. The field of engineering concerned with construction science and materials is known as civil engineering. This book discusses the fundamentals as well as modern approaches of construction science. A number of latest researches have been included to keep the readers up-to-date with the global concepts in this area of study.

This book has been the outcome of endless efforts put in by authors and researchers on various issues and topics within the field. The book is a comprehensive collection of significant researches that are addressed in a variety of chapters. It will surely enhance the knowledge of the field among readers across the globe.

It gives us an immense pleasure to thank our researchers and authors for their efforts to submit their piece of writing before the deadlines. Finally in the end, I would like to thank my family and colleagues who have been a great source of inspiration and support.

Editor

Assessing the seismic performance of rammed earth walls by using discrete elements

Quoc-Bao Bui[1]*, Tan-Trung Bui[2] and Ali Limam[2]

*Corresponding author: Quoc-Bao Bui, University of Savoie, Chambery 73000, France

E-mail: Quoc-Bao.BUI@univ-savoie.fr

Reviewing editor: Sanjay Kumar Shukla, Edith Cowan University, Australia

Abstract: Rammed earth (RE) is attracting renewed interest throughout the world because of its low embodied energy and its interesting hygric-thermal behavior. Several studies have recently been carried out to investigate this material. However, the seismic behavior of RE walls is still an important subject that needs to be more thoroughly investigated. The present study assesses the seismic performance of RE walls by using the discrete element modeling (DEM) and the nonlinear pushover method. Firstly, nonlinear "force–displacement" curves of the studied wall were obtained by DEM. Secondly, the standard "acceleration–displacement" curves were carried out following Eurocode 8. Thirdly, the above curves were superimposed to determine the intersection point (target point) which enabled to assess the seismic performance of the studied wall in the corresponding conditions (vertical load, seismic zone). The results show that the studied walls can have satisfactory resistance in seismicity zones ranging from "very low" to "moderate" (according to Eurocode 8). For "medium" seismicity zones, the studied structures should only be constructed on A-type soils (very good soil). For B-type soils, wall reinforcement techniques would be necessary. Without special reinforcements, studied RE structures seem unsuitable for "strong" seismicity zones, for all soil types.

Subjects: Geomechanics; Soil Mechanics; Structural Engineering; Waste & Recycling

Keywords: rammed earth; seismic performance; discrete element method; pushover; sustainable development

1. Introduction

Rammed earth (RE) is attracting interest in the context of sustainable development because of its low embodied energy and its interesting hygric-thermal behavior. Several studies have recently been conducted to investigate this material. However, the seismic behavior of RE walls requires more thorough investigation. The first exploratory study on the dynamic characteristics of RE

ABOUT THE AUTHOR

Dr Quoc-Bao Bui obtained his PhD degree in 2008 at School of Civil Engineering Lyon (ENTPE-INSA Lyon), France. From 2011, he has been associate professor at University Savoie Mont Blanc, Chambery, France. His research activities concentrate on the dynamic of structures (earthquake behavior, vibrational measurements, *in situ* non-destructive tests, beam-column joints) and non-conventional materials (rammed-earth, non-conventional concrete).

PUBLIC INTEREST STATEMENT

Rammed earth (RE) is attracting renewed interest throughout the world because of its "green" characteristics in the context of sustainable development. The present study assesses the seismic performance of RE walls. The results show that the studied walls can have satisfactory resistance in seismicity zones ranging from "very low" to "moderate" (according to Eurocode 8). For "medium" seismicity zones, the RE structures should only be constructed on A-type soils (very good soil). For B-type soils, wall reinforcement techniques would be necessary.

buildings was carried out by Bui, Hans, Morel, and Do (2011). The dynamic characteristics (natural frequencies, mode shapes and the damping) of the studied buildings were identified. These studies also showed that the analytical shear-beam model could reproduce the dynamic behavior of the buildings studied. Then Gomes, Lopes, and de Brito (2011) conducted a numerical study on the seismic resistance of RE constructions in Portugal, but there were still several limitations. Firstly, their model did not analyze the behavior of RE material in detail and there was not a validation step in their numerical model. Secondly, the seismic assessment was conducted using the classical elastic linear equivalent approach, which is less advantageous than the nonlinear approaches. Cheah, Walker, Heath, and Morgan (2012), Hamilton, McBride, and Grill (2006), Miccoli, Müller, and Fontana (2014), Miccoli, Oliveira, Silva, Muller, and Schueremans (2015), and Silva et al. (2013) present various experiments on the shear behavior of several RE walls (stabilized or unstabilized, unreinforced or reinforced). Ciancio and Augarde (2013) studied the out-out-plane of stabilized RE subject to lateral wind force. However, there has not yet been any quantitative study on the seismic performance of RE buildings on the structure scale.

From the view point of earthquake engineering, RE material does not seem favorable. Indeed, the material works essentially in compression and has a very low tensile strength; the walls' mass is high, which can cause considerable inertial forces during earthquakes. However, according to a number of post-seismic investigations, RE walls present acceptable behaviors. For example, in the Morris and Walker investigation (Morris & Walker, 2011) after the Darfield earthquake (2010, New Zealand, 7.1 on the Richter magnitude scale), only minor cracks were observed in RE walls. This means that when RE buildings are well designed and executed, they can have a satisfactory seismic performance. The present study assesses this performance quantitatively.

It is important to note that the seismic behavior of a RE building depends on several parameters: Earthquake action (seismicity zone, soil type, site factors), the structure's dynamic characteristics (natural frequencies, modal shapes, damping), and the material's characteristics (compressive, tensile strengths, Young's modulus, density). This is why for the same material (RE in this case), the seismic performance of each building may differ depending on its structural characteristics and the quality of its execution. This paper investigates three virtual RE buildings with current designs for RE houses in France and Europe: One story in RE walls, two stories in RE walls, the first story in RE walls and the second story a wooden structure. The RE walls are 50-cm-thick unstabilized RE, built by a pneumatic rammer and their seismic performance is investigated for an almost dry state, several months after their construction (about 2–3% of moisture content, Bui, Morel, Hans, & Meunier, 2009; Bui, Morel, Reddy, & Ghayad, 2009).

The assessment used the nonlinear pushover method, in a numerical model with the discrete element method. One of the main advantages of the numerical approach is the possibility to simulate several pushover tests on a full-scale wall. The wall modeled is in an actual RE house where *in situ* dynamic measurements were taken during the construction (Bui et al., 2011; Bui, Morel, Hans, et al., 2009). The dynamic characteristics of the wall were measured so that the relevancy of the numerical results could be checked.

2. Pushover method

Traditionally, earthquake-resistant design has been strength-based, using linear elastic analysis. Since inelastic behavior is usually allowed for strong earthquakes, this is not entirely rational. Strength-based design considers inelastic behavior only implicitly. Displacement-based (or deformation-based) design considers inelastic behavior explicitly, using nonlinear analysis (Chopra & Goel, 2002). Displacement-based design recognizes that in an earthquake, inelastic deformation can be greater than that in strength-based. The present paper uses the displacement-based design with the pushover method. It is important to note that the pushover method has not been included in previous French earthquake regulations. It is presented in Eurocode 8 (EN 1998-1:2004, 2004), which has been applied in France and Europe for several years.

The pushover method is a static nonlinear analysis. Firstly, a capacity curve is established (by experiments or numerical models, which enable nonlinear analysis). The seismic force (represented by the shear force at the wall base V_b) is transformed to spectral acceleration S_a (Figure 1(a)):

$$S_a = V_b/m \tag{1}$$

where m is the mass supported by the wall, and the displacement of the wall top u_t is replaced by the spectral displacement S_D ($S_D = u_t$).

Secondly, the seismic elastic spectrum S_a, which is given in the seismic design code and a function of the structure natural period T, is also transformed in the spectrum $S_a - S_d$ (Figure 1(b)) by the following relationship:

$$S_d = S_a/(2\pi/T)^2 \tag{2}$$

When the two above curves ($S_a - S_d$) are superposed (Figure 1(c)), the intersection point (D_1 for elastic spectrum or D_2 for inelastic spectrum) indicates the performance point (or "target point") that can give information on the damage state of the studied structure. More information on the pushover method can be found in Chopra and Goel (2002) and EN 1998-1:2004 (2004).

As explained above, the pushover method is a nonlinear analysis, so this approach recognizes that inelastic deformation should be taken into account. In general, the design (inelastic) spectrum is obtained by dividing the elastic spectrum by a "behavior factor" q. This is an important parameter in the pushover method; it accounts implicitly for inelastic response, the presence of damping and other force reduce effects, such as period elongation and soil–structure interaction. The behavior factor is defined as the ratio of the elastic acceleration response spectrum expected at a site to that of an inelastic spectrum used for design of a structure (Salvitti & Elnashai, 1996):

$$q = S_a^{elastic}/S_a^{inelastic} \tag{3}$$

The procedure to determine the "rational" q is quite complex because it depends not only on the material, but also on the structural configuration. This procedure will not be presented in detail here; only typical values proposed in Eurocode 8 will be discussed. q can be:

- 1 for structures with essentially elastic behavior.
- 1.5 for structures with limited ductility.
- 3 for structures with ductile behavior. Some ductile structures can have $q = 5$.

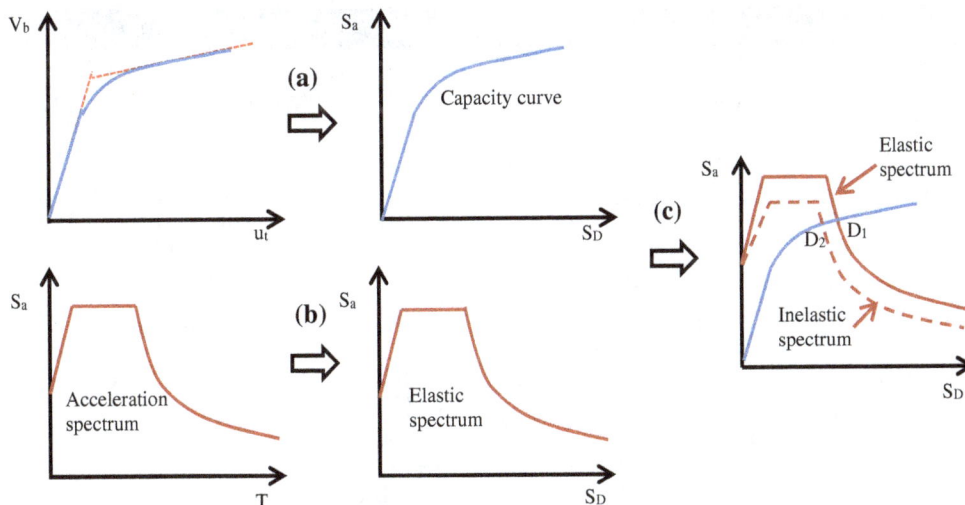

Figure 1. Synthesis of the pushover method, according to Eurocode 8.

There is not yet a specific value of q for RE structures, but for unreinforced masonry structures, Eurocode 8 authorizes that q is taken to be at least 1.5. If the inelastic spectrum is used, the performance point will be D_2 (Figure 1(c)).

3. Discrete element method and parametric studies

3.1. Discrete element method

The explicit discrete element modeling (DEM) based on finite difference principles originated in the early 1970s as the result of landmark work on the progressive movements of rock masses such as 2D rigid block assemblages (Cundall, 1971). This technique was then extended to the modeling of masonry and concrete (Bui, Limam, & Bui, 2014; Lourenço, Oliveira, Roca, & Orduña, 2005). However, to our knowledge, DEM has not yet been used to study RE structures.

The 3DEC code (Itasca, 2011) was used in the present study. The RE wall was modeled as an assemblage of discrete blocks (earthen layers), and the interfaces between earthen layers were modeled by introducing an interface law.

Earthen layers were assumed to be homogeneous and isotropic and were modeled by blocks that were further divided into a finite number of internal elements for stress, strain, and displacement calculations. The failure envelope used in this study was the Mohr–Coulomb criterion with a tension cut-off (Bui, Bui, Limam, & Morel, 2015).

Interfaces between earthen layers were modeled by an interface law between the blocks according to the Mohr–Coulomb interface model with a tension cut-off. This interface constitutive model considers both shear and tensile failure, and interface dilation is included. Further details of the constitutive behaviors of blocks and interfaces can be found in Itasca (2011).

3.2. Parametric studies

In a previous study by the authors (Bui et al., 2015), parametric studies identified 13 parameters necessary for the DEM; the summary of recommendations is presented in Table 1. In that table, f_c is the compressive strength and Young's modulus was calculated in the elastic part which was from 0 to 20% of the ultimate compressive stress (Bui, Morel, Hans, & Walker, 2014). These recommendations are strengthened with the results presented in other studies that used the finite element method (El Nabouche, Bui, Perrotin, Plé, & Plassiard, 2015; Miccoli, Oliveira, et al., 2015).

Table 1. Recommended parameters for earthen blocks and interfaces in DEM

Earthen blocks

Density d (kg/m³)	Young's modulus (E_b)	Poisson's ratio (ν)	Tensile strength (f_t)	Cohesion (c)	Friction angle (φ)	Dilatancy angle (Ψ)
~2,000	$(450-500) \times f_c$	0.22	$(0.07-0.1) \times f_c$	$(0.07-0.1) \times f_c$	45–51°	~12°
Bui, Morel, Hans, et al. (2009)	Bui, Morel, Hans, et al. (2009)	Bui, Morel, et al. (2014)	Bui, Bui, Limam, and Maximilien (2014)	Bui, Bui, et al. (2014), Cheah et al. (2012)	Bui et al. (2015), Bui, Bui, et al. (2014), Cheah et al. (2012), El Nabouche et al. (2015), Miccoli, Oliveira, et al. (2015)	Bui et al. (2015), Cundall (1971)

Interfaces

Normal stiffness (k_n)	Shear stiffness (k_s)	Tensile strength ($f_{t,interface}$)	Cohesion ($c_{interface}$)	Friction angle ($\varphi_{interface}$)	Dilatancy angle (Ψ)
$\dfrac{E_{wall} \cdot E_{block}}{(E_{block} - E_{wall}) \cdot h_{block}}$	$k_n/[2(1+\nu)]$	110–150 kPa	110–150 kPa	25–45°	~12°
Lourenço et al. (2005)	Lourenço et al. (2005)	Bui et al. (2015)	Bui et al. (2015)	Bui et al. (2015)	Bui et al. (2015)

According to the proposed values, Bui et al. (2015) obtained useful numerical results by comparing with experiments in two cases: Loading perpendicular to earthen layers and loading in the diagonal direction. The DEM application for the lateral loads will be presented in this paper.

4. Seismic capacity of RE walls

4.1. Wall description

The studied wall was constructed in a new RE house in the Rhone-Alpine region, France. It is 50 cm thick and is an unstabilized RE wall. *In situ* dynamic measurements were taken on this wall (during the construction phase, Bui, Morel, Hans, et al., 2009) and on the complete house (after the construction Bui et al., 2011), so the wall's dynamic characteristics were determined. A numerical model was constructed using DEM (Figure 2). The compressive strength determined from the compression tests on cylindrical samples was used (f_c = 1.9 MPa). Young's modulus, density and Poisson's ratio were also measured: 470 MPa (Bui, Morel, Hans, et al., 2009), 20 kN/m³ (Bui & Morel, 2009; Bui, Morel, Hans, et al., 2009) and 0.22 (Bui, Morel, et al., 2014), respectively.

A parametric study was conducted to identify other characteristics of the earthen blocks and the interfaces (Bui et al., 2015). The parameters identified are presented in Table 2, which reproduce the natural frequencies measured on site (Table 3).

4.2. Pushover curve

The wall was simplified but on the safety side: Although the wall has some resistances in the direction perpendicular to its plane, this resistance is generally ignored for a structural design against

Figure 2. Wall studied and its mesh in DEM.

Table 2. Parameters of earthen blocks and interfaces used in the DEM model

Earthen blocks

Tensile strength f_t	Cohesion c		Friction angle ϕ
133 kPa	133 kPa		45°
Bui, Bui, et al. (2014)	Bui, Bui, et al. (2014)		Bui, Bui, et al. (2014)

Interfaces

Normal stiffness k_n	Tensile strength $f_{t, interface}$	Cohesion $c_{interface}$	Friction angle $\phi_{interface}$
60 GPa/m	113 kPa	113 kPa	38°
Bui et al. (2015)	0	Bui et al. (2015)	Identified

Table 3. Comparison of frequencies obtained by DEM and by measurements			
Modes	f_{model} (Hz)	$f_{measured}$ (Hz)	$f_{model}/f_{measured}$
1	10.8	10.8	1.00
2	17.3	18.2	0.95
3	23.1	24.0	0.96
4	36.8	36.5	1.01

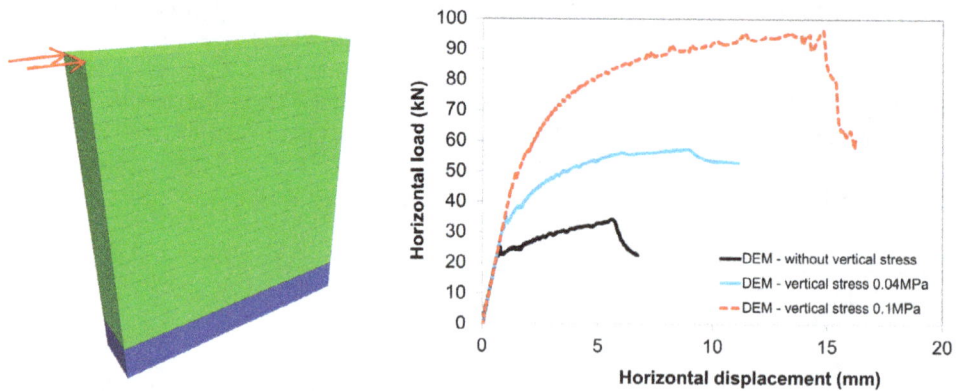

Figure 3. Discrete element model (left); load–displacement curves of the wall with different vertical stresses (right).

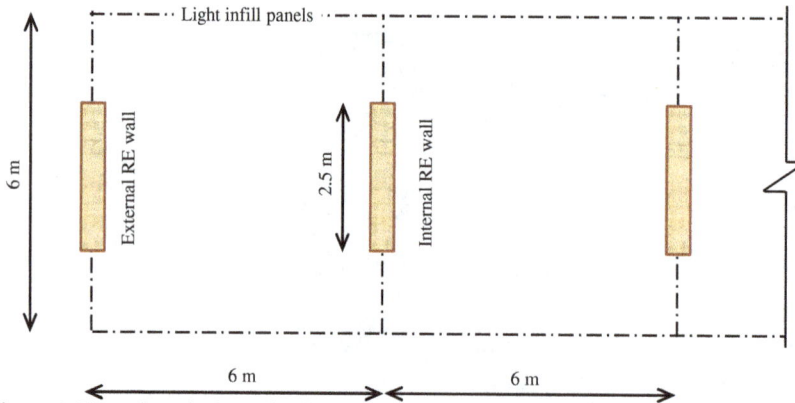

Figure 4. Plan of vertical studied houses.

lateral loadings, which is why the L-form wall can be simplified to a line form (Figure 3, left). To obtain the pushover curve, a horizontal force was applied to the wall top and incremented until the wall failed (post-peak).

In building construction, RE walls bear vertical stresses that come from the horizontal elements: The floors or roof (dead loads + live loads). The values of these loads in case of an earthquake will be presented in detail in the next section. Here, to assess the influence of the vertical stresses on the wall's behavior, three cases of uniform vertical stress on the wall top were simulated: 0, 0.04, and 0.1 MPa. The results are presented in Figure 3, right. It is logical that as the vertical stress increases, the shear performance of the wall increases. However, in a building, when a wall has greater vertical stress due to the horizontal elements (floors, roof), it must also bear a greater mass due to these elements. Therefore, from the earthquake engineering point of view, a wall bearing high vertical loads may be an unfavorable wall.

4.3. Seismic capacity

Because the seismic performance of a RE wall depends on both the vertical stress (favorable role) and the corresponding mass (unfavorable role), several cases should be investigated: Interior or exterior walls; one story, two stories or more; and the value of the loads from the horizontal elements (floor, roof), which depend on the dimensions of these elements between the bearing walls (influenced zones). For a general assessment of the seismic performance of the RE buildings in France (and Europe), three virtual RE buildings are considered:

- Only one story and all bearing walls are in RE (this is the case of the measured house).
- Two stories in RE walls and all bearing walls are in RE.
- First story with all bearing walls in RE and second story in a wooden structure.

The floor (or roof) space between bearing walls measures 6 m × 6 m, but the wall length is always 2.5 m, like the real house studied. The vertical elements between the walls and on the facades are light elements (wooden and glass infill). In general, the bearing RE walls have lengths greater than or equal to 2.5 m, so the length chosen also tends toward a greater safety. The wall height is 2.7 m, the same as the house studied.

In a seismic design, in addition to the earthquake load, other loads should also be taken into account; they are given according to Eurocode 8:

$$\sum \text{Dead load} + \varphi \cdot \Psi_{2i} \cdot \sum \text{Live load} \tag{4}$$

where $\Psi_{2i} = 0.3$; $\varphi = 1$ for the roof and 0.5 for the floors (for residential, office, and commercial buildings).

In this paper, only residential buildings are studied; therefore, according to Eurocode 1, the live loads composed of an exploitation load of 1.5 kN/m² and the light partitions of 0.5 kN/m². The usual self weight of the wooden floor was $g = 1.1$ kN/m². The synthesis of the mass and vertical stress due to the loads on the floors and roofs is given in Table 4.

Pushover simulations were performed in the DEM model for each wall to obtain the capacity curves. Then the demand spectrum was constructed for important class II (current buildings) and the soil type A under the building (Figure 5). Soil type A corresponds to rock or a very good soil (mean velocity of the shear waves $v_{s,30} > 800$ m/s). Indeed, RE buildings in France are usually constructed on soil A or B. Soil type B will be discussed below.

From the capacity curves and the demand spectra, the intersection target points can be determined for each case (wall type and seismicity zone). Then the inter-story drift can also be determined:

$$\text{Inter-storey drift} = \text{target displacement/storey height} \tag{5}$$

Table 4. Added vertical stresses and masses due to the loads on the floors and roofs

Building type	Wall	Added vertical stress (MPa)	Added mass (ton)
One story in RE	Exterior wall	0.025	9.8
	Interior wall	0.05	12.9
Two stories in RE	Exterior wall	0.04	18.5
	Interior wall	0.08	23.6
First RE + second wood	Exterior wall	0.04	12.0
	Interior wall	0.08	17.0

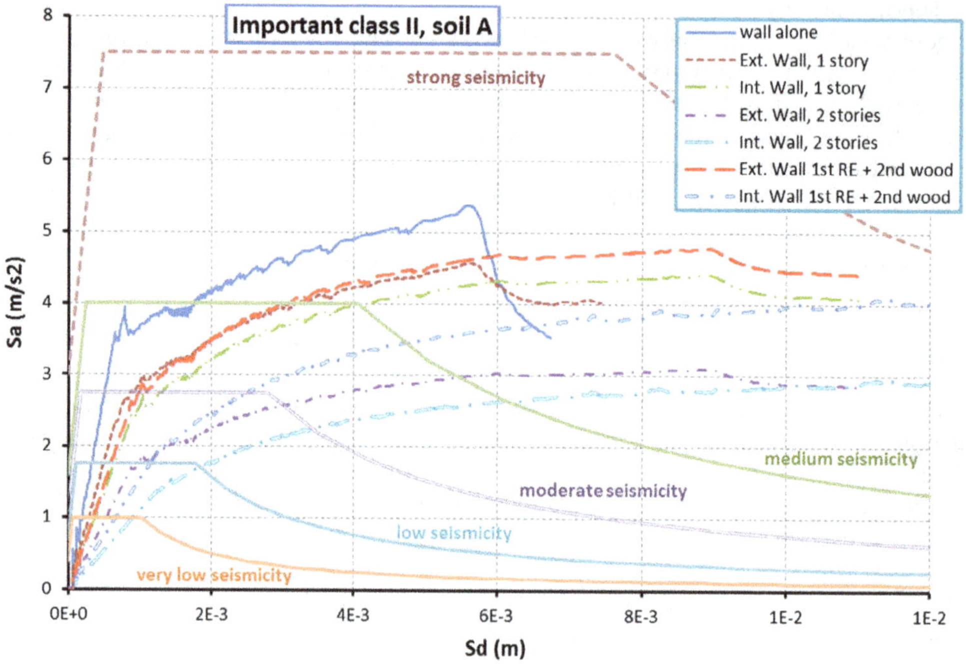

Figure 5. Capacity curves and demand spectra of different seismicity zones, for soil A.

The inter-story drift values determined are presented in Table 5.

To assess the seismic performance of the walls studied, a criterion must be chosen. For inter-story drift, Calvi (1999) proposed three damage limit states (LS) for masonry structures (Figure 6):

- LS2-Minor structural damage and/or moderate nonstructural damage; the building can be utilized after the earthquake, without any need for significant strengthening and repair to structural elements. The suggested drift limit is 0.1%.

- LS3-Significant structural damage and extensive nonstructural damage. The building cannot be used after the earthquake without significant repair. Still, repair and strengthening are feasible. The suggested drift limit is 0.3%.

- LS4-Collapse; repairing the building is neither possible nor economically reasonable. The structure will have to be demolished after the earthquake. Beyond this, LS global collapse with danger for human life has to be expected. The suggested drift limit is 0.5%.

Table 5. Damage assessment (with Inter-story drift, %) for soil A							
Seismicity zone	Wall alone	One story RE		First RE + Second wood		Two stories RE	
		Ext. wall	Int. wall	Ext. wall	Int. wall	Ext. wall	Int. wall
Very low	Slight (0.009)	Slight (0.013)	Slight (0.012)	Slight (0.013)	Slight (0.037)	Slight (0.023)	Slight (0.031)
Low	Slight (0.011)	Slight (0.016)	Slight (0.022)	Slight (0.023)	Slight (0.074)	Slight (0.033)	Slight (0.065)
Moderate	Slight (0.018)	Slight (0.027)	Slight (0.044)	Slight (0.038)	Slight (0.088)	Moderate (0.111)	Moderate (0.125)
Medium	Slight (0.062)	Moderate (0.101)	Moderate (0.147)	Moderate (0.108)	Moderate (0.185)	Moderate (0.202)	Moderate (0.227)
Strong	Complete (∞)	Complete (∞)	Complete (∞)	Complete (∞)	Complete (∞)	Complete (∞)	Complete (∞)

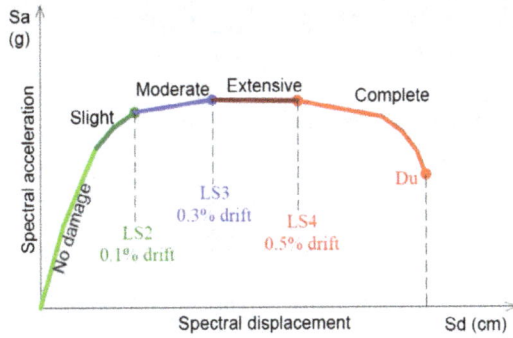

Figure 6. Performance levels on the pushover curve, according to Calvi (1999).

Following the above description, it is suggested that depending on the seismic demand, if a building has a behavior that does not exceed LS3, it can be considered satisfactory. Using the above criteria, the damage states of the wall studied for each corresponding seismicity zone can be determined and are presented in Table 5. It can be observed that except for the strong seismicity zones, RE buildings on an A-type soil can present satisfactory seismic performance.

Following the well-known equation (2) in the dynamic of structures ($S_a = \omega^2 S_d$), the initial slope of the capacity curve corresponds to ω^2 (with $\omega = 2\pi f$). That is why from this curve, the frequency f can be determined; the result is presented in Figure 7, noted "$f_{numerical}$". On this figure, h is the height of the structure. From this figure, it is interesting to note that the obtained inelastic frequencies have the same order of magnitude as the frequency obtained using other approaches: Empirical formula (Eurocode 8), noted "f_{1_EC8}"; analytical with shear beam theory (Bui et al., 2011), noted "$f_{1\_shear\ beam}$"; and *in situ* measurement (Bui et al., 2011), noted "$f_{experimental}$". This confirms the relevancy of the models used. It is logical that the inelastic frequencies have values slightly lower than the elastic frequency values, except the case of first story in RE and the second story in wooden structure. The reason may be that this mixed structure does not follow the relationships established for the current structures (one material for all walls).

In the same procedure, RE buildings on a B-type soil were also assessed. The B-type soil following Eurocode 8 corresponds to good soil (shear wave velocity $v_{s,30}$ = 360–800 m/s). The results are presented in Table 6.

It can be observed that for soil B, without special reinforcement, RE buildings in the investigated configuration do not seem adapted for medium and strong seismicity zones.

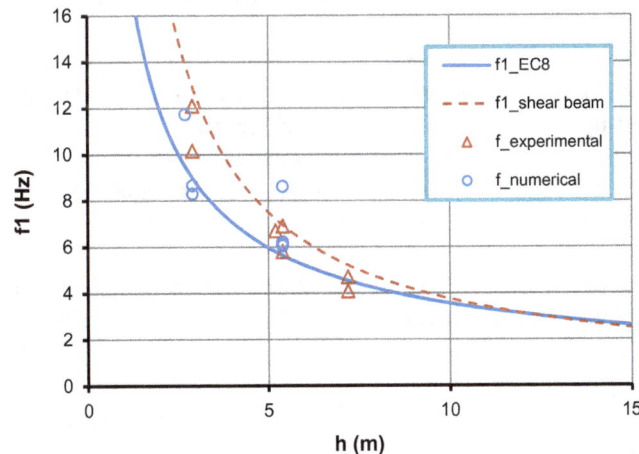

Figure 7. First frequencies obtained from different approaches: empirical (EC8), analytical (shear beam), experimental and numerical.

Seismicity zone	Wall alone	One story RE		First RE + Second wood		Two stories RE	
		Ext. wall	Int. wall	Ext. wall	Int. wall	Ext. wall	Int. wall
Very low	*Slight*	*Slight*	*Slight*	*Slight*	*Slight*	*Slight*	*Slight*
Low	*Slight*	*Slight*	*Slight*	*Slight*	*Slight*	*Slight*	Moderate
Moderate	*Slight*	*Slight*	*Slight*	*Slight*	Moderate	Moderate	Moderate
Medium	Moderate	Complete	*Extensive*	*Extensive*	*Extensive*	Complete	Complete
Strong	Complete	Complete	Complete	Complete	Complete	Complete	Complete

Table 6. Damage assessment (for soil B)

4.4. Discussion

For the configurations studied in this paper, the results show that for "very low" to "moderate" seismicity zones: Unreinforced RE buildings can be constructed on A and B soils. For "medium" seismicity zones, only soil A is acceptable. For other seismicity zones with other soil types, RE constructions are not authorized without appropriate reinforcements. This result is similar to a condition imposed in Eurocode 8 for unreinforced masonry:

$$a_g \cdot S < 2m/s^2 \tag{6}$$

where a_g is the design horizontal acceleration and S depends on soil types (Appendix A).

Indeed, from Equation (6), for "very low" to "moderate" seismicity zones, unreinforced masonry buildings can be constructed on soils A and B. For "medium" seismicity zones, only soil A is appropriate. For other seismicity zones with other soil types, unreinforced masonries are not authorized. This condition is similar to the results above obtained in this study. Therefore, this condition in Eurocode 8 for unreinforced masonry seems applicable also for RE buildings studied in this paper. However, in practice, if the RE wall length increases (more than 2.5 m for a 6 m span like the house studied in this study), the wall's seismic performance may also increase.

5. Conclusions and prospects

From a general point of view, the static nonlinear method gives more pertinent information than the static elastic method. For structures in which the dynamic behavior is dominated by the first vibrational mode (e.g. low-rise buildings), the pushover technique gives good estimations of displacements and can give useful information to assess the building's seismic vulnerability.

The results confirm that the mass of the upper stories does not play a favorable role because on the one hand it increases the vertical stress on the walls but on the other hand, it increases also the seismic action (inertia effect).

For the configurations studied in this paper, the results show that for "very low" to "moderate" seismicity zones: Unreinforced RE buildings can be constructed on A and B soils. For "medium" seismicity zones, only soil A is acceptable. For other seismicity zones with other soil types, RE constructions are not authorized without appropriate reinforcements. This result shows that the condition imposed in Eurocode 8 for unreinforced masonry seems applicable also for RE buildings.

This paper concentrated on the in-plane behavior of RE walls. Their out-of-plane behavior will also be studied in future studies. Other reinforcement techniques will also be tried to check their relevancy for RE buildings.

Funding
The authors wish to thank the French National Research Agency (ANR) for the funding of the PRIMATERRE project [project number ANR-12-Villes et Bâtiments Durables]. The Indo French Centre for the Promotion of Advanced Research (CEFIPRA) is also warmly thanked for its support.

Author details
Quoc-Bao Bui[1]
E-mail: Quoc-Bao.BUI@univ-savoie.fr
Tan-Trung Bui[2]
E-mail: tan-trung.bui@insa-lyon.fr
Ali Liman[2]
E-mail: ali.limam@insa-lyon.fr
[1] University Savoie Mont Blanc, LOCIE, CNRS, POLYTECH Annecy-Chambéry, Chambery 73000, France.
[2] University of Lyon, LGCIE, INSA Lyon, 69621 Villeurbanne Cedex, France.

References
Bui, Q. B., & Morel, J. C. (2009). Assessing the anisotropy of rammed earth. *Construction and Building Materials, 23*, 3005–3011. http://dx.doi.org/10.1016/j.conbuildmat.2009.04.011
Bui, Q. B., Morel, J. C., Hans, S., & Meunier, N. (2009). Compression behaviour of non-industrial materials in civil engineering by three scale experiments: The case of rammed earth. *Materials and Structures, 42*, 1101–1116. http://dx.doi.org/10.1617/s11527-008-9446-y
Bui, Q. B., Morel, J. C., Reddy, B. V. V., & Ghayad, W. (2009). Durability of rammed earth walls exposed for 20 years to natural weathering. *Building and Environment, 44*, 912–919. http://dx.doi.org/10.1016/j.buildenv.2008.07.001
Bui, Q. B., Hans, S., Morel, J.-C., & Do, A.-P. (2011). First exploratory study on dynamic characteristics of rammed earth buildings, *Engineering Structures, 33*, 3690–3695. http://dx.doi.org/10.1016/j.engstruct.2011.08.004
Bui, T. T., Bui, Q. B., Limam, A., & Maximilien, S. (2014). Failure of rammed earth walls: From observations to quantifications. *Construction and Building Materials, 51*, 295–302. http://dx.doi.org/10.1016/j.conbuildmat.2013.10.053
Bui, T. T., Limam, A., & Bui, Q. B. (2014). Characterization of vibration and damage in masonry structures: Experimental and numerical analysis. *European Journal of Environmental and Civil Engineering, 18*, 1118–1129.
Bui, Q. B., Morel, J. C., Hans, S., & Walker, P. (2014). Effect of moisture content on the mechanical characteristics of rammed earth. *Construction and Building Materials, 54*, 163–169. http://dx.doi.org/10.1016/j.conbuildmat.2013.12.067
Bui, T. T., Bui, Q. B., Limam, A., & Morel, J. C. (2015). Modeling rammed earth wall using discrete element method. *Continuum Mechanics and Thermodynamics, 28*, 523–538.
Calvi, G. M. (1999). A displacement-based approach for vulnerability evaluation of classes of buildings. *Journal on Earthquake Engineering, 3*, 411–438.

Cheah, J. S. J., Walker, P., Heath, A., & Morgan, T. K. K. B. (2012). Evaluating shear test methods for stabilised rammed earth. *Proceedings of the ICE – Construction Materials, 165*, 325–334. http://dx.doi.org/10.1680/coma.10.00061
Chopra, A. K., & Goel, R. K. (2002). A modal pushover analysis procedure for estimating seismic demands for buildings. *Earthquake Engineering & Structural Dynamics, 31*, 561–582.
Ciancio, D., & Augarde, C. (2013). Capacity of unreinforced rammed earth walls subject to lateral wind force: Elastic analysis versus ultimate strength analysis. *Materials and Structures, 46*, 1569–1585. http://dx.doi.org/10.1617/s11527-012-9998-8
Cundall, P. A. (1971). A computer model for simulating progressive, large scale movements in blocky rock systems. *Proceeding of International Symposium Rock Fracture, ISRM, Nancy, 1*, 2–8.
El Nabouche, R., Bui, Q. B., Perrotin, P., Plé, O., Plassiard, J.-P. (2015, June 21–24). Numerical modeling of rammed earth constructions: Analysis and recommendations. In *1st International Conference on Bio-based Building Materials*, Clermont Ferrand.
EN 1998-1:2004. (2004). *Design of structures for earthquake resistance–part 1* (157 p.). European Committee for Standardization, Brussels.
Gomes, M. I., Lopes, M., & de Brito, J. (2011). Seismic resistance of earth construction in Portugal. *Engineering Structures, 33*, 932–941. http://dx.doi.org/10.1016/j.engstruct.2010.12.014
Hamilton, III, H. R., McBride, J., & Grill, J. (2006). Cyclic testing of rammed-earth walls containing post-tensioned reinforcement. *Earthquake Spectra, 22*, 937–959. http://dx.doi.org/10.1193/1.2358382
Itasca. (2011). 3DEC–three dimensional distinct element code, version 4.1. Minneapolis, MN: Author.
Lourenço, P. B., Oliveira, D. V., Roca, P., & Orduña, A. (2005). Dry joint stone masonry walls subjected to in-plane combined loading. *Journal of Structural Engineering, 131*, 1665–1673. http://dx.doi.org/10.1061/(ASCE)0733-9445(2005)131:11(1665)
Miccoli, L., Müller, U., & Fontana, P. (2014). Mechanical behaviour of earthen materials: A comparison between earth block masonry, rammed earth and cob. *Construction and Building Materials, 61*, 327–339. http://dx.doi.org/10.1016/j.conbuildmat.2014.03.009
Miccoli, L., Oliveira, D. V., Silva, R. A., Muller, U., & Schueremans, L. (October, 2015, September). Static behaviour of rammed earth: Experimental testing and finite element modeling. *Materials and Structures, 48*, 3443–3456.
Morris, H., & Walker, R. (2011). Observations of the performance of earth buildings following the February 2011 Christchurch earthquake. *Bulletin of the New Zealand Society for Earthquake Engineering, 44*, 358–367.
Salvitti, L. M., & Elnashai, A. S. (1996). *Evaluation of bahaviour factors for RC buildings by nonlinear dynamic analysis* (Paper No. 1820). 11th World Conference on Earthquake Engineering. Elsevier, Acapulco, Mexico. ISBN: 0 08 042822 3.
Silva, R. A., Oliveira, D. V., Miranda, T., Cristelo, N., Escobar, M. C., & Soares, E. (2013). Rammed earth construction with granitic residual soils: The case study of northern Portugal. *Construction and Building Materials, 47*, 181–191.

Appendix A

For unreinforced masonry, Eurocode 8 imposes the following condition: $a_g. S < 2$ m/s^2

where

- a_g: Design horizontal acceleration: $a_g = g_I . a_{gr}$

where a_{gr}: Ground acceleration referenced according to the seismicity zones

Seismicity zones	a_{gr} (m/s^2)
1 (very low)	0.4
2 (low)	0.7
3 (moderate)	1.1
4 (medium)	1.6
5 (strong)	3

g_I: Important factor. For category II (current buildings, $h < 28$ m): $g_I = 1$; category III (schools, meeting halls, etc.): $g_I = 1.2$

- S: Depends on soil types

Soil types	S (zones 1–4)
A (rock)	1
B (stiff deposit of sand, gravel or over consolidated clay)	1.35
C (deposit of medium-density sand)	1.5
D (deposit of noncohesive soils from low to medium density)	1.6
E (alluvium)	1.8

2

Sensitivity of crumb rubber particle sizes on electrical resistance of rubberised concrete

Sakdirat Kaewunruen[1]* and Ratthaphong Meesit[1]

*Corresponding author: Sakdirat Kaewunruen, Birmingham Centre for Railway Research and Education, School of Civil Engineering, University of Birmingham, Birmingham, UK

E-mail: s.kaewunruen@bham.ac.uk

Reviewing editor: Raja Rizwan Hussain, King Saud University, Saudi Arabia

Abstract: Railway track components often suffer from high aggressive loading and vibrating conditions of railway environment, causing high maintenance costs due to impact damage, rail seat abrasion and excessive noise and vibration to surrounding equipment. Thus, it is essential to have novel improvement of material capabilities in order to solve or reduce these problems. A nanoengineered improvement method for concrete material using crumb rubber has been recently introduced to railway applications. However, for modern electrified railway tracks, structural materials will need to provide electrical and signal insulation for effective operations of track circuits and electrification. This paper firstly highlights the importance of the particle sizes of crumb rubbers on the electrical resistivity of the concrete modified by crumbed rubbers. It shows that microscale crumb rubbers induce lesser electrical conduction capacity than nanoscale crumb rubber.

Subjects: Built Environment; Civil, Environmental and Geotechnical Engineering; Materials Science

Keywords: crumb rubber; concrete; electrical resistance; sensitivity; particle size; engineered concrete; railway track; insulation

ABOUT THE AUTHORS

Sakdirat Kaewunruen is Senior Lecturer in Railway and Civil Engineering with over 220 technical publications, and Ratthaphong Meesit is formerly a graduate student at The University of Birmingham. The University of Birmingham is host to the world-leading Birmingham Centre for Railway Research and Education (BCRRE), a multi-disciplinary group of staff from the Schools of Civil Engineering, Electronic, Electrical & Systems Engineering, Mechanical Engineering and Materials & Metallurgy. This study is part of a collaborative research project between the University of Birmingham (UOB) and The University of Illinois at Urbana Champaign (UIUC) through BRIDGE grant. The project will build on research strengths of both UoB and UIUC to perform investigations on dynamic damping, dielectric property and dynamic resistance of micro and nanoengineered crumbed rubber concrete (CRC). The applications of the advanced materials such as modified CRC in railway tracks could potentially improve safety, resilience and reliability of rail infrastructure.

PUBLIC INTEREST STATEMENT

This research is the world first to deal with electrical resistance of innovative nano and microengineered concrete, of which this electrical property is crucial. The investigation will lead the way on the practical applications of the new materials to railway industry. An important focus on this research is the recycling of crumb rubbers, which are derived from wasted rubber tires and plastics. Its wide spread applications in railway construction and maintenance will strengthen environmental impact to societies. Insights into the sensitivity of the crumb rubber particle sizes on the dielectric property of modified concrete will assure better material design and its compliance and compatibility with systems requirements.

1. Introduction

Railway infrastructure owners all over the world are suffering from many of wheel and rail irregularities, resulting in increased maintenance, more frequent monitoring and track patrol, and the rapid track degradation that can lead to poor ride quality. This mechanism also undermines structural capacity of materials used in railway environments. As such, railway track components often wear and tear rapidly from the aggressive loading and vibrating conditions. Some of the problems are impact damage, rail seat abrasion and excessive noise and vibration to surrounding equipment. These problems play an important role in sustainable and effective railway operations. Therefore, it is essential to develop new techniques to mitigate these problems. Nowadays, several innovative methods are developed in order to suppress the impact load and vibration before and after passing through sleepers to other track components (ballast, subgrade and ground). For example, rail and sleeper pads are installed underneath the rail and sleepers, respectively, to increase the damping of the track system (Carrascal, Casado, Polanco, & Gutiérrez-Solana, 2007; Connolly, Kouroussis, Laghrouche, Ho, & Forde, 2015; Kaewunruen & Remennikov, 2009, 2010; Remennikov & Kaewunruen, 2008). However, these two composite products can wear and come off from the main structure due to various factors such as looseness of fastening system, climate change and so on (Griffin, Mirza, Kwok, & Kaewunruen, 2014; Kernes, Edwards, Dersch, Lange, & Barkan, 2011). Therefore, finding the sustainable way to reduce the impact load and vibration problem is still a big challenge in railway industry.

A method for improving dynamic resistance and vibration energy absorption of railway concrete material has recently been introduced for manufacturing railway sleepers. The dynamic damping of concretes can be improved using waste rubber tyres as microfiller. Not only can the crumb rubber improve dynamic damping property while maintain structural strength at acceptable level, such the method provides an environment-friendly avenue by recycling synthetic wastes such as wheel tires, plastic and other industry rubbers. Thanks to significant and popular utilisation of cement and concrete composites in railway infrastructure, the total environmental benefit of such method in railway systems is potentially substantial.

This paper highlights the evaluation of electrical resistance of crumb rubber concrete (CRC) and focuses on the sensitivity of particle sizes of crumb rubber on the electrical conductivity. Monitoring electrical resistance of structural concrete is critical for effectiveness of modernised railway operations. Electrification, signalling control, train communication and detection, track circuits, automatic train control, and level crossing safety control within railway corridor are among the subsystems that require the ability of structural concrete for electrical insulation at certain extent. For instance, rail/sleeper fastening assemblies and sleepers shall ensure a minimum electrical resistance between the running rails of 10 Ohms per track kilometre (Kaewunruen & Vaughan, 2009; RailCorp, 2012). As a result, material design must comply with systemic requirements in order to avoid multi-million dollars penalty when the railway track malfunctions. This study is the first to address the sensitivity of particle characteristics on the dielectric property of engineered concrete. The outcome of this research will help material engineers to develop and improve construction materials for real-world applications.

2. CRC

In this development, CRC has been designed as structural material for manufacturing concrete sleepers. The fundamental characteristics of CRC include:

2.1. Properties of fresh concrete

The fresh density of CRC with rubber aggregate was observed by Siddique and Naik (2004) and Su, Yang, Ling, Ghataora, and Dirar (2015). They illustrated that fresh CRC has lower density than normal concrete, and increase in percentage of crumb rubber affects the reduction in fresh density. This is because the crumb rubber has low specific gravity. In addition, non-polar nature of rubber particles may repel water and attract air on rubber surface, which would cause air void increase. Workability of CRC was evaluated by measuring slump of fresh concrete. Various researches have stated that increase in crumb rubber replacement reduces the workability of fresh CRC (Najim & Hall, 2011, 2012; Pacheco-Torgal, Ding, & Jalali, 2012). In addition, the size of crumb rubber also has an impact to the workability of concrete. Su et al. (2015) found that there is a reduction of slump, once the size of rubber particle is decreased. However, this workability issue can be solved by adding proper amount of plasticizer admixture (about 2–3%) into the concrete (Aiello & Leuzzi, 2010; Topçu & Bilir, 2009).

2.2. Properties of hardened concrete

As well known, compressive strength is one of the main properties of concrete. For CRC, most of previous researches have obviously shown that increase in rubber content leads to reduction in compressive strength (Aiello & Leuzzi, 2010; Chou, Lin, Lu, Lee, & Lee, 2010; Eldin & Senouci, 1993; Güneyisi, Gesoğlu, & Özturan, 2004; Issa & Salem, 2013). In addition, a smaller reduction in compressive strength was observed when only fine aggregate was replaced by crumb rubber. Su et al. (2015) studied about the effect of different sizes of rubber particles (3, 0.5, 0.3 mm) to the compressive strength of CRC. They revealed that at 28 days, cube compressive strength of concrete increases with a decrease in the rubber particle size due to a better void filling ability of finer crumb rubber. Moreover, the types of crumb rubber also have an significant impact to the compressive strength.

According to the reduction of compressive strength, tensile strength of CRC is also influenced on the amount of rubber content. Increase in rubber content decreases both splitting and flexural strength (Güneyisi, Gesoğlu, & Özturan, 2004). However, reduction of tensile strength seems to be less impact than in the case of compressive strength (Pacheco-Torgal et al., 2012). The reduction of tensile strength also depends on the size of rubber particle. Greater rubber particle will negatively impact to the tensile strength (Ganjian, Khorami, & Maghsoudi, 2009; Su et al., 2015). Furthermore, the types of aggregate that is substituted also have an effect to tensile strength. Larger decrease in tensile strength occurs when replacing coarse aggregate rather than fine aggregate with rubber waste (Aiello & Leuzzi, 2010). Elastic modulus of the concrete tends to reduce when increasing the percentage replacement of coarse or fine aggregate with waste rubber (Ganjian et al., 2009; Kumar, Sandeep, & Sudharani, 2014; Onuaguluchi & Panesar, 2014). Even though the elastic modulus of CRC decreases, the strain rate property increases considerably when increasing rubber content. This makes CRC has lower brittleness index than plain concrete, which causes CRC has greater ductility performance (Snelson, Kinuthia, Davies, & Chang, 2009; Zheng, Huo, & Yuan, 2008). In addition, it was revealed that CRC also has higher toughness and better energy absorbing ability compared to normal concrete (Aliabdo, Abd Elmoaty, & AbdElbaset, 2015; Li et al., 2004). Therefore, this property could be advantageous for railway application.

For CRC, very few researches have been dedicated to dielectric property. In principle, rubber content in concrete performs as an electrical insulator which influences the resistivity of CRC becomes higher than that of plain concrete (Issa & Salem, 2013). Sukontasukkul (2009) has conducted the experiment to investigate noise absorption ability of CRC. Two size of crumb rubber were considered: passing sieve No. 6 (3.36 mm) and 26 (0.707 mm). These groups of crumb rubber were used to replace fine aggregate. The noise absorption ability of each concrete mix was compared using noise absorption coefficient (α). As a result, CRC seemed to have better noise absorption ability compared to reference concrete. Even though temperature is different (low, normal and high), CRC still performed well in terms of sound absorption than plain concrete (Holmes, Browne, & Montague, 2014).

Thermal resistance of CRC has been studied by Kaloush, Way, and Zhu (2005). The result illustrated that coefficient of thermal expansion of CRC will decrease when crumb rubber content is increased both heating (expansion) and cooling (contraction) cycles. This means CRC can be more resistant in thermal changing than normal concrete. In addition, Sukontasukkul (2009) also found the relationship between the size of crumb rubber and thermal resistivity. Smaller rubber particles provided a better thermal resistance to concrete. Sukontasukkul and Chaikaew (2006) have found that concrete with smaller rubber particles seemed to have less per cent weight loss than concrete contained bigger rubber particles as shown in Figure 1.

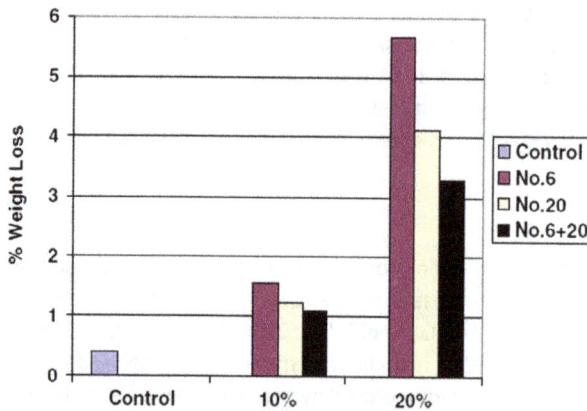

Figure 1. Abrasion test results (Sukontasukkul & Chaikaew, 2006).

Table 1. Mixture proportions of concrete(kg/m³)							
No.	Mixes	Cement	Gravel	Sand	Silica fume	425 μ rubber	75 μ rubber
1	RFC	530	986	630	–	–	–
2	SFC	477	986	630	53	–	–
3	SFRC-425-5	477	986	599	53	31.5	–
4	SFRC-425-10	477	986	567	53	63	–
5	SFRC-75-5	477	986	599	53	–	31.5
6	SFRC-75-10	477	986	567	53	–	63

However, Kang, Zhang, and Li (2012) have found the opposite results when only replacing same volume of fine aggregate with crumb rubber. They found that CRC seemed to have a better abrasion resistance than plain and silica fume concrete (lower per cent mass loss). The abrasion resistance of CRC tended to increase as rubber content increase. This may be because the CRC has excellent dynamic performances in terms of energy absorption, toughness and cracking resistance. Therefore, theses parameters results in better abrasion resistance in CRC.

After reviewing the characteristics of CRC, it can be obviously seen that CRC has a lot of advantages. However, there are still some aspects of mechanical properties requiring improvement such as compressive and tensile strength, and elastic modulus. The critical review suggests that mixing a smaller size of rubber waste particle at micro and nanoscales into the concrete can mitigate those problems. Therefore, this study is the first to investigate the electrical properties of high-strength CRC aimed for railway applications. The very small particles of rubber waste as microscale (75 μ) were selected to use in the experiment.

2.3. Design of engineered concrete

Six concrete mixes have been designed in accordance with British Standards (British Standards Institution, BS EN 197-1, 2011). The reference concrete (RFC) is designed using water-cement ratio of 0.44 and slump value of 60–180 mm, in order to achieve a target mean strength of 63 MPa at 28 days. The second mix is the reference concrete, which 10 weight per cent (wt%) of cement is replaced by silica fume (SFC). For the remaining concrete mixes, they are modified from the second mix by replacing 5 and 10 wt% of fine aggregate with 425 and 75 μ rubber powders, respectively. All of mixture portions are presented in Table 1.

3. Measurement of electrical resistance

In general, two methods used for measuring the electrical resistivity of concrete are two-pole and four-pole method (Konsta-Gdoutos & Aza, 2014). Both methods are very similar in terms of

Figure 2. Four poles resistivity testing, after Konsta-Gdoutos and Aza (2014).

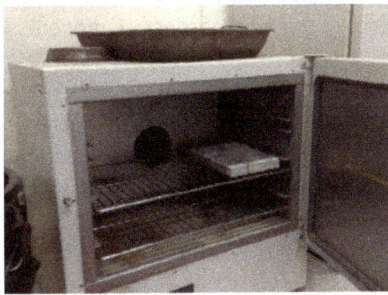

(a) Moisture Controlling   *(b) Experimental Set Up*

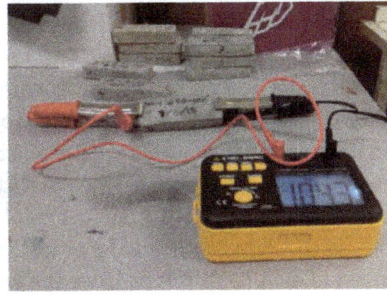

Figure 3. Arrangement of electrical resistivity.

application of current and voltage passing concrete. However, two-pole method applies two electrodes at both ends of the concrete, and they are used to measure the current intensity and voltage difference. While in four-pole method, there are four poles of electrodes applied on the concrete sample. Two outer poles are used for measuring the current and other two poles are used for determining the voltage as shown in Figure 2.

In this study, two poles method was selected. The type of concrete specimen used in the test was W45 × H20 × L120 mm prism. The equipment used was insulation tester (Victor VC60B$^+$) which can be applied to determine electrical resistance (Ohms) of concrete samples. Before the test, the concrete samples were removed from the curing tank at 28 days. Then, they were cleaned and put into the oven where the temperature was controlled to be constant at 30°C (see Figure 3(a)). After 24 hours, the samples were taken out from the oven, and the experiment was set up by connecting the lead wires to both ends of the concrete as shown in Figure 3(b). The insulation tester was turned on, and the voltage was kept constant at 1,000 V. After that, the electrical resistance of concrete specimen was measured, and the value at 60 sec was recorded as an electrical resistance of that sample.

4. Particle-based sensitivity on dielectric property

In this study, the electrical resistivity experiment was conducted when the concrete samples had age of 28 days. The data obtained from the test was the electrical resistance (Ohms). Therefore, to convert it into electrical resistivity, the common equation was used as presented below:

$$\rho = \frac{RA}{L} \tag{1}$$

where ρ is the electrical resistivity (Ohm-m), R is the electrical resistance (Ohms), A is the cross-sectional area of specimen and L is the length of specimen.

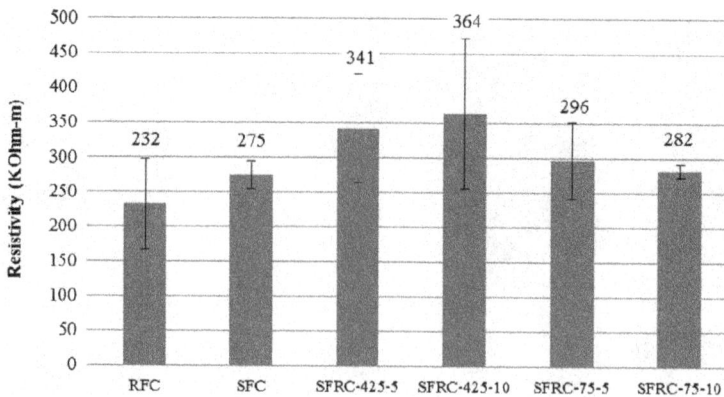

Figure 4. Electrical resistivity of each concrete mixes.

As shown in Figure 4, the RFC had the lowest electrical resistivity (232 KOhm-m) when compared to other mixes. The silica fume tended to have a positive impact to this kind of concrete properties (Onuaguluchi & Panesar, 2014). It made the electrical resistivity increase approximately 18.53% in this study. Replacing very fine aggregate with microcrumb rubber either 5 wt% or 10 wt% could significantly enhance the electrical resistivity compared to RFC and SFC. This is because the rubber is dielectric material, thus, crumb rubber inside concrete acted as an insulator preventing the current transfer between two measuring electrodes. Small particle of rubbers can fills the void better and replace tapped moisture inside concrete, resulting in higher electrical resistance. Moreover, this study also found that the concrete contained the bigger size of rubber particles has a higher electrical resistivity than that of finer one. This is because the insulated spherical dimension of the rubber particles is larger. Accordingly, when there is more amount of the larger crumb rubber, the electrical resistivity becomes higher. In contrast, nanoparticle rubbers do not improve the electrical performance even there is more amount of such particles.

Overall, the electrical resistivity of rubberised concrete mixes tends to be relatively higher than ordinary high-strength concrete. Interestingly, the results are quite consistent when compared to the results obtained by previous researches (Issa & Salem, 2013).

5. Concluding remarks

Railway track components are usually subjected to high aggressive loading and vibrating conditions within railway corridor. Novel improvement of material capabilities to mitigate such problems becomes a new research agenda. This study focuses on a nanoengineered improvement method for concrete material using crumb rubber, which has been recently introduced to railway applications. This paper firstly highlights the importance of the particle sizes of crumb rubbers on the electrical resistivity of the concrete modified by crumbed rubbers.

This research establishes the development of the environment-friendly concrete using waste rubber tire (425 and 75 μ) as microfiller to comply with the requirements for making railway concrete sleeper. Six different types of concrete have been designed and manufactured in accordance with British standards. The experiments have been conducted to determine electrical resistance of the specimens at 28 days. CRC concrete specimens have exhibited a better characteristic in electrical resistance in comparison with ordinary high-strength concrete. Replacing very fine aggregate with microscale crumb rubber can significantly enhance the ability to resist the current flow inside the concrete. This property is very meaningful and highly practical in modern railway track applications because better electrical resistance of concrete sleeper can ensure the reliable and safe railway operations using track circuits.

Acknowledgement

The first author would like to thank Australian Government Department of Education, Research and Innovation for his Endeavour Executive Fellowships at Massachusetts Institute of Technology, at Harvard University, at Chalmers University of Technology, and at Railway Technical Research Institute. Kaewunruen

Funding

The authors would like to gratefully acknowledge the University of Birmingham's BRIDGE Grant, which financially supports this study as a part of the project "Improving damping and dynamic resistance in concrete through Micro and Nano engineering for sustainable and environmental-friendly applications in railway and other civil construction". This project is part of a collaborative BRIDGE program between the University of Birmingham and the University of Illinois at Urbana Champaign.

Author details

Sakdirat Kaewunruen[1]
E-mail: s.kaewunruen@bham.ac.uk
Ratthaphong Meesit[1]
E-mail: RXM496@student.bham.ac.uk
[1] Birmingham Centre for Railway Research and Education, School of Civil Engineering, University of Birmingham, Birmingham, UK.

References

Aiello, M., & Leuzzi, F. (2010). Waste tyre rubberized concrete: Properties at fresh and hardened state. *Waste Management, 30*, 1696–1704. http://dx.doi.org/10.1016/j.wasman.2010.02.005

Aliabdo, A. A., Abd Elmoaty, A. E., & AbdElbaset, M. M. (2015). Utilization of waste rubber in non-structural applications. *Construction and Building Materials, 91*, 195–207. http://dx.doi.org/10.1016/j.conbuildmat.2015.05.080

British Standards Institution, BS EN 197-1. (2011). *Cement. Composition, specifications and conformity criteria for common cements*. London: British Standards Institution.

Carrascal, I., Casado, J., Polanco, J., & Gutiérrez-Solana, F. (2007). Dynamic behaviour of railway fastening setting pads. *Engineering Failure Analysis, 14*, 364–373. http://dx.doi.org/10.1016/j.engfailanal.2006.02.003

Chou, L.-H., Lin, C.-N., Lu, C.-K., Lee, C.-H., & Lee, M.-T. (2010). Improving rubber concrete by waste organic sulfur compounds. *Waste Management & Research, 28*, 29–35.

Connolly, D., Kouroussis, G., Laghrouche, O., Ho, C., & Forde, M. (2015). Benchmarking railway vibrations—Track, vehicle, ground and building effects. *Construction and Building Materials, 92*, 64–81. http://dx.doi.org/10.1016/j.conbuildmat.2014.07.042

Eldin, N. N., & Senouci, A. B. (1993). Rubber-tire particles as concrete aggregate. *Journal of Materials in Civil Engineering, 5*, 478–496. http://dx.doi.org/10.1061/(ASCE)0899-1561(1993)5:4(478)

Ganjian, E., Khorami, M., & Maghsoudi, A. (2009). Scrap-tyre-rubber replacement for aggregate and filler in concrete. *Construction and Building Materials, 23*, 1828–1836. http://dx.doi.org/10.1016/j.conbuildmat.2008.09.020

Griffin, D. W. P., Mirza, O., Kwok, K., & Kaewunruen, S. (2014), Composite slabs for railway construction and maintenance: A mechanistic review. *The IES Journal of Civil and Structural Engineering, 7*, 243–262. doi:10.1080/19373260.2014.947909

Güneyisi, E., Gesoğlu, M., & Özturan, T. (2004). Properties of rubberized concretes containing silica fume. *Cement and Concrete Research, 34*, 2309–2317. http://dx.doi.org/10.1016/j.cemconres.2004.04.005

Holmes, N., Browne, A., & Montague, C. (2014). Acoustic properties of concrete panels with crumb rubber as a fine aggregate replacement. *Construction and Building Materials, 73*, 195–204. http://dx.doi.org/10.1016/j.conbuildmat.2014.09.107

Issa, C. A., & Salem, G. (2013). Utilization of recycled crumb rubber as fine aggregates in concrete mix design. *Construction and Building Materials, 42*, 48–52. http://dx.doi.org/10.1016/j.conbuildmat.2012.12.054

Kaewunruen, S., & Remennikov, A. M. (2009). Progressive failure of prestressed concrete sleepers under multiple high-intensity impact loads. *Engineering Structures, 31*, 2460–2473. http://dx.doi.org/10.1016/j.engstruct.2009.06.002

Kaewunruen, S., & Remennikov, A. M. (2010). Dynamic properties of railway track and its components: Recent findings and future research direction. *Insight - Non-Destructive Testing and Condition Monitoring, 52*, 20–22. http://dx.doi.org/10.1784/insi.2010.52.1.20

Kaewunruen, S., & Vaughan, S.. (2009, April). *Electrical conductivity of concrete turnout bearers* (Technical Report No. TR158, 22 p.). Sydney: Track Services RailCorp.

Kaloush, K., Way, G., & Zhu, H. (2005). Properties of crumb rubber concrete. *Transportation Research Record: Journal of the Transportation Research Board, 1914*, 8–14. http://dx.doi.org/10.3141/1914-02

Kang, J., Zhang, B., & Li, G. (2012). The abrasion-resistance investigation of rubberized concrete. *Journal of Wuhan University of Technology-Mater. Sci. Ed, 27*, 1144–1148. http://dx.doi.org/10.1007/s11595-012-0619-8

Kernes, R. G., Edwards, J. R., Dersch, M. S., Lange, D. A., & Barkan, C. P. (2011). Investigation of the impact of abrasion as a concrete crosstie rail seat deterioration (RSD) mechanism. *AREMA (2011). Annual Conference in Conjunction with Railway Interchange* (pp. 1–24). Minnesota: AREMA.

Konsta-Gdoutos, M. S., & Aza, C. A. (2014). Self sensing carbon nanotube (CNT) and nanofiber (CNF) cementitious composites for real time damage assessment in smart structures. *Cement & Concrete Composites, 53*, 162–169.

Kumar, G. N., Sandeep, V., & Sudharani, C. (2014). Using tyres waste as aggregate in concrete to form rubcrete-mix for engineering application. *International Journal of Research in Engineering and Technology, 3*, 500–509.

Li, G., Garrick, G., Eggers, J., Abadie, C., Stubblefield, M. A., & Pang, S.-S. (2004). Waste tire fiber modified concrete. *Composites Part B: Engineering, 35*, 305–312. http://dx.doi.org/10.1016/j.compositesb.2004.01.002

Najim, K. B., & Hall, M. R. (2011). Workability and mechanical properties of crumb-rubber concrete. *Proceedings of the Institution of Civil Engineers, 166*, 7–17.

Najim, K. B., & Hall, M. R. (2012). Mechanical and dynamic properties of self-compacting crumb rubber modified concrete. *Construction and Building Materials, 27*, 521–530. http://dx.doi.org/10.1016/j.conbuildmat.2011.07.013

Onuaguluchi, O., & Panesar, D. K. (2014). Hardened properties of concrete mixtures containing pre-coated crumb rubber and silica fume. *Journal of Cleaner Production, 82*, 125–131. http://dx.doi.org/10.1016/j.jclepro.2014.06.068

Pacheco-Torgal, F., Ding, Y., & Jalali, S. (2012). Properties and durability of concrete containing polymeric wastes (tyre rubber and polyethylene terephthalate bottles): An overview. *Construction and Building Materials, 30*, 714–724. http://dx.doi.org/10.1016/j.conbuildmat.2011.11.047

RailCorp. (2012). *Track engineering specification SPC 232 concrete sleeper*. Sydney: Asset Standards Authority.

Remennikov, A. M., & Kaewunruen, S. (2008). A review of loading conditions for railway track structures due to train and track vertical interaction. *Structural Control and Health Monitoring, 15,* 207–234. doi:10.1002/stc.227

Siddique, R., & Naik, T. R. (2004). Properties of concrete containing scrap-tire rubber—An overview. *Waste Management, 24,* 563–569. http://dx.doi.org/10.1016/j.wasman.2004.01.006

Snelson, D., Kinuthia, J., Davies, P., & Chang, S. (2009). Sustainable construction: Composite use of tyres and ash in concrete. *Waste Management, 29,* 360–367. http://dx.doi.org/10.1016/j.wasman.2008.06.007

Su, H., Yang, J., Ling, T.-C., Ghataora, G. S., & Dirar, S. (2015). Properties of concrete prepared with waste tyre rubber particles of uniform and varying sizes. *Journal of Cleaner Production, 91,* 288–296. http://dx.doi.org/10.1016/j.jclepro.2014.12.022

Sukontasukkul, P. (2009). Use of crumb rubber to improve thermal and sound properties of pre-cast concrete panel. *Construction and Building Materials, 23,* 1084–1092. http://dx.doi.org/10.1016/j.conbuildmat.2008.05.021

Sukontasukkul, P., & Chaikaew, C. (2006). Properties of concrete pedestrian block mixed with crumb rubber. *Construction and Building Materials, 20,* 450–457. http://dx.doi.org/10.1016/j.conbuildmat.2005.01.040

Topçu, İ. B., & Bilir, T. (2009). Experimental investigation of some fresh and hardened properties of rubberized self-compacting concrete. *Materials and Design, 30,* 3056–3065. http://dx.doi.org/10.1016/j.matdes.2008.12.011

Zheng, L., Huo, X., & Yuan, Y. (2008). Strength, modulus of elasticity, and brittleness index of rubberized concrete. *Journal of Materials in Civil Engineering, 20,* 692–699. http://dx.doi.org/10.1061/(ASCE)0899-1561(2008)20:11(692)

3

Design of a durable roof slab insulation system for tropical climatic conditions

Kasun Nandapala[1]* and Rangika Halwatura[1]

*Corresponding author: Kasun Nandapala, Department of Civil Engineering, University of Moratuwa, Moratuwa, Sri Lanka

E-mails: mpkcnandapala@gmail.com, kasuncn@uom.lk

Reviewing editor: Raja Rizwan Hussain, King Saud University, Saudi Arabia

Abstract: Flat roof slabs become popular day-by-day due to the advantages like cyclonic resistance, the possibility of future vertical extension, and the possibility of utilizing as an additional working space. However, a serious matter of concern is its thermal discomfort, for which air conditioning is used as the most common remedy. This has led to extensive use of energy, increasing the operational cost of the buildings, and contributing to global warming. Hence, the current trend is to go for passive techniques. Insulating roof slabs is identified as a better passive way to make buildings thermally comfortable. In this study, several existing roof slab insulation systems and their performances were investigated, and the most effective system for tropical climates was identified. Since that system had an issue in durability, a new system was developed with a discontinued-stripped supporting arrangement, which withstood a 4MT-point load. Further, it was proven by comparing literature that the newly designed system has a heat gain reduction of more than 75%.

Subjects: Built Environment; Engineering & Technology; Sustainable Development

Keywords: roof slabs; insulation; thermal performance; structural performance; durability

1. Introduction

It is evident that deforestation in the world has reached an intolerable level and has become the primary cause for the imbalance that the world is facing today. This phenomenon has aggravated mainly because of the unplanned construction and the extensive land consumption caused due to the

ABOUT THE AUTHORS

Kasun Nandapala is a full-time PhD candidate at the Department of Civil Engineering in University of Moratuwa, who is carrying out his core research on thermal insulation in roof slabs. Rangika Halwatura is a senior lecturer at the Department of Civil Engineering in University of Moratuwa. The research interests of this team include but not limited to thermal optimization of buildings, development of sustainable construction materials, invention of techniques to promote sustainable built environment.

PUBLIC INTEREST STATEMENT

Global warming and the associated climate change are the biggest issues that the current world is facing. Consequently, the natural disasters have increased in intensity and severity. Therefore, structures need to be made robust, for which the use of flat concrete roof slabs is a good strategy. It provides some additional advantages like possibility of future vertical extension and possibility of using as a working space. However, the thermal discomfort in the immediate underneath floor is its major drawback. Even though air conditioning solves the problem, it is at a higher cost in the form of energy consumption. Insulating slabs is a better strategy to prevent this. In this study, a roof slab insulation has been developed which is structurally sound and durable. It was proven that this system reduces more than 75% heat gain into a building, which eventually will either eliminate the necessity of air conditioning or will reduce the necessary capacity.

increased population. Consequently, due to the scarcity created, "land" has become one of the most expensive commodities, particularly in urban areas. In this context, multi-storey construction has become popular, as it produces a higher floor-area ratio (Dareeju, Meegahage, & Halwatura, 2011).

Consequently, use of flat concrete roof slabs has begun to be popular as it provides the flexibility of using either as a working space, as a rooftop garden or as a temporary shelter till a future vertical extension is taken place (Banting, 2005; Berardi, GhaffarianHoseini, & GhaffarianHoseini, 2014; Halwatura & Jayasinghe, 2007). Further, roof slabs increase the robustness of the structures and provide an additional cyclonic resistance which is very handy against the climate change taking place in the world (Halwatura, 2013).

However, the use of roof slabs has not been sufficiently penetrated to the middle-class population due to a variety of reasons. Thermal discomfort in the immediate underneath floor has been found to be the primary reason for that (Nandapala & Halwatura, 2014). The concrete roof slabs get heated in the daytime and emit long-wave radiation to the underneath space, causing the discomfort (Halwatura & Jayasinghe, 2008).

Air conditioning is the most common remedy to overcome this issue. Even though air conditioning resolves this issue, a higher operational cost has to be incurred. Further, this increases the energy usage which leads to the biggest problem in the world, Global Warming (Halawa et al., 2014; Lean & Rind, 2001; Macilwain, 2000).

Elaborating the impact of this in quantitative terms, in Singapore, buildings use up to 57% of the total energy usage of the country (Kwong, Adam, & Sahari, 2014). In Malaysia, a country with similar tropical climatic conditions, more than 30% of the total energy in buildings is used for making them thermally comfortable (Dong, Lee, & Sapar, 2005). Those findings sum up that around 15–20% of the total energy usage in tropical countries is used for enhancing thermal comfort in buildings. It is evident that the amount of energy used for comfort is much more than what the world can afford.

Hence, active cooling (in the form of Air Conditioning) is not a preferred remedy to achieve thermal comfort. Therefore, passive cooling techniques have emerged (Al-Obaidi, Ismail, & Abdul Rahman, 2014; Alvarado, Terrell, & Johnson, 2009; Sadineni, Madala, & Boehm, 2011), insulation in particular (Al-Homoud, 2005; Brito Filho & Santos, 2014; Dylewski & Adamczyk, 2014). Insulating the roof has been identified to be the best option as it is the element that contributes to about 70% of the total heat gain inside buildings (Halwatura, 2014; Vijaykumar, Srinivasan, & Dhandapani, 2007).

There are several roof slab insulation techniques tried out in the world. A research work carried out in Florida, USA, has obtained a 38% energy saving by applying a cool paint (Parker & Barkaszi, 1997), and another study in Italy, with the same technique, has achieved a 54% energy saving (Romeo & Zinzi, 2013). A daily heat gain reduction of 56% has been observed in Greece, using a 60 mm ventilated air gap as an insulator (Dimoudi, Androutsopoulos, & Lykoudis, 2006). A system developed in Sri Lanka, a tropical country, has achieved a heat reduction of 75% using a 25 mm polystyrene layer (Halwatura & Jayasinghe, 2008). Some laboratory experiments have obtained similar results as well (Alvarado & Martínez, 2008; Megri, Achard, & Haghighat, 1998).

Above figures incontrovertibly suggest that insulation can be effective in any climatic condition. Since this study focuses on tropics, the system developed in Sri Lanka was identified to be the most recent and the most effective to suit the conditions.

A further study on the system suggested that this particular system has been developed as an alternative to one of the most common insulation systems used in tropical countries, shown in Figure 1. It has an insulation layer on the structural slab and a protective screed on top of it.

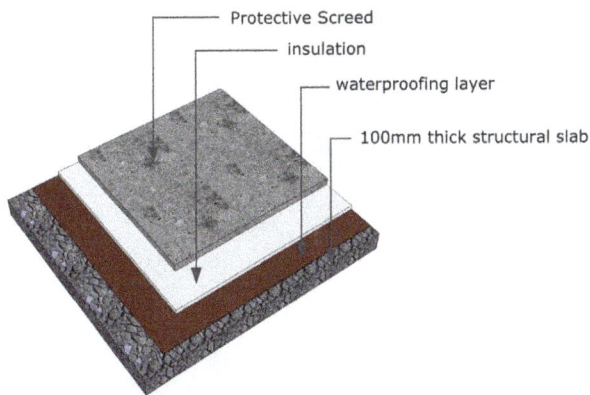

Figure 1. The most common roof insulation system used in tropical countries.

This system had been tested under practical conditions and its thermal performance is emphasized. However, this has imposed a restriction on loading, since a layer of weak material (insulation material) is placed between two layers of concrete, and the load path from the slab passes through it.

The system by Halwatura and Jayasinghe (shown in Figure 2) has been developed as a remedy for this. It has a set of concrete strips within the insulation layer to support the top screed. Hence, the load path passes through those strips, without transferring the load to the insulation layer. Further, a 50 mm × 50 mm steel mesh has been introduced to distribute the load.

However, a field study suggested that there is a concern about the durability of the system as some water patches were observed in slab soffits in the long run. A thorough study resulted in finding that water was stagnant in the polystyrene layer. This phenomenon can further be elaborated as follows:

The arrangement of supporting strips in plan view is shown in Figure 3. It is apparent that there is no drainage path for the penetrated water to flow out, as the insulation material is enclosed by a set of continuous concrete strips. Hence, a water head is developed in the system, resulting in a reduction in the lifespan of the waterproofing layer.

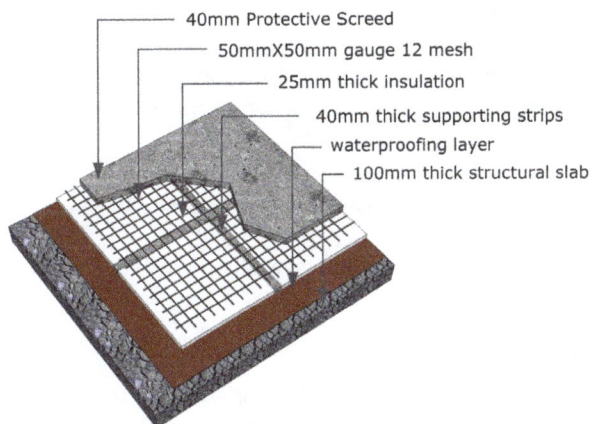

Figure 2. The insulation system with continuous concrete strips by Halwatura and Jayasinghe (2008).

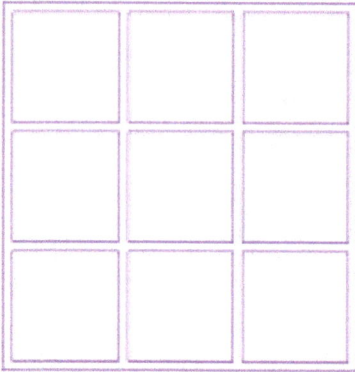

Figure 3. Plan view of the supporting arrangement of the system with continuous concrete strips.

2. Objectives

The primary goal of this study is to develop a new system that is structurally sound with a proper drainage path. The objectives of this study are as enlisted below:

- To investigate the importance of roof slab insulation and performance of existing techniques It was expected to study the importance of roof slab insulation in general, importance of it in tropical climates, existing techniques, their advantages and disadvantages by means of a literature survey. (This was presented in Section 1).
- To develop a new system that is structurally sound with an optimum structural arrangement The structural arrangement was optimized in such a way that it has no restriction for loading. Further, a proper drainage path was provided for the penetrated water to flow out. Most importantly, the thermal performance of the system was studied as well.

3. Methodology

3.1. Overall methodology

A literature survey was carried out to figure out the significance of roof slab insulation in tropical climates, the existing techniques, and their strengths and weaknesses.

A field study was carried out to check and verify the data in literature and to find out the practical issues of those techniques in the operational stage.

Finite element modeling by SAP 2000 software (verified by manual calculations) was used to optimize the structural arrangement. The procedure followed in this optimization process is described in Section 3.2.

Then, actual scale testing was performed to verify the data obtained by computer simulations.

Finally, the possible thermal gain reduction is predicted based on the results of a similar technique in literature.

3.2. The method followed to optimize the strip arrangement

It was decided to provide a proper drainage path along the insulation layer, as it was the major drawback identified in the system with continuous concrete strips. First of the options considered was to remove the supporting concrete strips in one direction, of which a typical plan view is shown in Figure 4. The objective of providing a support was to eliminate the restriction for loading. Hence, the system had to be designed in such a way that it can withstand any practical load applied on that. Therefore, a structural analysis was carried out assuming that the imposed load applied on the system is 5 kN/m², which is the maximum specified in BS 6399-1: 1996 (British Standards Institution, 1996).

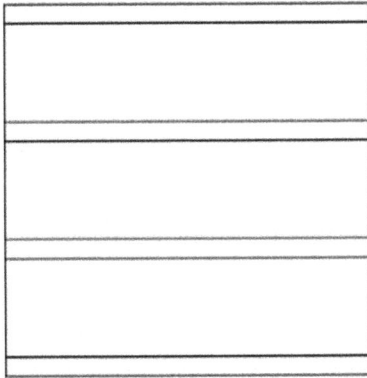

Figure 4. Plan view of the supporting arrangement after removing strips in one direction.

In this context, there were four variables to be considered:

(1) Spacing between strips

(2) Size of the strips

(3) Strength and the mix proportion of the concrete to be used

(4) Reinforcing arrangement in the protective screedAn optimum spacing for the system was to be found out. Because the system would have structurally failed if the strips are placed too far apart to each other, and the thermal performance of the system would have depleted, if they were placed too closer to each other as it increases the concrete area within the insulation layer.

It was intended to optimize the system by minimizing the concrete area within the insulation. Hence, a minimum size of the strips that would bear the load applied on that had to be determined.

The concrete used had to be strong enough to carry the load, and had to be able to be compacted in an area of a width of 40 mm. Hence, concrete with a lower maximum aggregate size (chip concrete) was used. A suitable proportioning was found out by laboratory experiments.

The protective screed had to be designed as a load bearing slab itself. Since concrete is a material which is weak in tension, some arrangement of reinforcement had to be incorporated into the system (Min, Yao, & Jiang, 2014). The bottom reinforcement was fixed to be a 50 mm × 50 mm gauge 12 mesh, due to the convenience of construction. Four options were considered for the top reinforcing arrangement: no reinforcement, 6 mm mild steel bars near supports, 10 mm tor steel bars near supports, and a similar continuous 50 mm × 50 mm mesh (double nets). Then the optimum arrangement was found by computer simulations.

Thereafter, the system was further optimized by varying the spacing between strips while discontinuing them. A typical arrangement of this case is shown in Figure 5. In this case, three more variables were added;

(1) Spacing between strips (Number 1 in Figure 5)

(2) Longitudinal spacing between supports (Number 2 in Figure 5)

(3) Length of the supports (Number 3 in Figure 5)

Different finite element models were developed by varying the spacing of strips, and the optimum arrangement for each value of spacing was found out. This process went on until the top screed becomes a flat slab with a set of blocks as supports.

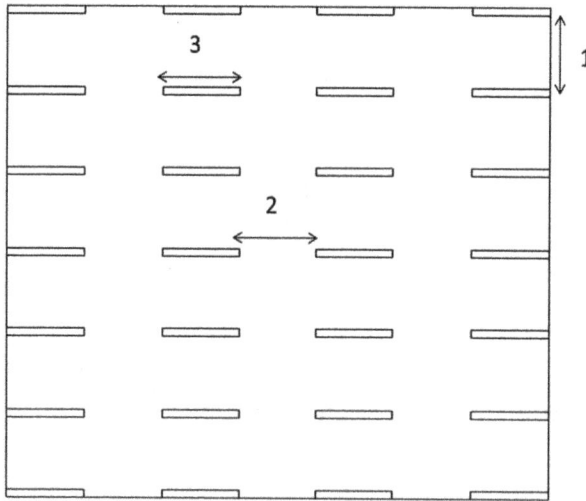

Figure 5. Variables to be considered in optimizing the strips in longitudinal direction.

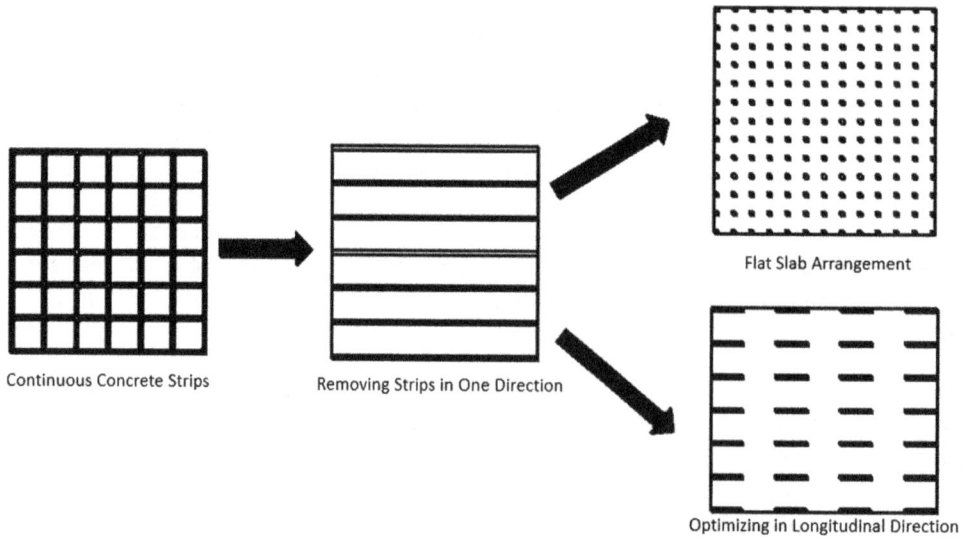

Continuous Concrete Strips

Removing Strips in One Direction

Flat Slab Arrangement

Optimizing in Longitudinal Direction

Figure 6. The process followed to optimize the system.

Then the system with a minimum concrete area within the insulation layer was selected as the best arrangement.

A graphical elaboration on the process followed to optimize the system is depicted in Figure 6.

4. Results

4.1. Step 1: Removing strips in one direction

Initially, the size of strips was fixed to 50 mm, and a concrete strength of 15 N/mm² was assumed for initial trial calculations. Then, the system was analyzed for different values of spacing, and the bearable loadings were calculated for each spacing, and each reinforcing arrangement by a reverse calculation of the procedure explained in BS 8110 part 1:1997 (British Standards Institution, 1997). A typical model developed by SAP 2000 software (the model developed for the arrangement shown in Figure 3) is shown in Figure 7.

Figure 8 shows the results obtained by computer simulations (Only hogging bending moment is shown here as it was the critical parameter). It shows the moment capacities for different top

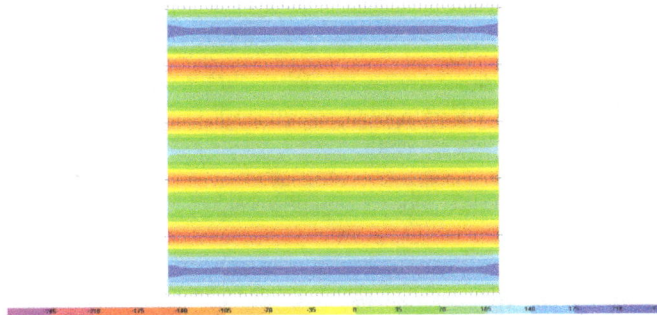

Figure 7. A typical model obtained by computer simulations with strips only in one direction.

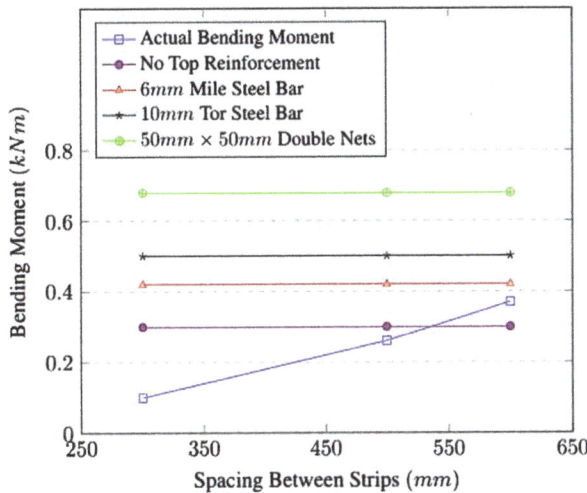

Figure 8. Actual bending moments and moment capacities of different reinforcing arrangements when the strips in one direction are removed.

reinforcing arrangements described in Section 3, and the actual bending moments for different strip-spacings.

Those results suggest that any form of the selected top reinforcement can satisfy the moment resistance required. However, the strips should be spaced in less than 540 mm if no top reinforcement is provided.

This last option was selected for further analysis due to the convenience in construction.

4.2. Step 2: Discontinuing the strips

As it has been mentioned in Section 3, the next step was to find out the optimum arrangements by varying the spacing between strips. A set of computer models were developed for various options of strip-spacing (number 1 in Figure 5), spacing between supports (number 2 in Figure 5) and length of the supports (number 3 in Figure 5). The optimum arrangements obtained for three different values of strip-spacings are presented in Table 1.

Table 1. The optimum arrangements obtained for different values of strip-spacing

Spacing between strips (mm)	Spacing between supports (mm)	Length of the supports (mm)
("1" in Figure 5)	("2" in Figure 5)	("3" in Figure 5)
300	400	300
400	300	300
500	100	200

4.3. Flat slab arrangement

The next option was to minimize the size of the supports and to support the system by a set of blocks. In this case, the protective screed layer behaves as a flat slab. Figure 9 shows the actual bending moments and the moment capacities for different block-spacings with a 50 mm × 50 mm gauge 12 mesh as reinforcement.

The results suggest that it is possible to achieve the required structural capacity, if the blocks are spaced at 150 mm or less in both directions.

4.4. Step 4: Selecting a suitable width of the strips/supports

Since the height of the supporting strips is small in comparison with its cross-sectional area, the buckling failure was ruled out. Hence, the minimum width required was calculated by a compressive strength calculation. The results obtained are shown in Table 2.

Results show that a minute width is sufficient to carry the load. However, a minimum width of 25 mm is selected owing to the practicality of construction.

4.5. Step 5: Selecting the best system

The next step was to single out a system out of the four options short-listed (shown in Table 3). Since the intention is to optimize the system, it was intended to minimize the concrete are in the layer, since concrete increases the composite conductivity of the layer as it is not an insulation material.

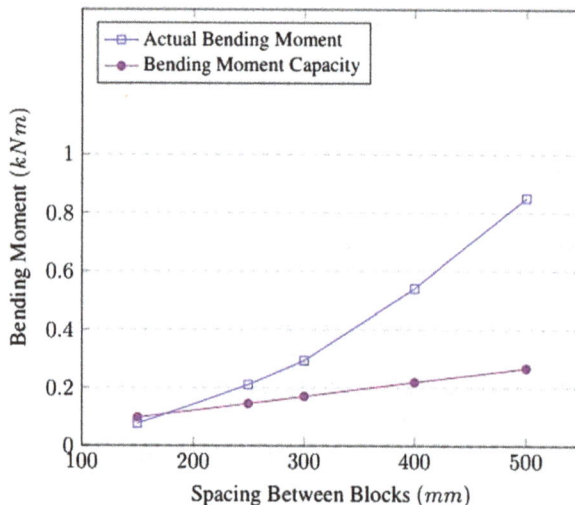

Figure 9. Bending moments and moment capacities of the protective screed with a 50 mm × 50 mm gauge 12 Mesh for a flat slab arrangement with different support spacings.

Table 2. Calculations for finding minimum width of strips				
Calculation	300 mm spacing between strips	400 mm spacing between strips	500 mm spacing between strips	Flat slab arrangement
Effective area (m²)	0.21	0.24	0.15	0.04
Dead load (kN)	0.21	0.24	0.15	0.04
Live load (kN)	1.05	1.20	0.75	0.20
Design load (kN)	1.97	2.26	1.41	0.38
Minimum area of support (mm²)	131	150	94	25
Minimum required width (mm)	0.44	0.50	0.47	5.00

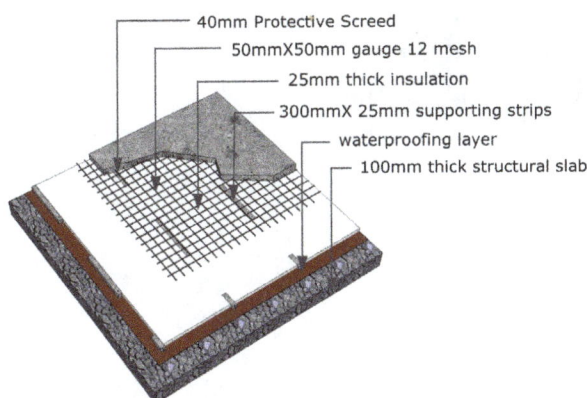

Figure 10. Isometric view of the derived system.

The concrete/total area ratio of the existing system is shown in Table 3 for comparison purpose. It clearly shows that all the systems selected have a much lower concrete area than the existing system. The system with the lowest concrete area (the system with 400 mm strip-spacing) was selected as the best system. An isometric view of this system is shown in Figure 10.

4.6. Step 6: Selecting a suitable concrete mix
The other variable fixed in Section 3 was the mix proportion of the concrete used. Since the supporting strips of the selected system are only 25 mm thick, it was necessary to specify a lower maximum aggregate size for the concrete. As chipped metal (with a maximum size of 10 mm) is a common construction material, a mix design was performed to achieve the assumed strength of 15 N/mm^2. Several options were considered by varying the Water–Cement ratio from 0.6 - 0.8.

All the tested mixes gained the required strength of 15 N/mm^2. Hence, the mix tried out for a water–cement ratio of 0.7, with 1:2:3 volume proportion of cement, sand and metal respectively was selected as the suitable mix proportion due to the convenience of specifying in practice.

4.7. Physical model testing
The next step was to check the strength of the system by physical model testing. The system was loaded with a calibrated proving ring to measure the applied load, and the deflection was measured with a dial gauge. The experimental setup used is shown in Figure 11. Both the readings were continuously taken down till the system failed. The obtained load-deflection curve is shown in Figure 12.

The graph in Figure 12 shows that the system can be loaded up to about 30 kN without cracking, and the system can be loaded up to 37 kN (approximately 4MT) without failing structurally. This value is higher than any practical load specified in BS 6399-1: 1996 (British Standards Institution, 1996). Therefore, it is proven that this system is structurally sound.

Table 3. Calculations for finding minimum width of strips					
Calculation	300 mm Spacing	400 mm Spacing	500 mm Spacing	Flat slab	Existing system (Halwatura & Jayasinghe, 2008)
Concrete area (m^2)	3.71	3.51	3.57	5.61	16.16
Total area (m^2)	97.5	105.5	101.5	99.0	100.8
Concrete ratio	3.8%	3.3%	3.5%	5.7%	16.0%

Figure 11. Experimental setup of the actual scale testing.

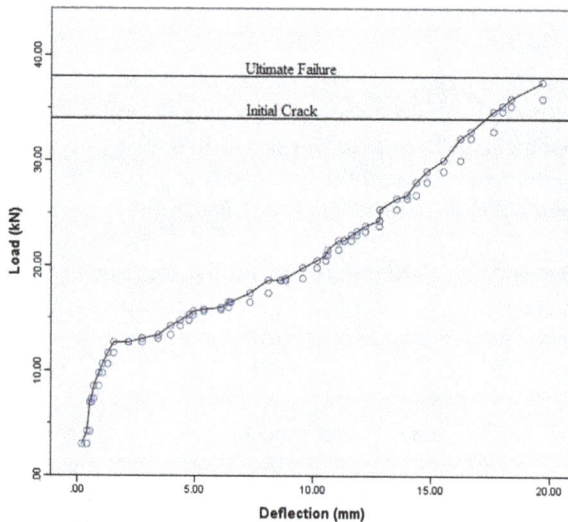

Figure 12. The graph of load vs. deflection obtained by actual scale testing.

Even though the deflection showed higher values, it was observed that the top screed sags independently without affecting the structural slab. Hence, it was not considered as a serviceability failure of the system.

5. Discussion

The necessity of a supporting arrangement for the top screed was to make the insulation system structurally sound, eliminating the restriction for loading the roof slab. The system designed addressing that issue had a durability issue since a drainage path was not provided for the penetrated water. The major aim of developing a new system was to address those two issues.

The actual scale testing in Section 4.7 proved that it is structurally sound, by withstanding a 4 MT point load on the system. Since the supporting strips are discontinued, a drainage path is provided within the insulation layer, addressing the issue of durability.

However, since this is a thermal insulation system, it is necessary to compare the thermal performance of the system with existing techniques.

Table 4. Comparison of thermal conductivities of the new system and the existing system

System	Composite conductivity of the insulation layer $(Wm^{-1}K^{-1})$	Composite conductivity of the system $(Wm^{-2}K^{-1})$
The existing system (Halwatura & Jayasinghe, 2008)	0.039	1.1
Newly designed system	0.034	1.0

An adjustment for thermal conductivity values was necessary to be made since the insulation layer consists of a set of concrete supports within the insulation layer. The adjustment was made according to Equation (1) (Progelhof, Throne, & Ruetsch, 1976).

$$\frac{1}{K_I} = \frac{1-\phi}{K_p} + \frac{\phi}{K_c} \tag{1}$$

where K_I is the thermal conductivity of the composite insulation layer $(Wm^{-1} K^{-1})$; K_p is the thermal conductivity of the insulation material – polystyrene $(Wm^{-1} K^{-1})$; K_c is the thermal conductivity of concrete $(Wm^{-1} K^{-1})$; ϕ is the volume fraction of concrete.

A comparison of the conductivity values between the newly designed system and the system by Halwatura and Jayasinghe is shown in Table 4 (Please see Appendix A for detailed calculations). It shows a 9% reduction of heat transfer in the new system.

The system by Halwatura and Jayasinghe has proven to achieve a 75% heat reduction into the buildings (Halwatura & Jayasinghe, 2008). Table 4 shows that the newly designed system has a lower thermal conductivity. Hence, the newly designed system should theoretically have a heat gain reduction of more than 75%.

A few limitations of this study can be identified. First, the optimization technique used is simple as it is performed by varying the structural arrangement and mix proportions. A further optimization may be performed by replacing the materials with newly invented, more effective materials, either as the insulator or as a structural element.

In this study, the thickness of the insulation material was taken to be 25 mm. Thus, the only mode of failure of the supporting strips considered was crushing. However, if a researcher intends to vary the thickness of the insulation, other modes of failure like buckling should be considered depending on the thickness considered.

6. Conclusions

Roof slab insulation is very significant to mitigate and adapt to global warming. Developing an insulation system that is thermally effective, structurally sound and durable was the main objective of this study. A system that contains an insulation layer on top of the structural slab and a protective screed on top of it, which is supported by a set of discontinuous concrete strips was selected as the option to develop. The discontinuity of the strips provides a drainage path, making the system more durable in comparison with similar existing techniques, while the concrete strips provide the structural stability. The optimization was performed by finite element modeling. It was found out that a set of 300 mm × 25 mm strips in 300 mm longitudinal clear spacing and 400 mm transverse spacing cast by 1:2:3 chip–concrete with a water–cement ratio of 0.7 can withstand an imposed load of 5 kN/m², which is the maximum specified for a roof. An actual scale physical model withstood a 4MT-point load, emphasizing that the system is structurally sound. A calculation of composite conductivities and a comparison with an existing system has proven that this system can reduce the heat gain into a building by more than 75%.

Funding

This work was financially supported by Senate Research Committee, University of Moratuwa [grant number SRC/LT/2011/17].

Author details

Kasun Nandapala[1]

E-mails: mpkcnandapala@gmail.com, kasuncn@uom.lk

ORCID ID: http://orcid.org/0000-0003-4088-2987

Rangika Halwatura[1]

E-mails: rangikauh@gmail.com, rangika@uom.lk

ORCID ID: http://orcid.org/0000-0002-9206-5607

[1] Department of Civil Engineering, University of Moratuwa, Moratuwa, Sri Lanka.

References

Al-Homoud, M. S. (2005). Performance characteristics and practical applications of common building thermal insulation materials. *Building and Environment, 40,* 353–366. Retrieved from http://www.sciencedirect.com/science/article/pii/S0360132304001878

Al-Obaidi, K. M., Ismail, M., & Abdul Rahman, A. M. (2014). Passive cooling techniques through reflective and radiative roofs in tropical houses in Southeast Asia: A literature review. *Frontiers of Architectural Research, 3,* 283–297. Retrieved from http://www.sciencedirect.com/science/article/pii/S2095263514000399

Alvarado, J. L., & Martínez, E. (2008). Passive cooling of cement-based roofs in tropical climates. *Energy and Buildings, 40,* 358–364. Retrieved from http://www.sciencedirect.com/science/article/pii/S0378778807000965

Alvarado, J. L., Terrell, Jr., W., & Johnson, M. D. (2009). Passive cooling systems for cement-based roofs. *Building and Environment, 44,* 1869–1875. Retrieved from http://www.sciencedirect.com/science/article/pii/S0360132308003028

Banting, D. (2005). *Report on the environmental benefits and costs of green roof technology for the city of Toronto.* Retrieved from http://www.torontopubliclibrary.ca/detail.jsp?Entt=RDM396307\ &amp;R=396307

Berardi, U., GhaffarianHoseini, A., & GhaffarianHoseini, A. (2014). State-of-the-art analysis of the environmental benefits of green roofs. *Applied Energy, 115,* 411–428. Retrieved from http://www.sciencedirect.com/science/article/pii/S0306261913008775

British Standards Institution. (1996). *Loading for buildings, Part 1.* London: Author.

British Standards Institution. (1997). *Structural use of concrete.* London: Author.

Brito Filho, J. P., & Oliveira Santos, T. V. (2014). Thermal analysis of roofs with thermal insulation layer and reflective coatings in subtropical and equatorial climate regions in Brazil. *Energy and Buildings, 84,* 466–474. Retrieved from http://www.sciencedirect.com/science/article/pii/S0378778814006859

Dareeju, B. S. S. S., Meegahage, J. N., & Halwatura, R. U. (2011). Indoor thermal performance of green roof in a tropical climate. In *International Conference on Building Resilience.* Retrieved from https://www.researchgate.net/profile/Rangika_Halwatura/publication/278032545_97_impact_of_green_roofs_on_urban_heat

Dimoudi, A., Androutsopoulos, A., & Lykoudis, S. (2006). Summer performance of a ventilated roof component. *Energy and Buildings, 38,* 610–617. Retrieved from http://www.sciencedirect.com/science/article/pii/S0378778805001957

Dong, B., Lee, S. E., & Sapar, M.H. (2005). A holistic utility bill analysis method for baselining whole commercial building energy consumption in Singapore. *Energy and Buildings, 37,* 167–174. Retrieved from http://www.sciencedirect.com/science/article/pii/S0378778804001811

Dylewski, R., & Adamczyk, J. (2014). 12 - Life cycle assessment (LCA) of building thermal insulation materials. In F. Pacheco-Torgal, L. F. Cabeza, J. Labrincha, & A. de Magalhães (Eds.), *Eco-efficient construction and building materials* (pp. 267–286). Woodhead Publishing. Retrieved from http://www.sciencedirect.com/science/article/pii/B9780857097675500121

Halawa, E., van Hoof, J., & Soebarto, V. (2014). The impacts of the thermal radiation field on thermal comfort, energy consumption and control---A critical overview. *Renewable and Sustainable Energy Reviews, 37,* 907–918. Retrieved from http://www.sciencedirect.com/science/article/pii/S1364032114003657

Halwatura, R. U. (2013). Effect of turf roof slabs on indoor thermal performance in tropical climates: A life cycle cost approach. *Journal of Construction Engineering, 2013,* e845158. Retrieved from http://www.hindawi.com/journals/jcen/2013/845158/abs/

Halwatura, R. U., & Jayasinghe, M. T. R. (2007). Strategies for improved micro-climates in high-density residential developments in tropical climates. *Energy for Sustainable Development, 11,* 54–65. Retrieved from http://www.sciencedirect.com/science/article/pii/S097308260860410X

Halwatura, R. U., & Jayasinghe, M. T. R. (2008). Thermal performance of insulated roof slabs in tropical climates. *Energy and Buildings, 40,* 1153–1160. Retrieved from http://www.sciencedirect.com/science/article/pii/S0378778807002411

Halwatura, R. U. (2014). Performance of insulated roofs with elevated outdoor conditions due to global warming. *Journal of Environmental Treatment Techniques, 2,* 134–142.

Kwong, Q. J., Adam, N. M., & Sahari, B. B. (2014). Thermal comfort assessment and potential for energy efficiency enhancement in modern tropical buildings: A review. *Energy and Buildings, 68,* 547–557. Retrieved from http://www.sciencedirect.com/science/article/pii/S0378778813006166

Lean, J., & Rind, D. (2001). Sun-climate connections. Earth's response to a variable sun. *Science, 292,* 234–236.

Macilwain, C. (2000). Global-warming sceptics left out in the cold. *Nature, 403,* 233–233. Retrieved from http://www.nature.com/nature/journal/v403/n6767/full/403233a0.html

Megri, A. C., Achard, G., & Haghighat, F. (1998). Using plastic waste as thermal insulation for the slab-on-grade floor and basement of a building. *Building and Environment, 33,* 97–104. Retrieved from http://www.sciencedirect.com/science/article/pii/S0360132397000292

Min, F., Yao, Z., & Jiang, T. (2014). Experimental and numerical study on tensile strength of concrete under different strain rates. *The Scientific World Journal, 2014,* e173531. Retrieved from http://www.hindawi.com/journals/tswj/2014/173531/abs/

Nandapala, K., & Halwatura, R. (2014). Prioratizing effective means of retrofitting flat slabs to meet public demands in order to promote sustainable built environment. *Proceedings of the Special Session on Sustainable Buildings and Infrastructure* (Vol. 1174--1180). Kandy.

Parker, D. S., & Barkaszi, Jr., S. F., (1997). Roof solar reflectance and cooling energy use: Field research results from Florida. *Energy and Buildings, 25,* 105–115. Retrieved from http://www.sciencedirect.com/science/article/pii/S0378778896010006

Progelhof, R. C., Throne, J. L., & Ruetsch, R. R. (1976). Methods for predicting the thermal conductivity of composite systems: A review. *Polymer Engineering & Science, 16,* 615–625. Retrieved from http://onlinelibrary.wiley.com/doi/10.1002/pen.760160905/abstract

Romeo, C., & Zinzi, M. (2013). Impact of a cool roof application on the energy and comfort performance in an existing non-residential building. A Sicilian case study. *Energy and Buildings, 67*, 647–657. Retrieved from http://www.sciencedirect.com/science/article/pii/S0378778811003288

Sadineni, S. B., Madala, S., & Boehm, R. F. (2011). Passive building energy savings: A review of building envelope components. *Renewable and Sustainable Energy Reviews,* *15*, 3617–3631. Retrieved from http://www.sciencedirect.com/science/article/pii/S1364032111002504

Vijaykumar, K. C. K., Srinivasan, P. S. S., & Dhandapani, S. (2007). A performance of hollow clay tile (HCT) laid reinforced cement concrete (RCC) roof for tropical summer climates. *Energy and Buildings, 39*, 886–892. Retrieved from http://www.sciencedirect.com/science/article/pii/S0378778806002507

Appendix A

Calculating thermal conductivity of the insulation layer

For the Newly Designed System, (The assumed thermal conductivities are shown in Table A1)

Table A1. Thermal conductivities of the materials used	
Material	**Thermal conductivity (Halwatura & Jayasinghe, 2007)**
Concrete	$1.7\,W/m^{-1}K^{-1}$
Polystyrene	$0.033\,W/m^{-1}K^{-1}$

$$\frac{1}{K_I} = \frac{1-\phi}{K_p} + \frac{\phi}{K_c}$$
$$= \frac{1-3.3\%}{0.033} + \frac{3.3\%}{1.7}; \quad (\phi = 3.3\% \text{ by Table 3})$$
$$K_I = 0.034 Wm^{-1}K^{-1}$$

For the Existing System,

$$\frac{1}{K_I} = \frac{1-\phi}{K_p} + \frac{\phi}{K_c}$$
$$= \frac{1-16\%}{0.033} + \frac{16\%}{1.7}; \quad (\phi = 16\% \text{ by Table 3})$$
$$K_I = 0.039 Wm^{-1}K^{-1}$$

Calculating thermal conductivities of the systems themselves

Table A2. Surface resistances of roof slab (Halwatura & Jayasinghe, 2008)		
Location	**Symbol**	**Surface resistance**
Top surface	R_T	0.04
Soffit	R_S	0.14
Insulation system	R_I	(calculated above)

Thermal Resistance of the New System $= \dfrac{T_1}{K_1} + \dfrac{T_2}{K_2} + \dfrac{T_3}{K_3}$; $(T_i -$ Thickness of the layer$)$

$$= \dfrac{0.04}{1.7} + \dfrac{0.025}{0.034} + \dfrac{0.1}{1.7}$$

$$= 0.82 m^2 KW^{-1}$$

Air-to-Air Resistance of the New System $= R_T + R_I + R_S$

$$= 0.04 + 0.82 + 0.14$$

$$= 1.0 m^2 KW^{-1}$$

Hence, the Composite Conductivity of the newly designed system $= \dfrac{1}{1.0}$

$$= 1.0 Wm^{-2}K{-}1$$

Thermal Resistance of the Existing System $= \dfrac{T_1}{K_1} + \dfrac{T_2}{K_2} + \dfrac{T_3}{K_3}$

$$= \dfrac{0.04}{1.7} + \dfrac{0.025}{0.039} + \dfrac{0.1}{1.7}$$

$$= 0.72 m^2 KW^{-1}$$

Air-to-Air Resistance of the Existing System $= R_T + R_I + R_S$

$$= 0.04 + 0.72 + 0.14$$

$$= 0.90 m^2 KW^{-1}$$

Hence, the Composite Conductivity of the existing system $= \dfrac{1}{0.9}$

$$= 1.1 Wm^{-2}K{-}1$$

4

Estimating a municipal water supply reliability

O.G. Okeola[1]* and S.O. Balogun[2]

*Corresponding author: O.G. Okeola, Department of Water Resources & Environmental Engineering, University of Ilorin, Ilorin, Nigeria

E-mail: ogolayinka@unilorin.edu.ng

Reviewing editor: Sarah Bell, University College London, UK

Abstract: The availability and adequacy of water in a river basin determine the design of water resources projects such as water supply. There is a further need to regularly appraise availability of such resource for municipality at a distant future to help in articulating contingent plan to handle its vulnerability. This paper attempts to empirically determine the reliability of water resource for a municipal water supply. An approach was first developed to estimate municipality water demand that lack socioeconometric data using a purpose-specific model. Hydrological assessment of river Oyun basin was carried out using Markov model and sequent peak analysis to determine the reliability extent for the future demand need. The two models were then applied to Offa municipality in Kwara state, Nigeria. The finding revealed the reliability and adequacy of the resource up till year 2020. The need to start exploring a well-coordinated conjunctive use of resources is recommended. The study can serve as an organized baseline for future work that will consider physiographic characteristics of the basin and climatic dynamics. The findings can be a vital input into the demand management process for long-term sustainable water supply of the town and by extension to urban township with similar characteristic.

Subjects: Civil, Environmental and Geotechnical Engineering; Engineering Economics; Engineering Management

Keywords: catchment; water demand; model; empirical; population

1. Introduction and literature review

1.1. Water resources assessment

Reservoir system is a requisite requirement for most urban water supply. Reservoirs for water supply are generally operated by a set of predetermined rules formulated on the basis of historical inflow, design storage capacity, and safe yield criteria (Moy, Cohon, & ReVelle, 1986). Inflow hydrology is an

ABOUT THE AUTHOR

O.G. Okeola is a senior research fellow at the Department of Water Resources & Environmental Engineering, University of Ilorin, Nigeria. He holds a PhD in water resources management. His research interests focus on stochastic and deterministic hydrology, decision analysis, water supply management, and water policy. He has written instructional monograph and joint technical papers. He is a member of the American Society of Civil Engineer (ASCE), International water Association (IWA), Nigerian Society of Engineers (NSE), American Institute of Steel Construction (AISC) and a Registered Engineer with the Council for Regulation of Engineering in Nigeria (COREN).

PUBLIC INTEREST STATEMENT

This work looked in to the reliability of water resource for a municipal water supply usage in Kwara State of Nigeria. An approach was first developed to estimate municipality water demand. Thereafter, hydrological assessment of River Oyun was carried out using Markov model and sequent peak analysis to determine the reliability extent for the future demand need. The findings indicated the reliability and adequacy of the resource up till year 2020, but for domestic usage only. The need to start exploring a well-coordinated conjunctive use of resources is imperative.

important aspect of a reservoir operation studies. Seasonal demand may be relatively fixed, but in contrast, variation in natural streamflow between seasons may be highly variable. In evaluating reliability of water supply system, the availability of water must be compared with water demand. This is usually done by simulating water supply system using historic streamflows and anticipated demands. Fiering (Vogel, 1987) documented three principal shortcomings of strict use of the historical records, thus: (1) the analysis is based solely on the historical record and it is unlikely that some flow sequence will recur during the active life of the completed structure (2) the mass diagram does not help designer to establish or calculate the risk to be taken with regard to water shortages during period of low flow, and (3) the length of the historical record is likely to differ from the economic life of the proposed structure.

Sharma, Tarboton, and Lall (1997) concluded that it is very important to generate synthetic streamflow sequences to analyze alternative designs, operation policies, and rules for water resources systems and that the dependence structure of streamflow sequences is often assumed to be Markovian. In an earlier study, Itube, Mejia, and Dawdy (1972) and Yurekli and Ozturk (2003) noted that generating extreme values is most significant in design and planning. For general surface water resources systems, mathematical optimization techniques have been applied to the study of reservoir planning management as related to urban water supply in a single purpose reservoir and including irrigation, flood control, hydropower, etc. in a multipurpose reservoir with varying degree of success (Barnes & Chung, 1986; Guariso, Rinaldi, & Soncini-Sessa, 1986; Karamouz, Szidarovszky, & Zahraie, 2003; Martin, 1987; Moy et al. 1986; Strycharczyk & Stedinger, 1987).

Water resources infrastructure and policies are designed in part to reduce the risks inherent to climatically driven, variable water resources systems and are built for adaptation to climate variability (Kiparsky, Joyce, Purkey, & Young, 2014). The water supply planning and management philosophy has been based largely on the concept of firm yield which should exceed demand by some reasonable margin of safety (Cabezas & Wurbs, 1986). This work followed this philosophy in its water resources availability assessment approach and was based on the yield that can be achieved, and other reliabilities, given a streamflow sequence. The persistence of high flows and of low flows, often described by their correlation, affects the reliability with which a reservoir of a given size can provide a specified yield (Loucks, Stedinger, & Haith, 1981).

1.2. Water demand

Estimating and forecasting water demand become necessary as the urban population dependent on public water supplies increases rapidly and new demands for water are not easily met. Considerable efforts have been put into the development of urban water supply projections in the last four decades resulted in a wealth of understanding and sophisticated forecasting techniques in this field. Different kinds of data-sets have been used ranging from household data to aggregated data. The quantity and type of data available determine which forecasting method should be considered for application. There is no absolute level of accuracy that is appropriate in all demand forecasting situations. However, it is important to understand the key determinants of water usage. According to Nieswiadomy (1992), a consensus on the proper estimation methodology has not been reached.

Arbués, García-Valiñas, and Martínez-Espiñeira (2003) give a state-of-the-art review on estimation of residential water demand with distinct attention to variables, specification model, data-set, and most common econometric problems. Wurbs (1994) earlier made a general characterization of water use forecasting by (1) the level of complexity of the mathematical relationships between water use and explanatory variables and (2) the level of sectoral, spatial, seasonal, and other disaggregation of water users. The complexity of the relationships, however, depends primarily on how many and which explanatory variables are included in the equations. A large number of studies of the demand for urban water have appeared in the literature since the classic Howe and Linaweaver study of 1967 (Martin & Thomas, 1986). The approach most widely used for water forecasting is the per capita method, which assumes that the population is the single explanatory variable. It provides adequate explanation on water use and assumes other variables to be unimportant or perfectly correlated with population.

Other methods improve on this by considering many factors such as price, income, housing type, household size, and climate that are known to affect water use. The use of multivariate model reduces the degree of subjectivity in the analysis and makes better use of available data. The method includes variables observe to be significantly correlated with water usage and not necessarily those suggested by a priori economic reasoning. The disadvantages are that data requirements may be considerable and may be difficult to collect. These models reflect more of correlation rather than causation and consequently may omit potentially important relationship. The econometric demand model however differs in that they are based on economic reasoning and include only variables which are either expected to be causally related to or found to be significantly correlated with water usage. Econometric demand models are available mostly for residential water use and in the developed economies. A variety of econometric techniques which are based on semi-log equations are used to model residential water demand. These include ordinary least square, two and three stages least squares (2SLS, 3SLS), instrumental variables, generalized least squares, log–log model, double-log model, and discrete/continuous choice (DCS) (see e.g. Hussain, Thrikawala, & Barker, 2002; Mazzanti & Montini, 2006; Mitchell, 1999; Olmstead, Michael Hanemann, & Stavins, 2007; Schleich & Hillenbrand, 2009)

Stephenson and Randel (2003) estimate Rand water's demand in South Africa for 2000–2020 using statistical approach and obtained confidence limits for individual sectors. A decomposed demographic model was developed and used in predicting water demand until 2020. A wider uncertainty in future consumption was foreseen. Therefore, conservative planning was recommended for future water resource projects. This was to be achieved through low capital and operating intensive schemes. Cochran and Cotton (1985) used multiple regression models in a municipal water demand study for Oklahoma City and Tulsa, Oklahoma. The results indicate that price and per capita income were predictive variables for Oklahoma City's water demand, while only per capita income was found to be a predictor for consumption in Tulsa. Mimi and Smith (2000) and Khadam (1984) employed this approach in water demand studies for Rammallah and Khartoum, respectively. Both studies also found price and size of household significant, but the later was inversely (i.e. as household sizes increases, per capita water use decreases). Mylopoulos, Mentes, and Theodossiou (2004) applied a cubic functional form of an econometric model to study a residential water demand which allows the use of different price elasticities for different levels of water demand. The data used for the econometric analysis were obtained through a survey of consumers in the city of Thessaloniki, Greece. Panel estimation methods were then employed to estimate model parameters. The results showed that a cubic form of the demand equation can provide appropriate estimates of price elasticities for different "consumption groups" of residential customers.

The traditional method of forecasting industrial water use is the water requirement approach (Hanemann, 1998). This approach postulates that water use in an industrial establishment varies proportionately with the scale of production in that establishment. Scale is measured in terms of physical units of output, monetary value of output, or the size of labor force employed. There are two approaches. The first approach is a constant factor of proportionality which leads to the following forecasting Equations 1 and 2. In this approach, α_i and β_i are treated as constants. They vary by industry, i, but are fixed over all the establishments in an industry.

$$X_i = \alpha_i y_i$$

(1)

Alternatively,

$$X_i = \beta_i E_i$$

(2)

Table 1. Typical demand figure for commercial and institution establishments in urban areas	
Usage	Demand allowance
Small businesses, shops and offices	Up to 35 liter/capita/day (applied as per capita allowance to the whole urban population)
Offices	65 liters/day/employee*
Departmental stores	100–135 liters/day/employee*
Hospitals	350–500 liters/day/bed
Hotels	250 liters/day/bed
Schools	25–75 liters/day/pupil*

*Note: These figures should only be applied when the above are operating or open.
Source: Walliingford Ltd (2003).

where X_i is the water intake in an establishment in the ith type of industry, y_i is the production by the establishment, α_i is the water intake per unit of output in the ith type of industry, E_i is the number of employees in the establishment, β_i is the water intake per employee in the ith type of industry.

The second approach is more sophisticated. It relaxes the assumption of strict proportionality and postulates as follow:

$$X_i = \alpha_i y_i^{\gamma} \tag{3}$$

Alternatively,

$$X_i = \beta_i E_i^{\gamma} \tag{4}$$

where γ may or may not vary with industry i. Water use increases less than proportionately with scale of production if $\gamma < 1$, and more than proportionately if $\gamma > 1$. Dziegielewski (1988) found that a value of $\gamma = 0.7$ fit the data well for the US manufacturing industry. Several studies have reportedly found the number of employees to be highly correlated with water demand and therefore, in a unit use approach, may be used to estimate a water coefficient for a group of establishment (Cook, Urban, Maupin, Pratt, & Church, 2001).

For urban institutional and commercial water demand estimate where there is no available metered record, one estimation method is to apply a demand allowance on a per capita basis for various institutions and commercial buildings. Typical allowances for commercial and institutional establishment are as shown in Table 1. These allowances assume piped water connections and waterborne sanitation, and should be adjusted down where the establishments have a lower level of service for instance, standpipes, hand pumps, or VIP latrines in schools. Most of the sophisticated statistical and econometrical models are not entirely applicable in developing countries albeit the characteristic of developing economy. Oyegoke and Oyesina (1984) contend that in estimating design figures for water demand in developing countries, the philosophy should be to provided first for the basic needs and then incorporate various factors that may affect demand in the particular situation.

2. Water demand model formulation

The objective is to find simple relationship which accounts for as much of the variability of demand as possible. Hence, in the formulation of the model, water demand is addressed as basically non-irrigation demand which include the following principal determinant components typical of urban water requirement: residential, industrial, commercial, institution, and system losses. The forecasting relationship to estimate water demand is based on specific assumption reflecting the following local situation common in most developing countries:

- urban water uses are predominantly residential and commercial,
- water use is not metered, and
- fixed rates, independent on amount consumed, thus quantity has no correlation and causation with price.

The model is formulated to take into account the major uncertainty associated with water demand which is the population. Therefore, the per capita method is used and estimate of principal components incorporated. The justification for this are: (1) paucity of long-period socioeconomic data; and (2) no large contingents of seasonal residents. The total urban water demand U_{wd} forecast model is given in Equation 5.

$$U_{wd} = \gamma \left(x \left(P_t (1+r)^n \right) + \sum_{j=1}^{m} \sum_{i=1}^{k} q_{ji} b_i \right) \tag{5}$$

where P_t is the population at present time t; U_{wd} is the water demand in cubic meter per day in year n; r is the rate of growth of population; n is the length of time for which the projection is made; x is the per capita water requirement in cubic meter for domestic/residential use; γ is a factor greater than 1 for system losses and contingent usage; q_{ji} is the estimated water requirement of establishment i per employee, pupil, or bed space in category j; b_i is the number of employees, pupils, bed spaces, etc.; m is the number of principal determinant components ($m = 1$ for commercial, $m = 2$ for institution, $m = 3$ for industrial); k is the number of establishment in commercial/industrial/institution categories.

3. Hydrological model formulation

There are two key approaches to simulate the variability of water resources potential in a hydrosystem including the uncertainty resulting from this. The first is based on historical time series while the other stochastic based. In other, to stimulate Oyun River Water Resources System, a synthetic hydrological system was generated using Markov model. A number of stochastic models have been considered in the literature for synthetic generation and forecasting of hydrological processes (Fortin, Perreault, & Salas, 2004; Karamouz et al., 2003; Loucks et al., 1981; Philipose & Srinivasan, 1995).

Hydrologic processes such as streamflow are oftenly well represented by stationary linear models such as autoregressive (AR) and autoregressive moving average models. These models are usually capable of preserving the historical statistics such as mean, variance, skewness, and covariance (Fortin et al., 2004; Philipose & Srinivasan, 1995; Shaw, 1988). The basic form of the AR model of order p, (AR, p), with constant parameters is (Philipose & Srinivasan, 1995):

$$Z_{v,\tau} = \sum_{j=1}^{p} \phi_{j,\tau} Z_{v,(\tau-j)} + \varepsilon_{v,\tau} \tag{6}$$

where the subscript v and τ denote the year and the period, respectively, $\{Z_{v,\tau}\}$ is the time series suitably transformed and standardized and has an expected value equal to zero, $\phi_{j,\tau}$ are the AR parameters, and $\{\varepsilon_{v,\tau}\}$ is the error term (white noise) and assumed to be uncorrelated.

The adequacy of the flow in Oyun River and the impoundment to meet projected demand was assessed using hydrologic time series modeling approach. The model uses monthly step and has its stochastic component handled by representing the inflow as a Markov process and the fitting of probability distributions to the inflows. The statistical parameters of the underlying population of historical streamflow are first derived and subsequently used to develop a stochastic model of reservoir inflows. The stochastic streamflow model provides large number similar sequences that are used to estimate the reliability with which a storage reservoir can deliver prescheduled quantities of water (Philipose & Srinivasan, 1995). The monthly streamflow is predicted up to the year 2020 using

Markov. The 50 years inflows are used in the (1) estimation of the yield of the River Oyun (2) reliability of flow, and (3) the development of storage–yield function. The yield of the Oyun reservoir was subsequently determined using sequent peak method. The sequent peak model can be used to trade off capacity and yield given a mode of operation that does not allow failure (Moy et al., 1986).

4. Application of the models to study area

River Oyun is the source of water supply and the catchment (Figure 1) is oblong in shape and it is very long compared with its breadth. The climate of the catchment is the common type with the tropical savannah grassland of Africa. There is not much climatic variation and hence the hydrologic variation in the catchments is also insignificant (Adebosin, 1986). The water demand model was applied to water supply scheme in Offa Township, the headquarters of Offa local Government Area (LGA) in Kwara state. The Offa township is completely inside the catchment which lies entirely inside Kwara

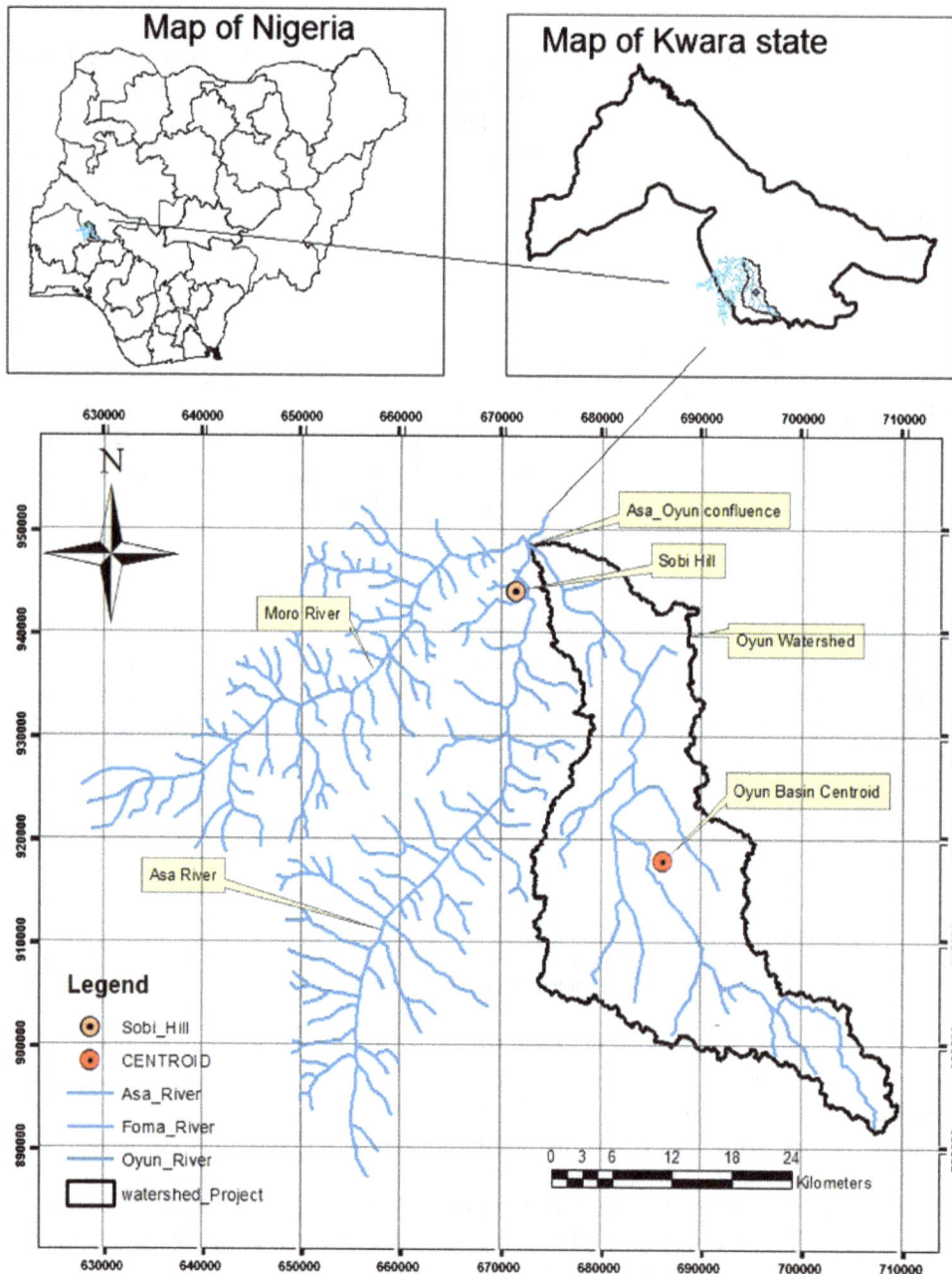

Figure 1. Map of Nigeria showing study area and delineated river Oyun catchment.

Table 2. Salient features of Oyun reservoir and dam	
Dam height	9.8 m
Location	6 km north of Offa
Catchment area (Adedayo, 2014)	830 km²
Spillway crest elevation	406.58 m.a.s.l
Spillway length	80 m
Embankment	409.5 m.a.s.l
Gross storage volume	3500000 m³
Net storage volume	2900000 m³

Source: KWWC (2009).

State of Federal Republic of Nigeria (Figure 1) between latitudes 8° 38′ and 9° 50′ N and between longitudes 8° 03′ and 8° 15′ E. Table 2 shows the salient features of Oyun reservoir and dam for a single purpose municipal water supply.

The yield of the Oyun reservoir was subsequently determined using sequent peak method. The sequent peak model can be used to trade off capacity and yield given a mode of operation that does not allow failure (Moy et al., 1986). Primary data on the customer categories connected to the public utility are collected from Kwara State Water Corporation (KWWC), the state-owned corporation responsible for municipal water supply. The standard water usages (Table 1) were used in the estimation of the principal components of water demand in the model based on the primary data and on the available numbers of individual components. The urban water demand was estimated as given in the model (Equation 5) and computed using C++ code.

5. Results and discussion

5.1. Water demand
The domestic, commercial, industrial, and institutional water demands between the periods 1996 and 2008 are estimated. The standard estimated water usage values in Table 1 are used in the estimation of principal components of water demand based on the available confirmed numbers in individual component. The domestic estimates are based on 120 lcpd and 2.83% annual population growth rate. The population projection was based on 1991 estimate (Okeola, 2000). Figure 2 shows the trend in the different categories of water demand for the period. Due to lack of reliable data on industry, commercial, and institutions for making future projections, the long-term water demand forecast for the year 2020 was limited to domestic demand (Figure 3). It was based on 2006 population census figures (National Population Commission, 2007).

5.2. Water resources
Using sequent peak analyses, the Oyun reservoir storage–yield function is: $y = 0.977s^{0.4037}$; where s is storage in Mm³. The storage–yield function is well correlated with coefficient 0.97. The probability of exceedance that corresponds to the 75, 80, 85, 90, and 95% reliabilities of flow is as shown in

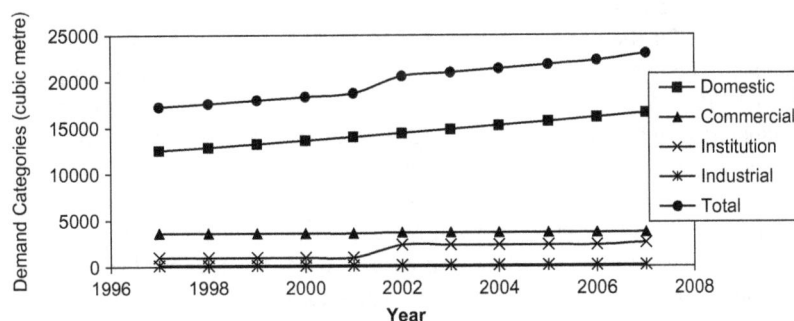

Figure 2. Offa water demand trend between 1996 and 2008.

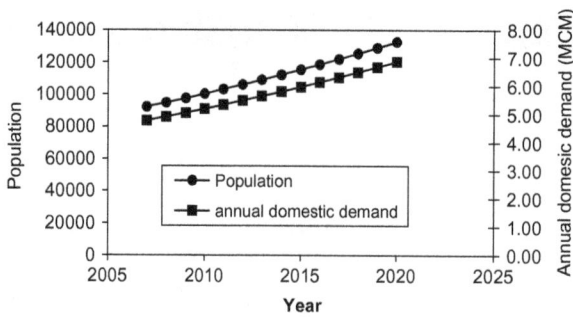

Table 3. Flow frequency distribution for Oyun River (Mm³)			
Exceedence probability (%)	Annual flow (Mm³)		
	Normal	Log normal	Log Pearson
99	1.38	7.59	7.82
95	6.83	9.86	9.23
90	9.79	11.37	11.42
85	11.73	12.30	12.83
80	13.09	13.02	13.36
75	14.61	14.15	13.91

Figure 3. Projection of Offa population and annual domestic demand.

Table 3 at normal, lognormal, and log Pearson distribution models. The probability of exceedance is the difference between the percentage and correspondence percentile. For example, the reliability of the flow with 10% percentile is 90%, which implies that it has 90% frequency of reoccurrence. A high value of flow has high return period (low probability of reoccurrence and therefore a low reliability), while a small value of flow has low return period (high probability of reoccurrence and therefore a high reliability).

From the result of the analysis on the 50-year record, the available safe annual flow for the river is 6.8 Mm³. This is the volume that is available 95% of the time. The sequent peak analysis result shows that the maximum monthly draft possible from Oyun reservoir is 1.67 Mm³, for a reservoir capacity of 3.07 Mm³. However, the net volume of the reservoir is 2.9 Mm³. Practically, the 1.67 Mm³ monthly draft will still be met. The monthly stream flows are predicted up to the year 2020 using Markov model. The water resources assessment shows that the average annual flow for the River Oyun at the Oyun dam is 6.9 Mm³. The highest flow occurs in the month of September and October. The current total annual estimated water demand is 8.2 Mm³ out of which 5.9 Mm³ is for domestic usage only. This water source can still meet the demand of the domestic consumers but will not be adequate beyond the year 2020. There is a steady growth in the population and water demand up till year 2020 as shown in Figure 3.

6. Conclusion and recommendation
There are numerous programs for the short term aimed at improving water supply in Offa using underground water sources. The agencies involved in this program are Rural Water Supply Agency of the State government, UNICEF, ADP, LNRBA, and the Offa and Oyun LGAs. These programs should be managed and coordinated such that they are integrated into centralize system for effective conjunctive usage and thus improved upon the short-term measure.

On the long term when the demand exceeds the resource capacity, planning for new resources is the best solution. There are two ways to go about it in this case. One is to increase the height of the dam so as to increase the impounding capacity. The other is interbasin water transfer. This is the physical transportation of water out of one river basin to another basin. They both require adequate technical feasibility studies covering hydrological, geotechnical, geological, and socioeconomic investigations.

The study can serve as an organized baseline for future work, particularly in obtaining improved estimate for industrial, commercial, institutional water use categories, and planning conjunctive uses of water resources. The findings of the study can be a vital input into the demand management process for long-term sustainable water supply of the town and by extension to urban township with similar characteristic in Nigeria and third world countries.

Funding
The authors received no direct funding for this research.

Author details
O.G. Okeola[1]
E-mail: ogolayinka@unilorin.edu.ng
S.O. Balogun[2]
E-mail: olasoulman@gmail.com
[1] Department of Water Resources & Environmental Engineering, University of Ilorin, Ilorin, Nigeria.
[2] National Centre for Hydropower Research & Development, Ilorin, Nigeria.

References
Adebosin, H. O. O. (1986). *Siting of reservoirs in river Oyun basin* (BE Project). Department of Civil Engineering, University of Ilorin, Ilorin.

Adedayo, I. T. (2014). *GIS application in the evaluation of peak runoff hydrograph for Oyun River watershed in Kwara state using remote sensing data* (MGIS thesis). Department of Geography and Environmental Sciences, University of Ilorin, Nigeria.

Arbués, F., García-Valiñas, M. A., & Martínez-Espiñeira, R. (2003). Estimation of residential water demand: A state-of-the-art review. *The Journal of Socio-Economics, 32*, 81–102. http://dx.doi.org/10.1016/S1053-5357(03)00005-2

Barnes, Jr., G. W., & Chung, F. I. (1986). Operational planning for California water system. *Journal of Water Resources Planning and Management, 112*, 71–86. http://dx.doi.org/10.1061/(ASCE)0733-9496(1986)112:1(71)

Cabezas, L. M., & Wurbs, R. A. (1986). Economic evaluation of urban water supply systems. *Journal of Urban Planning and Development, 112*, 46–59. http://dx.doi.org/10.1061/(ASCE)0733-9488(1986)112:2(46)

Cochran, R., & Cotton, A. W. (1985). Municipal water demand study, Oklahoma city and Tulsa, Oklahoma. *Water Resources Research, 21*, 941–943. http://dx.doi.org/10.1029/WR021i007p00941

Cook, Z., Urban, S., Maupin, M., Pratt, R., & Church, J. (2001). *Domestic, commercial, municipal and industrial water demand assessment and forecast in Ada and Canyon counties, Idaho*. Idaho, ID: Department of Water Resources.

Dziegielewski, B. (1988). Urban non residential water use and conservation. In M. Waterstone & R. John Burt (Eds.), *Proceedings of the Symposium on water-use data for water resources management* (pp. 371–380). Tucson, AZ: American Water Resources Association.

Fortin, V., Perreault, L., & Salas, J. D. (2004). Retrospective analysis and forecasting of streamflows using a shifting level model. *Journal of Hydrology, 296*, 135–163. http://dx.doi.org/10.1016/j.jhydrol.2004.03.016

Guariso, G., Rinaldi, S., & Soncini-Sessa, R. (1986). The management of lake Como: A multiobjective analysis. *Water Resources Research, 22*, 109–120. http://dx.doi.org/10.1029/WR022i002p00109

Hanemann, W. M. (1998) *Determinants of urban water use*. Retrieved May 26, 2003, from http://are.berkeley.edu/courses/are262/spring98/water.pdf

Hussain, I., Thrikawala, S., & Barker, R. (2002). Economic analysis of residential, commercial, and industrial uses of water in Sri Lanka. *Water International, 27*, 183–193. http://dx.doi.org/10.1080/02508060208686991

Itube, I. R., Mejia, J. M., & Dawdy, D. R. (1972). Streamflow simulation: 1. A new look at Markovian models, fractional Gaussian noise, and crossing theory. *Water Resources Research, 8*, 921–930.

Karamouz, M., Szidarovszky, F., & Zahraie, B. (2003). *Water resources systems analysis*. Boca Raton, FL: Lewis Publishers.

Khadam, M. O. (1984). Managing water demand in developing countries. In *Proceeding of International Seminar on water resources management practices* (pp. 07–215). Ilorin.

Kiparsky, M., Joyce, B., Purkey, D., & Young, C. (2014). Potential impacts of climate warming on water supply reliability in the Tuolumne and Merced river basins, California. *PLoS ONE, 9*, e84946. doi:10.1371/journal.pone.0084946

KWWC. (2009). *Kwara State Water Corporation*. Ilorin, Kwara State, Nigeria.

Loucks, D. P., Stedinger, J. R., & Haith, D. A. (1981). *Water resource systems planning and analysis*. Englewood Cliffs, NJ: Prentice-Hall.

Martin, Q. W. (1987). Optimal daily operation of surface-water systems. *Journal of Water Resources Planning and Management, 113*, 453–470. http://dx.doi.org/10.1061/(ASCE)0733-9496(1987)113:4(453)

Martin, W. E., & Thomas, J. F. (1986). Policy relevance in studies of urban residential water demand. *Water Resources Research, 22*, 1735–1741. http://dx.doi.org/10.1029/WR022i013p01735

Mazzanti, M., & Montini, A. (2006). The determinants of residential water demand: Empirical evidence for a panel of Italian municipalities. *Applied Economics Letters, 13*, 107–111. http://dx.doi.org/10.1080/13504850500390788

Mimi, Z., & Smith, M. (2000). Statistical domestic water demand model for the west bank. *Water International, 25*, 464–468. http://dx.doi.org/10.1080/02508060008686854

Mitchell, G. (1999). Demand forecasting as a tool for sustainable water resource management. *International Journal of Sustainable Development and World Ecology, 6*, 231–241. http://dx.doi.org/10.1080/13504509909470014

Moy, W., Cohon, J. L., & ReVelle, C. S. (1986). A programming model for analysis of the reliability, resilience, and vulnerability of a water supply reservoir *Water Resources Research, 22*, 489–498. http://dx.doi.org/10.1029/WR022i004p00489

Mylopoulos, Y. A., Mentes, A. K., & Theodossiou, I. (2004). Modeling residential water demand using household data: A cubic approach. *Water International, 29*, 105–113. http://dx.doi.org/10.1080/02508060408691753

National Population Commission. (2007). *Federal Republic of Nigeria official gazette* (Vol. 94, No. 24). Abuja.

Nieswiadomy, M. L. (1992). Estimating urban residential water demand: Effects of price structure, conservation, and education. *Water Resources Research, 28,* 609–615. http://dx.doi.org/10.1029/91WR02852

Olmstead, S. M., Michael Hanemann, W., & Stavins, R. N. (2007). Water demand under alternative price structures. *Journal of Environmental Economics and Management, 54,* 181–198. http://dx.doi.org/10.1016/j.jeem.2007.03.002

Okeola, O. G. (2000). *Evaluation of the effectiveness of Oyun regional water supply scheme, Kwara State* (ME thesis). Department of Civil Engineering, University of Ilorin, Nigeria.

Oyegoke, S., & Oyesina, D. (1984). Determination of water demand in a developing economy. In *International Seminar on Water Resources Management Practices* (pp. 217–227). Ilorin, Nigeria.

Philipose, M. C., & Srinivasan, K. (1995). Construction of storage–performance–yield relationships for a reservoir using stochastic simulation. *International Journal of Water Resources Development, 11,* 289–302. http://dx.doi.org/10.1080/07900629550042245

Schleich, J., & Hillenbrand, T. (2009). Determinants of residential water demand in Germany. *Ecological Economics, 68,* 1756–1769.

Sharma, A., Tarboton, D. G., & Lall, U. (1997). Streamflow simulation: A nonparametric approach. *Water Resources Research, 33,* 291–308. http://dx.doi.org/10.1029/96WR02839

Shaw, E. M. (1988). *Hydrology in practice.* London: Chapman and Hall.

Stephenson, D., & Randel, B. (2003). Water demand theory and projections in South Africa. *Water International, 28,* 512–518.

Strycharczyk, J. B., & Stedinger, J. R. (1987). Evaluation of a "reliability programming" reservoir model. *Water Resources Research, 23,* 225–229. http://dx.doi.org/10.1029/WR023i002p00225

Vogel, R. M. (1987). Reliability indices for water supply systems. *Journal of Water Resources Planning and Management, 113,* 563–579. http://dx.doi.org/10.1061/(ASCE)0733-9496(1987)113:4(563)

Walliingford, H. R. (2003). *Handbook for the assessment of catchment water demand and use.* Oxon: HR Walliingford.

Wurbs, R. A. (1994). *Computer models for water resources planning and management* (IWR Report 94-NDS-7). Alexandria,VA: US Army Corps of Engineers, Institute of Water Resources.

Yurekli, K., & Ozturk, F. (2003). Stochastic modeling of annual maximum and minimum streamflow of Kelkit stream. *Water International, 28,* 433–441. http://dx.doi.org/10.1080/02508060308691721

Utilisation of iron ore tailings as aggregates in concrete

Francis Atta Kuranchie[1]*, Sanjay Kumar Shukla[1], Daryoush Habibi[1] and Alireza Mohyeddin[1]

*Corresponding author: Francis Atta Kuranchie, School of Engineering, Edith Cowan University, 270 Joondalup Drive, Joondalup, Western Australia 6027, Australia

E-mail: akuranchie@yahoo.com

Reviewing editor: Anand J. Puppala, University of Texas at Arlington, USA

Abstract: Sustainable handling of iron ore tailings is of prime concern to all stakeholders who are into iron ore mining. This study seeks to add value to the tailings by utilising them as a replacement for aggregates in concrete. A concrete mix of grade 40 MPa was prepared in the laboratory with water–cement ratio of 0.5. The concrete were cured for 1, 2, 3, 7, 14 and 28 days. The properties of the concrete such as workability, durability, density, compressive strength and indirect tensile strength were tested. A controlled mix of concrete was also prepared in similar way using conventional materials and the results were compared with the tailings concrete. It was found that the iron ore tailings may be utilised for complete replacement for conventional aggregates in concrete. The iron ore tailings aggregates concrete exhibited a good mechanical strength and even in the case of compressive strength, there was an improvement of 11.56% over conventional aggregates concrete. The indirect tensile strength did not improve against the control mix due high content of fines in the tailings aggregates but showed 4.8% improvement compared with the previous study where the conventional fine aggregates was partially replaced by 20% with iron ore tailings.

Subjects: **Civil, Environmental and Geotechnical Engineering; Engineering & Technology; Mining, Mineral & Petroleum Engineering**

Keywords: **iron ore tailings; concrete; aggregates; compressive strength; indirect tensile strength; slump**

1. Introduction

Western Australia (WA) is endowed with a large reserve of iron ore. This has attracted a lot of iron ore mining activities for many years in the state. Iron ore resource is a very significant player of the economy

ABOUT THE AUTHORS

The authors are affiliated to the Geo-technical and Geo-environmental research group of Edith Cowan University in the School of Engineering. Our research focus is in sustainability of the environment and quality in building and construction as well as geotechnical engineering. This work is part of the several research works in the focus subject.

PUBLIC INTEREST STATEMENT

This research has highlighted the possibility of introducing alternative and cheaper materials as aggregates for concrete technology. It promotes sustainability in construction by conserving conventional construction materials. It also gives positive impacts in the economy of construction projects. As an additional advantage, it has introduced an economical and environmentally friendly way to manage mine tailings. This has a potential of reducing the burden on mining industries in the management of their mine tailings.

of WA, with an average yearly production of 316 million tonnes which accounts for about 47% of the total value of all resources mined in the state (W. A. Department of Mines & Petroleum, 2009).

The long-time existence of large-scale iron ore mining in Western Australia has resulted in the accumulation of huge quantities of iron ore tailings in the state which needs to be handled, disposed off and monitored properly. It has been reported that in WA, on average, the production of 1 tonne of iron ore results in the generation of 2 tonnes of iron ore tailings (Price, 2004). It can therefore be estimated that about 632 million tonnes of iron ore tailings are generated yearly in WA. This quantity is so huge to the extent that if sustainable handling of these tailings is not found, it could lead to adverse effects on the environment and human health. One of the common environmental problems associated with these tailings is the formation of acid mine drainage which can be a potential source of surface and ground water pollution (Cassiano, Juarez, Pagnussat, Schneider, & Tubino, 2012). These tailings also consume a lot of land that could be used for other purposes, and compromise the good looks of the environment in these areas (Thomas, Damare, & Gupta, 2013; Yellishetty, Karpe, Reddy, Subhash, & Ranjith, 2008). There can be a potential of erosion from these tailings dumps into the environment (Yellishetty, Mudd, & Shukla, 2012). The current practice is that a smaller percentage of these tailings are returned to the mined-out areas as backfills to fill the void created, while majority of them is stockpiled in the environment or returned to a special tailings dam built for storage purposes.

In order to minimise the problems created by the mine tailings, one potential application area to explore is its utilisation in building and construction. This is because there is a greater potential in this sector where recycled waste products could be considered as construction and building materials. An example is the use of these mine tailings as aggregates in concrete (Shetty, Nayak, & Vijayan, 2014; Thomas et al., 2013; Thomas & Gupta, 2013; Yellishetty et al., 2008).

Aggregates make up about 70–80% of a concrete mix (Shetty et al., 2014; Thomas & Gupta, 2013). As the natural granite quarries for aggregates are gradually decreasing, there would be the need for alternative materials to be used as natural aggregates in concrete. If mine tailings are considered as a partial or complete replacement of natural aggregates in concrete, majority of these tailings could be recycled and used sustainably, by turning these mine tailings into useful resource and providing cheaper alternatives in concrete production (Ugama, Ejeh, & Amartey, 2014). This will eliminate the need to mine virgin materials such as concretes aggregates. The scarce resources could be conserved while at the same time providing sustainable solution to the handling of the tailings. This could make it possible for the mining industries concerned to generate extra income to defray their cost of production to maximise profit margin.

The processing activities associated with the iron ore beneficiation is such that it results in the tailings with particle sizes ranging from fine to coarse (Kuranchie, Shukla, & Habibi, 2014a). If these tailings are segregated properly, both fine and coarse aggregates for concrete can be obtained. Therefore, if the iron ore mining companies in WA could incorporate comprehensive utilisation of the tailings in their operation, it could lead to cleaner production and sustainable development in their operation. This could have a potential of reducing both cost and environmental impacts created by these tailings (Haibin & Zhenling, 2010).

The aim of this paper is to study the feasibility of using the iron ore tailings as both fine and coarse aggregates in making concrete. The paper evaluates the technical and environmental characteristics of the concrete with the main concern of recycling and adding economic value to the iron ore tailings to be used as alternative cheaper materials for concrete aggregates.

2. Experimental programme

In this section, the standard process to make concrete was followed. Several tests were conducted to measure physical properties of concrete at different ages and the results were compared with those of normal concrete made using conventional aggregates.

2.1. Materials

For the normal concrete, conventional materials such as sand, cement and granite aggregates were used. For the tailings concrete being studied, iron ore tailings were used together with the cement to make the new concrete.

The iron ore tailings were obtained from Mount Gibson Iron Extension Hill Operations in Perenjori of Western Australia (WA). The particle–size distribution curve of the original tailings as obtained is shown in Figure 1. The size of the tailings range from fines (<75 μm) to coarse (≤ 32 mm). Other physical and chemical properties of the tailings are presented in Tables 1 and 2, respectively (Kuranchie et al., 2014a).

The mine tailings were segregated using various sieve sizes. The particles ranging from 4.75 mm and below were used as fine aggregates while those ranging from 20 mm down to 10 mm were used as coarse aggregates in making the tailings concrete shown in Figure 2(a) and (b). Figure 3 shows the particle–size distribution curves of the tailings aggregates and they conformed to AS 2758.1: 2014 (Standards Australia, 2014a).

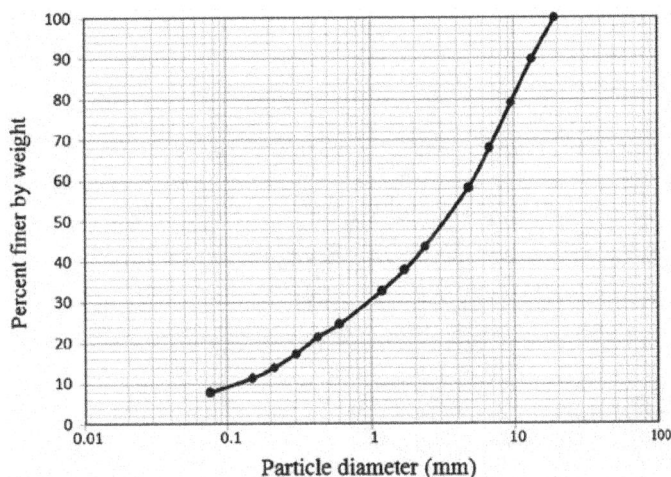

Figure 1. Particle–size distribution curve of tailings as received.

Table 1. Physical properties of iron ore tailings	
Property	**Value**
Fines (clay and silt) (%)	6.5
Sand content (%)	69.5
Gravel content (%)	24.0
Specific gravity	2.65
Minimum dry density, ρ_{dmin} (kg/m³)	1,685.00
Maximum dry density, ρ_{dmax} (kg/m³)	1,860.00
Effective grain size, D_{10} (mm)	0.13
D_{60} (mm)	2.67
D_{30} (mm)	0.70
Coefficient of uniformity, C_U	20.54
Coefficient of curvature, C_c	1.41
Soil classification as per USCS	SW-SM (Well graded sand-silty sand)

Source: Kuranchie et al. (2014a).

Table 2. Chemical composition of iron ore mine tailings	
Constituents	Amount (%)
CaO	0.03
SiO_2	57.31
Al_2O_3	9.58
Fe_2O_3	25.13
MgO	0.08
SO_3	0.16
Na_2O	0.04
K_2O	0.04
TiO_2	0.61
As	–
Loss on ignition	6.67

Source: Kuranchie et al. (2014a).

(a) Mine tailings fine aggregates

(b) Mine tailings coarse aggregates

(c) Natural fine sand aggregates

(d) Granite coarse aggregates

Figure 2. Normal and tailings aggregates.

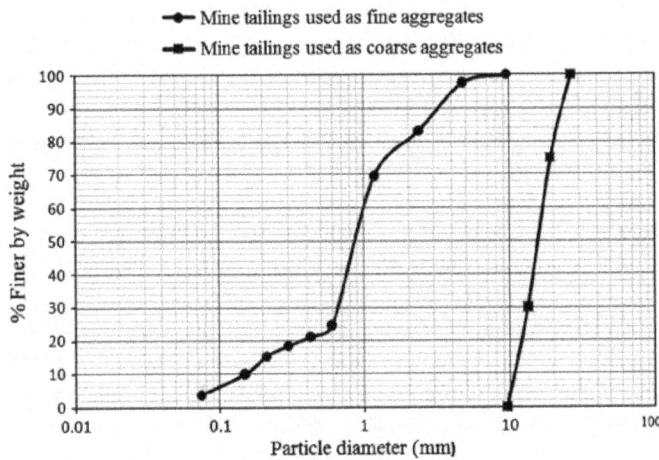

Figure 3. Particle–size distribution curves of the tailings used as aggregates.

Ordinary general purpose Portland cement of grade of 43 which conforms to AS 3972 (Standards Australia, 2010) was used. This was obtained from Cockburn Cement Limited in Western Australia. The cement had a normal consistency of 29.5%, soundness of 1 mm with initial setting time of 2 h and final setting time of 3.15 h. The specific gravity of the cement was 3.1.

Commonly used sand for various construction purposes in Perth known as "brickies sand" was obtained from a quarry site, 40 km North of Perth city. This was used as fine aggregates in the normal concrete as a control mix. This can be seen in Figure 2(c). The physical properties of the sand are given in Table 3 (Kuranchie, Shukla, & Habibi, 2014b). The sand was classified as poorly graded sand as per the Unified Soil Classification System (USCS).

Natural granite quarry of sizes 10 and 20 mm, which were obtained from commercial quarry, were mixed and used as the coarse aggregates in the control mix. This is shown in Figure 2(d).

2.2. Mix design and preparation of specimens

A concrete mix design of grade 40 MPa with a water–cement ratio of 0.5 and a mix design ratio of 1:2:3 (cement:fine aggregates:coarse aggregates) was prepared in the laboratory. The mix design recommended by Marsh (1997) was followed and it conforms to AS 1012.2 (Standards Australia, 2014d).

Table 3. Physical properties of sand	
Property	**Value**
Fines (%)	3.85
Sand (%)	96.15
Specific gravity	2.66
Minimum dry density, ρ_{dmin} (kg/m^3)	1,392
Maximum dry density, ρ_{dmax} (kg/m^3)	1,598
Effective grain size, D_{10} (mm)	0.18
D_{60} (mm)	0.38
D_{30} (mm)	0.26
Coefficient of uniformity, C_U	2.11
Coefficient of curvature, C_c	0.99
Soil classification as per USCS	Poorly graded sand (SP)

Source: Kuranchie et al. (2014b).

Ordinary general purpose Portland cement of grade of 43 which conforms to the Standard AS 3972 (Standards Australia, 2010) was used. This was obtained from Cockburn Cement Limited in Western Australia. The cement had a normal consistency of 29.5%, soundness of 1 mm with initial setting time of 2 h and final setting time of 3.15 h. The specific gravity of the cement was 3.1. For the concrete made from the mine tailings, the particle sizes of the mine tailings as described above were used as fine and coarse aggregates.

In the tailings aggregates concrete, i.e. where 100% of the aggregates consisted of iron ore tailings as described above, 18 cylindrical mould specimens of size 100 mm internal diameter and 200 mm length were prepared for 1, 2, 3, 7, 14 and 28 days curing duration for compressive strength test. Other 18 cylindrical moulds of size 150 mm internal diameter and 300 mm length were prepared from the same mix with the same curing duration for the indirect tensile strength test. The mixes were prepared at a controlled indoor temperature range of 25 to 30°C. The process was repeated using conventional aggregates to prepare the control mix which was used for comparison purposes.

After preparation, the small moulds were covered with their lids and plastic sheets were used to cover the big moulds. The specimens were kept in their moulds for 24 h before de-moulding. The specimens were then placed in the concrete curing chamber for the various curing durations explained above. Some quality assessment tests such as the density and the slump tests were performed on the fresh concrete, while after curing, other quality assessment tests on hardened concrete such as compressive and split tensile strength were also conducted.

2.3. Tests for fresh concrete

Density and slump tests were conducted on the fresh concrete for both the control mix and the tailings aggregates mix. Density of both the fresh tailings aggregate concrete and the normal concrete were determined following usual laboratory procedures such that it conformed to AS 1012.5 for determining the density of a fresh concrete (Standards Australia, 1999).

In determining the slump, a standard slump cone with dimensions height 310 mm, lower diameter 100 mm and upper diameter 198 mm was used. The concrete was filled in the cone in three layers. Each layer was compacted using a metal rod for 25 times to ensure good compaction. The concrete was carefully de-moulded from the cone. The cone was then placed upside down close to the concrete and the difference in heights was measured as the slump using a ruler. The procedure according to AS 1012.3.1 was followed (Standards Australia, 2014b). Figure 4 shows the procedure for the slump test.

Figure 4. Determination of the slump of a fresh concrete.

2.4. Tests for hardened concrete

Compressive strength test, indirect tensile strength test and acid resistance and alkalinity test were performed after the concrete have been cured for the various curing duration of the concrete specified. The acid resistance and alkalinity test was limited to the tailings aggregate concrete while all other test were conducted on both the normal aggregate and tailings aggregate concretes. Figures 5(a) and (b) show some examples of how the tailings and the normal aggregates concretes specimens look respectively after they have been cured. The tailings aggregates concretes have a brownish colour and this was due to the brown colour of the iron ore tailings.

Compressive strength was tested for the smaller cylindrical concrete specimens following AS 1012.9 (Standards Australia, 2014c) with a compression testing machine. Three specimens were tested for each curing scheme and the average of the values was used. The compressive strength was calculated by dividing the maximum load (force) observed from the compression machine by the cross-sectional area of the specimen tested and the unit was expressed in megapascals (MPa).

Indirect tensile strength was tested for the beams in the bigger cylindrical moulds. This test was consistent with AS 1012.10 (Standards Australia, 2000). Similarly, three specimens were tested for the indirect tensile strength in each curing duration schemes; 1, 3, 7, 14 and 28 days and the average of the three values was used. The test was done using a compression machine in the laboratory. The indirect tensile strength of the specimen was calculated using the formula

$$T = \frac{2000P}{\pi LD} \qquad (1)$$

where T is the indirect tensile strength in MPa, P is the maximum applied load indicated by the compression machine in kN, L is the length of the cylinder in mm and D is the diameter of the specimen.

Mine tailings may contain some sulphide minerals and heavy metals, and there is a possibility that these minerals may come into contact with other oxidising agents such as oxygen and this will lead to the production of some acidic contents (Hitch, Ballantyne, Hindle, & Norman, 2010; Thomas & Gupta, 2013). When the iron ore tailings aggregates concrete are exposed to these extreme conditions, disintegration of the concrete can occur due to acid attack. Also, the ferrous content of the iron ore tailings is subject to corrosion which could also have negative effects on the long-term durability of the concrete. Alkalinity and acid resistance test is therefore a means to check the long-term durability of the tailings concrete against acid attack and corrosion.

In testing for the alkalinity and acid resistance of the concrete, the pH of the resulting solution from the concrete was tested. In doing this, the concrete cubes after the various curing days were dried for 24 h in the oven at a temperature of 105°C. The specimen was cooled down to room temperature and was broken with a hammer to separate the mortar from the concrete. The dried mortar was ground

(a) Tailings aggregate concretes   (b) Normal aggregate concretes

Figure 5. Tailings and normal aggregates concrete.

using a ball mill and sieved to less than 150 μm size. Ten grams of the undersize was then mixed with distilled water and stirred. The pH of the resulting solution was taken using a digital pH meter. If the pH of the resulting solution is above 7, it indicates an alkaline condition which will have the potential of low acid attack. This also indicates that corrosion will be very low. This methodology was recommended by Thomas and Gupta (2013).

3. Results and discussion

The results for the various tests on both fresh and cured concrete have been summarised in this section and it has been discussed in detail. The results include both the normal aggregates and tailings aggregates concrete.

In the absence of experimental results, the density of a fresh concrete may be considered as 2,400 kg/m^3 as a recommended by AS 3600 (Standards Australia, 2009). In this work, the density of the normal concrete is found to be 2,345 kg/m^3 and the density of the tailings concrete is 2,362 kg/m^3. The difference is due to the slightly higher specific gravity of the tailings. The two densities are comparable with those reported in Thomas and Gupta (2013).

Slump test is used to measure the workability and consistency of a fresh concrete (Marsh, 1997). In this work, the slump for the normal concrete was calculated as 80 mm and that of the tailings concrete was 20 mm. Slump ranging from 0 to 25 mm is considered as having low workability.

Compressive strength is the main mechanical property of a concrete that is normally specified in supply of concrete. Figure 6 shows the compressive strength of both normal and tailings aggregates concrete. It was observed that the compressive strength for all the curing ages were higher in iron ore tailings aggregates concrete than in the normal concrete as the control mix. At age of 28 days, the compressive strength of the tailings aggregates concrete was 36.95 MPa while that of the control mix was 33.12 MPa, which shows an improvement of 11.56% of tailings aggregates concrete over normal concrete. This improvement of strength of the iron ore tailings aggregates concrete may be attributed to the chemistry associated with the chemical composition of the iron ore tailings. It has been reported that iron compounds have the potential to accelerate cement hydration (Yellishetty et al., 2008) and this could be the main factor for the improvement of compressive strength observed in the iron ore tailings aggregates concrete which has higher percentage of iron compounds.

Tensile strength of concrete is another crucial mechanical property of concrete. This shows the strength of how the aggregates are bonded to the other materials in the concrete. Concrete structures are susceptible to tensile cracking and it becomes an important factor, especially when the concrete is intended to be used for the design of highway and airfield slabs (Neville, 2012). Therefore,

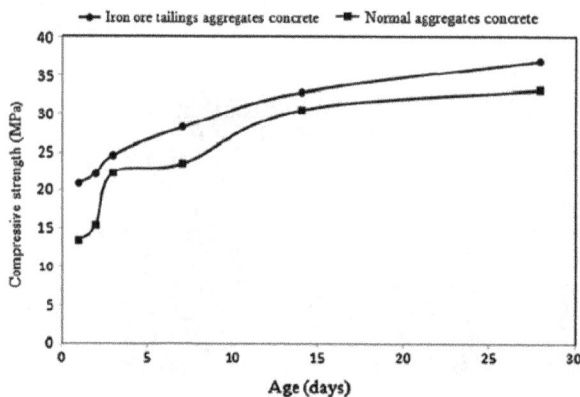

Figure 6. Compressive strength of normal and tailings aggregate concretes.

the results from the indirect tensile strength test will investigate the influence of the aggregates structure on adhesion and the strength of the bond between the concrete materials.

Figure 7 shows indirect tensile strength of both normal and tailings aggregates concretes for various curing ages from 1 to 28 days. It can be observed that the indirect tensile strength favourably increases with the ageing for both normal (control mix) and tailings aggregates concrete. From Figure 7, it is also observed that tensile strength for one day curing for tailings aggregates concrete was slightly higher than the control mix. As the concrete continues to age, the tensile strength of the normal aggregates concrete became higher than the tailings aggregates concrete. At age 28 days, the control mix achieved indirect tensile strength of 3.36 MPa, while that of the tailings aggregates concrete was 2.82 MPa.

It is noted that while the compressive strength improved, the tensile strength did not improve in the iron ore tailings aggregates concrete as compared with the control mix, although the compressive strength is directly proportional to the tensile strength. The reason may be attributed to the differences in the quantity of fines found in both materials used as fine aggregates. From Tables 1 and 3, it can be seen that iron ore tailings contain 6.5% fines, while the natural sand used in this work contains only 3.85% fines. It should be noted that in this work, 100% of the iron ore tailings were used as fine aggregates in the tailings aggregates concrete. Therefore, this particular concrete will have more fines as compared with the control mix. The higher content of fines in the iron ore tailings aggregates concrete will increase the demand for water in the mix and this will reduce the bond strength existing in the aggregate–cement paste, leading to a reduction in tensile strength (Adedayo & Onitiri, 2012; Ugama et al., 2014). However, the tensile strength exhibited by the iron ore tailings aggregates concrete in this work increased favourably with ageing and there is 4.8% improvement as compared with similar work where only 20% of natural sand fine aggregates were replaced by iron ore tailings as reported earlier by Ugama et al. (2014).

Figure 8 shows the pH values of resulting solution of the iron ore tailings aggregates concrete for the various curing ages. The pH values for the various ages of the concrete ranges from 11.8 to 12.5. These values can be termed as having high alkaline conditions and will, therefore, make acid attack and potential for corrosion very low. This is a good sign of long-term durability of the iron ore tailings concrete.

4. Economic and environmental implication

The cost of managing and monitoring mine tailings is very important to the life of every mine. This is because of some stringent regulations and legislations put in place by the appropriate governmental authorities for the mining companies to adhere to in dealing with their tailings. As part of this, mining companies set aside specific budget for this purpose which is normally included in their cost of production. Comprehensive utilisation of mine tailings in this scenario will provide extra income to

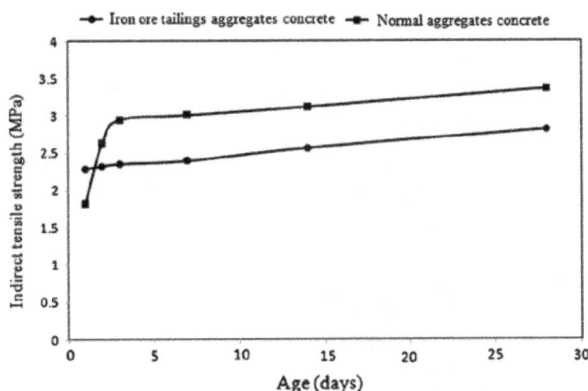

Figure 7. Indirect tensile strength of normal and tailings aggregate concretes.

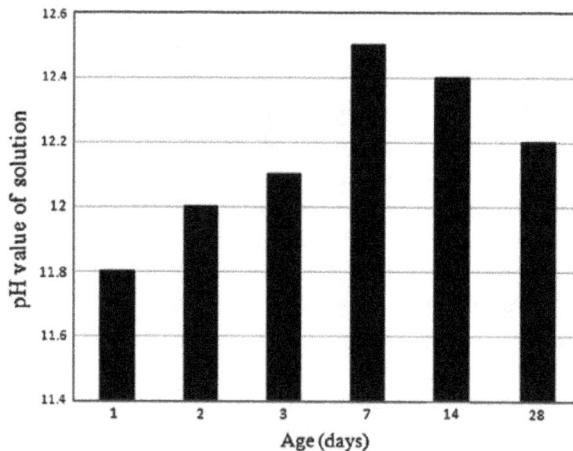

Figure 8. pH values showing the alkalinity of iron ore tailing aggregates concrete.

offset this cost for the mining companies upon mine closure. According to Packey (2012), the incorporation of utilising mine tailings in the operations of the mining companies could reduce rehabilitation costs and minimise the effects of mine closures on mining communities and bring about new economic activity.

Based on the results from this study, and the fact that aggregates forms about 70–80% of concrete, iron ore mine tailings in Western Australia could be considered as aggregates materials for concrete production where huge volumes of iron ore tailings could be utilised. The utilisation of the iron ore tailings could have positive effects on the environment because the quantity of iron ore tailings could be reduced tremendously. This could also lead to the reduction of adverse environmental effects and render the operations of the iron ore mine companies more sustainable in state.

5. Conclusions

This study evaluated the possibility of completely replacing conventional aggregates in concrete with iron ore mine tailings produced in Western Australia to attain the same or better outputs for technical specifications. In the study, both conventional and tailings aggregates were used for comparison. Concrete mix design of 40 MPa and water–cement ratio of 0.5 was used. Based on the results from this study, the following general conclusions could be made:

(1) The compressive strength of the concrete with iron ore tailings aggregates at 28 days was 36.95 MPa which shows an improvement of 11.56% over the concrete with conventional aggregates. This is mainly because of favourable chemical composition of the iron ore tailings.

(2) The split tensile strength exhibited by the concrete with tailings aggregates was 2.82 MPa at 28 days and this is slightly lower than concrete with conventional aggregates by 16% due to higher quantity of fines in the iron ore tailings as compared with the natural sand in the control mix. However, the tensile strength increased favourably with ageing and there was still 4.8% improvement on the tensile strength as compared with similar study reported earlier.

(3) The concrete with tailings aggregates has a low potential of corrosion and a low potential to acid attack due to high pH values of their resulting solution.

(4) The utilisation of iron ore tailings as aggregates in concrete could have positive environmental implications to the mining companies and the mining communities and will provide cheaper alternative materials to bring about economy in concrete production.

Funding
The authors received no direct funding for this research.

Author details
Francis Atta Kuranchie[1]
E-mail: akuranchie@yahoo.com
Sanjay Kumar Shukla[1]
E-mail: s.shukla@ecu.edu.au
Daryoush Habibi[1]
E-mail: d.habibi@ecu.edu.au

Alireza Mohyeddin[1]
E-mail: a.mohyeddin@ecu.edu.au
[1] School of Engineering, Edith Cowan University, 270
Joondalup Drive, Joondalup, Western Australia 6027,
Australia.

References

Adedayo, S. M., & Onitiri, M. A. (2012). Tensile properties of iron ore tailings filled epoxy composites. *The West Indian Journal of Engineering, 35*, 51–59.

Cassiano, R. D. S., Juarez, R. D. A. F., Pagnussat, D., Schneider, I. A. H., & Tubino, R. M. C. (2012). *Use of coal waste as fine aggregates in concrete blocks for paving.* Paper presented at the 10th International Conference on Concrete Block Paving, Shanghai, China.

Haibin, L., & Zhenling, L. (2010). Recycling utilization patterns of coal mining waste in China. *Resources, Conservation and Recycling, 54*, 1331–1340. doi:10.1016/j.resconrec.2010.05.005

Hitch, M., Ballantyne, S. M., Hindle, S. R., & Norman, B. (2010). Revaluing mine waste rock for carbon capture and storage. *International Journal of Mining, Reclamation and Environment, 24*, 64–79. doi:10.1080/17480930902843102

Kuranchie, F. A., Shukla, S. K., & Habibi, D. (2014a). Utilisation of iron ore mine tailings for the production of geopolymer bricks. *International Journal of Mining, Reclamation and Environment*, 1–44. doi:10.1080/17480930.2014.993834

Kuranchie, F. A., Shukla, S. K., & Habibi, D. (2014b). Studies on electrical resistivity of Perth sand. *International Journal of Geotechnical Engineering, 8*, 449–457. doi:10.1179/19397 87913Y.0000000033

Marsh, B. K. (1997). *Design of normal concrete mixes* (2nd ed., pp. 1–30). Watford: Construction Research and Communications.

Neville, A. M. (2012). *Properties of concrete* (5th ed., pp. 310–311). Harlow: Pearson Education.

Packey, D. J. (2012). Multiproduct mine output and the case of mining waste utilization. *Resources Policy, 37*, 104–108. doi:10.1016/j.resourpol.2011.11.002

Price, P. A. (2004). *Description of operating procedures of BHP Billiton iron ore* (Document number 12622268, pp. 1–31).

Shetty, K. K., Nayak, G., & Vijayan, V. (2014). Use of red mud and iron tailings in self compacting concrete. *International

Journal of Research in Engineering and Technology, 3*, 111–114.

Standards Australia. (1999). *Methods of testing concrete, determination of mass per unit volume of freshly mixed concrete* (No. AS 1012.5, p. 2). Sydney: Author.

Standards Australia. (2000). *Methods of testing concrete, determination of indirect tensile strength of concrete cylinders* (No. AS 1012.10, pp. 1–8). Sydney: Author.

Standards Australia. (2009). *Concrete structures* (No. AS 3600). Sydney: Author.

Standards Australia. (2010). *General purpose and blended cements* (No. AS 3972). Sydney: Author.

Standards Australia. (2014a). *Aggregates and rock for engineering purposes, concrete aggregates* (No. AS 2758.1, pp. 1–32). Sydney: Author.

Standards Australia. (2014b). *Methods of testing concrete, determination of properties related to the consistency of concrete—Slump test* (No. AS 1012.3.1, pp. 1–8). Sydney: Author.

Standards Australia. (2014c). *Methods of testing concrete, compressive strength tests—Concrete, mortar and grout specimens* (No. AS 1012.9, pp. 1–12). Sydney: Author.

Standards Australia. (2014d). *Methods of testing concrete, preparing concrete mixes in the laboratory* (No. AS 1012.2, p. 20). Sydney: Author.

Thomas, B. S., Damare, A., & Gupta, R. C. (2013). Strength and durability characteristics of copper tailing concrete. *Construction and Building Materials, 48*, 894–900. doi:10.1016/j.conbuildmat.2013.07.075

Thomas, B. S., & Gupta, R. C. (2013). Mechanical properties and durability characteristics of concrete containing solid waste materials. *Journal of Cleaner Production, 48*, 1–6. doi:10.1016/j.jclepro.2013.11.019

Ugama, T. I., Ejeh, S. P., & Amartey, D. Y. (2014). Effect of iron ore tailing on the properties of concrete. *Civil and Environmental Research, 6*, 7–13.

W. A. Department of Mines and Petroleum. (2009). *Mineral and petroleum statistics digest 2008/2009*. Perth: Author.

Yellishetty, M., Karpe, V., Reddy, E., Subhash, K. N., & Ranjith, P. G. (2008). Reuse of iron ore mineral wastes in civil engineering constructions: A case study. *Resources, Conservation and Recycling, 52*, 1283–1289. doi:10.1016/j. resconrec.2008.07.007

Yellishetty, M., Mudd, G. M., & Shukla, R. (2012). Prediction of soil erosion from waste dumps of opencast mines and evaluation of their impacts on the environment. *International Journal of Mining, Reclamation and Environment, 27*, 88–102. doi:10.1080/17480930.2012.6551

6

Calibration of safety performance function for crashes on inter-city four lane highways in India

Naveen Kumar ChikkaKrishna[1]*, Manoranjan Parida[1] and Sukhvir Singh Jain[1]

*Corresponding author: Naveen Kumar ChikkaKrishna, Department of Civil Engineering, Transportation Engineering Group, Indian Institute of Technology Roorkee, Roorkee 247667, Uttarakhand, India
E-mail: cnaveenkh@gmail.com

Reviewing editor: Filippo G. Pratico, University Mediterranea of Reggio, Italy

Abstract: There is a significant need to improve the highway safety during roadway planning, design and operations in developing countries like India. To receive appropriate consideration, safety needs to be dealt objectively within the transportation planning and highway design processes. Lack of available tools is a deterrent to quantify safety of a transportation facility during the planning or highway design process. The objective of this paper is to develop safety performance functions considering various elements involved in the planning, design and operation of a section on four-lane National Highway (NH)-58 located in the state of Uttarakhand, India. The mixed traffic on Indian multilane highways comes with a lot of variability within, ranging from different vehicle types to different driver characteristics. This could result in variability in the effect of explanatory variables on crashes across locations. Hence, explanatory variables for highway segment safety analysis considered were geometric characteristics like curvature change rate, slope change rate, transverse slope and traffic characteristics in the form of average daily traffic, light vehicle traffic, light commercial vehicle traffic, heavy vehicle traffic, two-wheelers, non-motorised traffic volume and operating speed were analysed against dependent variable as crash count per 200 m per year. Safety performance functions involving the explanatory variables are calibrated to predict

ABOUT THE AUTHORS

Naveen Kumar ChikkaKrishna holds a Master's degree in Traffic and Transportation Engineering from National Institute of Technology Calicut. He is pursuing his doctoral research in the Department of Civil Engineering, Indian Institute of Technology Roorkee. His research interests include road safety analysis, highway geometric design, traffic impact analysis and transportation planning.

Manoranjan Parida is a professor in Civil Engineering Department and dean (SRIC) at IIT Roorkee. He is also holding the chair—MoRTH. His areas of specialisation include urban transportation planning, traffic safety, etc. He is recipient of Jawaharlal Nehru Birth Centenary Award, Medals and Prizes of Institution of Engineers (India).

Sukhvir Singh Jain is a professor in Civil Engineering Department at IIT Roorkee. His main areas of specialisation include pavement management system, transport infrastructure systems, urban transport planning and design, transport environment interaction, intelligent transport system, integrated development of public transport system and road traffic safety.

PUBLIC INTEREST STATEMENT

In this study, we develop safety performance functions considering various elements involved in the planning, design and operation of four-lane national highways in India. Explanatory variables for highway segment safety analysis considered were geometric characteristics and traffic characteristics analysed against dependent variable as crash count and collision type. Safety performance functions involving the explanatory variables are calibrated to predict crash frequency using Poisson Weibull technique and crash types are predicted using ordered logit model. Model results suggest that increase in traffic volume leads to higher probability of crash risk and traffic safety is significantly distorted by higher curvature change rate values.

crash frequency using Poisson Weibull technique and crash types are predicted using ordered logit model. Model results suggest that increase in traffic volume leads to higher probability of crash risk and traffic safety is significantly distorted by higher curvature change rate values.

Subjects: Civil, Environmental and Geotechnical Engineering; Technology; Transportation Engineering; Highway Safety; Bayesian Statistics

Keywords: crash frequency; crash type; Poisson Weibull; Bayesian inference; ordered logit model

1. Introduction

Over the years due to growth in Indian economy, there has been an unprecedented increase in road transportation and vehicular traffic load on the existing highway network in India, which has led to unsafe conditions on our highways. Casualties due to crashes on these roads are increasing year after year. The road safety is further deteriorated by poor maintenance of vehicle, bad driving practice, lack of enforcement, casual attitude of road users and least concern towards the basic road safety measures by road maintaining agencies.

At present in India, more than 486,476 crashes occur annually. About 137,572 people lose their lives in such crashes. Additionally, nearly 494,893 people get injured in road crashes that lead to lifelong misery for the victims and family (Road Accidents in India, 2014). Road Safety Study can ensure that various safety deficiencies in road are reviewed so that these can be taken care at appropriate stage of road design or operation and maintenance of road in a cost-effective way. In recent years, significant effort and investment have been made to enhance highway safety. In the backdrop of resource constraints, the allocation of resources for safety improvement projects must yield maximum possible return on investment. Identifying highway locations that have the potential for crash reduction with the implementation of effective safety counter measures is therefore an important step in achieving the maximum return on safety investment.

Considerable amount of research has been done in recent years for developed countries to establish relationships between crashes and various traffic flow characteristics, geometric characteristics at highway segments and intersections. Understanding of crash-contributing factors on highway system using recent scientific approaches is yet to take off in developing countries.

Miaou and Lum (1993) investigated statistical of four regression models—two conventional linear regression models and two Poisson regression models considering highway geometric design parameters. Study revealed that Poisson regression models overestimate or underestimate the crash frequency. Miaou (1994) evaluated the performance of Poisson, zero-inflated Poisson and negative binomial regression models through maximum likelihood method to predict truck crashes in relation with road geometric parameters. Hauer (1997) describes various conventional approaches for statistical analysis of road safety engineering. Empirical Bayes (EB) approach to the analysis of road accident data is explained extensively. EB approach was adopted to eliminate the bias in estimated accident rates that arise from selection criteria. Persaud, Lord, and Palmisano (2002) studied the transferability of safety performance functions or crash prediction models to other jurisdictions. Study suggested that a single calibration factor is inappropriate and that a disaggregation by traffic volume is preferred.

Miaou and Lord (2003) analysed traffic crashes with respect to traffic flows at intersections. They challenged the assumption of fixed dispersion parameter and worked with various dispersion parameter relationships and functional forms. This study also indicated the advantages of full Bayes versus EB method. Mitra and Washington (2007) developed eight different models with explanatory factors as traffic flow and geometric factors to estimate crashes. Study suggested that model specification may be improved by testing extra variation functions for significance.

Geedipally and Lord (2008) evaluated the safety performance functions using a varying dispersion parameter which precisely estimated crashes with smaller confidence intervals. Geedipally and Lord (2010) investigated crashes as per single- and multi-vehicle crashes separately versus modelling total crash frequency. Cheng, Geedipally, and Lord (2013) evaluated the application of Poisson Weibull (PW) and Poisson Gamma (PG) models and results revealed both the techniques are competitive.

Numerous studies have been performed by many researchers on road safety analysis in India. Landge (2006) reviewed different modelling approaches adopted worldwide. Dinu and Veeraragavan (2011) implemented random parameters count model. The results suggested that the model coefficients for traffic volume, proportion of cars, motorised two-wheelers (TW) and trucks in traffic, and driveway density and horizontal and vertical curvatures are randomly distributed across the locations. Krishnan, Anjana, and Anjaneyulu (2013) applied hierarchical modelling approach to estimate crash frequency and severity of single and dual carriageway roads. Research review illustrated that the safety performance of non-urban four-lane highways was seldom investigated on Indian highways. The review indicated that there is a need to develop models for estimating the safety performance of non-urban highways using recent statistical techniques adopted in developed countries.

The study aims to (1) identify the crash contributing factors and (2) to develop safety performance functions using Bayesian inference to predict crash frequency and different crash types on a section of divided four-lane National Highway-58 in India.

2. Study approach
Real-world crash data that are properly defined can identify the key contributing factors to traffic crashes in terms of crash frequency (number of crashes per segment), crash type (direct impact collision, rear-end collision, sideswipe collision, rollover collision and skid-related collisions) and crash severity (fatal, incapacitating, non-incapacitating, animal-related crashes and property damage only). Hence, for scientific research on crashes, one first needs to have a reliable crash database. In India, there is no organised crash database maintained for in-depth research on crashes. The absence of such nationwide systematic data, seriously impede the scientific research and analysis of road crashes in India. To address this issue and to identify the root cause of crashes in India, it is necessary to fully understand the traffic and crash affecting parameters. Hence, in this study, an attempt is made to create the crash database for scientific research on crashes on four-lane divided national highway.

2.1. Study area description
The National Highway-58 connects Indian capital New Delhi to Mana, near China border in Uttrakhand state. It serves as a lifeline to the hilly part of the state. The road is strategically important being the shortest route from Delhi to international China border. The highway has length of 536 km of which 230-km length in plain and rest in the hilly terrain. The highway connects important religious destinations which attract tourists from all over the country and world throughout the year. The highway has two-lane and four-lane stretches. Traffic on the highway is mixed in nature and comprises heavy and light vehicles. Most of the highway study segment falls in rural areas (approximately 85%).

2.2. Site selection
This national highway is maintained and operated by National Highway Authority of India (NHAI) and concessionaire Western Uttar Pradesh Toll Ltd (WUPTL). The study has been done for four-lane road between km 52.00 and 130.00 to identify all safety deficiencies responsible for road crashes. Route map of study section of National Highway-58 is shown in Figure 1.

2.3. Details of road geometrics
Figure 2 shows typical cross-section of the candidate stretch under analysis. Table 1 shows the road infrastructure details for the study area. As per Indian Roads Congress (2000), the ruling design speed for National Highway-58 in plain terrain is 100 kmph.

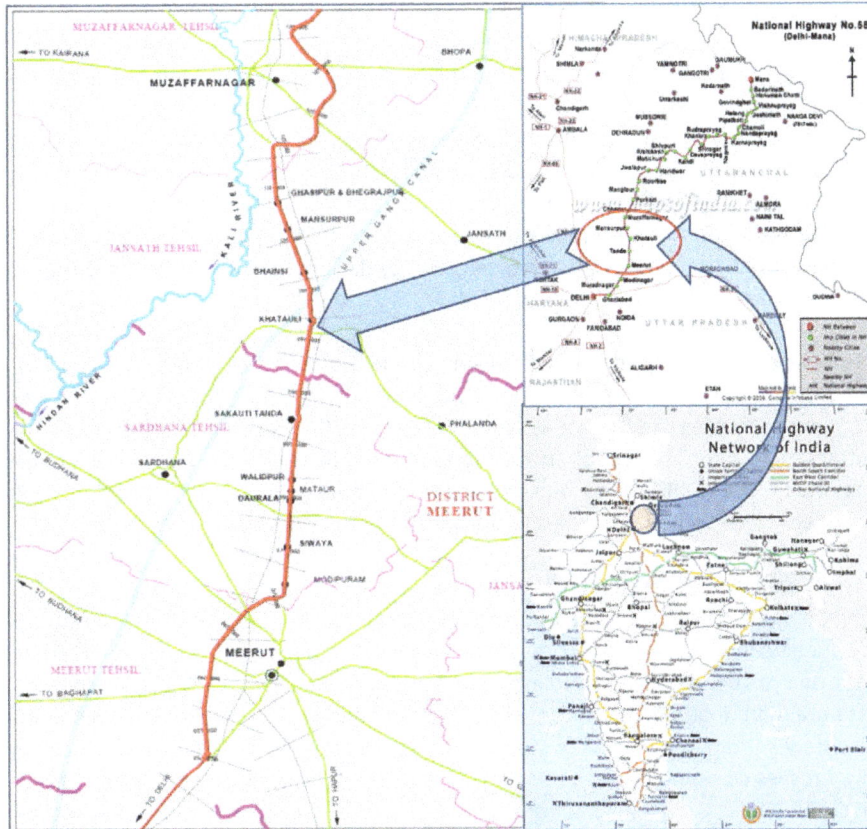

Figure 1. Study area route map of National Highway-58.

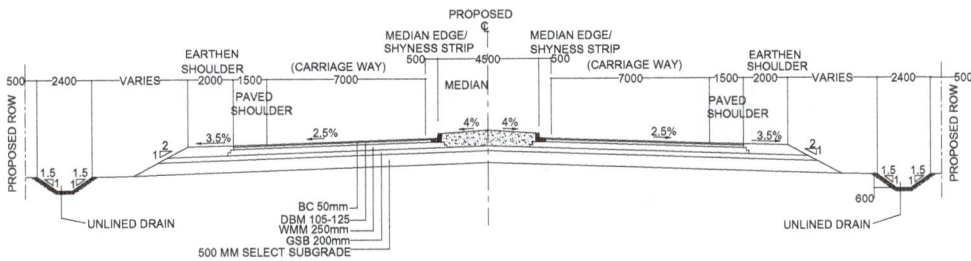

Figure 2. Typical cross-section for four-lane National Highway-58.

Note: All dimensions in millimeter.

Table 1. Road infrastructures details	
Highway features	**Description/Quantity**
Grade separated junctions	07 numbers
At-grade junctions	08 numbers
Major bridges	02 numbers
R O B	02 numbers at km 87.583 and 114.289
Underpasses	05 numbers at km 78.815, 87.400, 102.896, 118.550 and 122.175
Culverts	186 numbers
Truck lay byes	02 numbers
Bus lay bays	07 numbers
Toll plaza	1 number at km 75.990

2.4. Crash database description

From past studies, it is evident that any crash is a resultant of deficiency in any one of these factors, highway design, driver behaviour and vehicle defect. Hence, there are number of associated parameters for each of these aforementioned three factors leading to the occurrence of crashes and it is practically a challenging job to collect all these parameters. By considering the parameters applied in past crash prediction models and practical availability of data, data were collected for estimating the crash prediction models. Crash records for three years from May 2011 to April 2014 were collected from various police stations along the study section and WUPTL. Highway as-built drawings revealing the plan and profile of the study stretch and average daily traffic (ADT) for the study period was obtained from NHAI.

Classified traffic volume count survey was carried out manually at km 89.00 (near Dadri village) on NH-58 for 24 h on 6 June 2013. Later video graphic traffic volume count for morning and evening peak two hours was conducted at 15 major intersections. Assumption was made that there are no entry and exit of major traffic in between these intersections. Different traffic volumes like major highway traffic, minor road traffic, major road crossing traffic, merging and diverging traffic details were retrieved at each intersection from these video data using a C program.

2.5. Crash pattern and candidate segment

Total crash count (CC) per kilometre for the study period under consideration is as shown in Figure 3. Crash severity (for both intersection and segment crashes) and collision type (for segment crashes) statistics for the analysis period are revealed in Tables 2 and 3, respectively. From this, we can visualise that there are more than 20 crashes per kilometre throughout the analysis period along the study stretch. Safety performance functions were developed for crashes occurring on the highway segments only. Crashes occurring within a circle of 76 m (250 feet) were considered as intersection crash (Lord et al., 2008) and were excluded from the analysis data. Hence, there were 60 major segments (both directional) which were further divided into 200-m stretches. A minimum segment

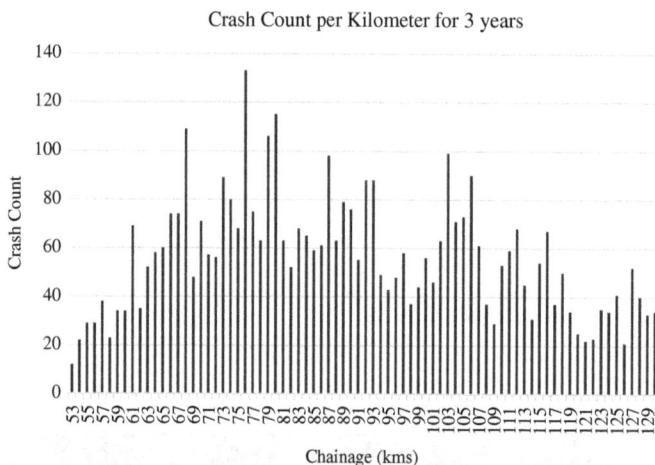

Figure 3. Total CC per kilometre along the study stretch.

Table 2. Crash severity statistics for the study stretch						
Time period	Fatal	Incapacitating injury	Non-incapacitating injury	Property damage	Animals killed	Total crashes
2011	66	237	293	365	378	1,339
2012	92	185	423	498	334	1,532
2013	81	198	361	451	399	1,490
Grand total	239	620	1,077	1,314	1,111	4,361
Share (%)	5.48	14.22	24.70	30.13	25.48	100

Time period	Direct impact collision (DI)	Rear end collision (REC)	Sideswipe collision (SSC)	Rollover collision (ROC)	Skidding collision (SKC)	Total segment crashes
2011	452	435	163	72	102	1,224
2012	460	478	157	80	202	1,377
2013	449	470	258	43	213	1,433
Grand total	*1,361*	*1,383*	*578*	*195*	*517*	4,034
Share (%)	31.21	31.71	13.25	4.47	11.86	100

Table 3. Collision-type statistics of segment crashes along the study stretch

length of 0.1 mile (≈161 m) (American Association of State Highway and Transportation Officials, 2010; Miaou, 1994) was considered to avoid low-exposure criteria and large statistical uncertainty of CC per short segment. Segments shorter than 162 m were combined with the neighbouring segments with comparably similar geometrical characteristics.

2.6. Influencing variables
CC per 200 m per year was taken as dependent variable in the crash prediction models. The safety parameters included in study were geometric characteristics like curvature change rate (CCR), slope change rate (SCR), transverse slope (TS) and traffic characteristics as ADT, traffic composition-like car (CAR) [composing cars, three- and four-wheeler autos], SUV, minibus, bus, light commercial vehicles (MTRUCK), heavy traffic volume (HTV) [two-axle trucks, multi-axle trucks, tractor-trailers], two-wheelers (TW) and non-motorised traffic volume (NMTV) [cycles, cycle rickshaws, animal-driven carts] and operating speed, V_{85} (American Association of State Highway and Transportation Officials, 2004; Jacob & Anjaneyulu, 2013) of vehicles in the traffic stream. For convenient calculations, natural log of traffic volume, speed, segment length and composition of vehicles was included in the models.

Curvature treated as CCR (Lamm, Wolhuter, Beck, & Rusher, 2001) of the segment, calculated as follows:

$$CCR \text{ [gon/m]} = \frac{\sum_i |\gamma_i|}{L} \qquad (1)$$

where γ_i is the deflection angle for a contiguous element (curve or tangent) i within a section of length L.

Tables 4 and 5 give the statistical summary of the variables selected to build the safety performance functions for crash frequency and crash-type prediction, respectively. The results in Table 4 revealed that for any subset of the independent variables, the CC exhibits overdispersion.

2.7. Model formulation and analysis
Two model forms were considered for analysis as per ease of access to required data, as in most of the situation in India, it is a challenging job to collect crash records, geometric design parameters and other variables for crash analysis.

The following generalized linear model functional form has been used in all Bayesian analyses:

Model A:

$$\mu_{ij} = \exp(Alpha + B[2]ADT_{ij} + Offset_i) \qquad (2)$$

Table 4. Statistical summaries of NH-58 crash data used in crash prediction models

Variable	Mean	Std. Dev.	Minimum	Maximum	Total
Total crashes	1.922	2.069	0	18	4,037
Segment length (m)	204.761	23.754	162	350	143,333
CCR (gon/m)	0.024	0.038	0	0.318	–
TS (%/m)	−2.048	2.353	−7	7	–
SCR (gon/m)	0.016	0.210	−1.589	1.543	–
Speed (kmph)	85.182	5.207	75.9	95	–
MO (no./segment)	0.087	0.282	0	1	61
ADT (veh/day)	23,027.662	6,006.542	13,563	40,936	–
CAR (veh/day)	10,276.515	2,935.922	5,056	19,264	–
SUV (veh/day)	3,063.441	921.151	1,409	6,838	–
MBUS (veh/day)	240.584	111.594	69	569	–
MTRUCK (veh/day)	6,034.901	2,241.029	2,664	12,208	–
BUS (veh/day)	730.959	287.526	357	1,793	–
HTV (veh/day)	1,538.635	1,106.875	693	6,693	–
TW (veh/day)	534.711	169.224	231	1,146	–
NMTV (veh/day)	608.405	357.652	35	1,523	–

Notes: Total Crashes (CC), Segment Length (Offset, meters), Curvature Change Rate (CCR), Slope Change Rate (SCR), Transverse Slope (TS), Operating Speed of Traffic (Speed), Median opening (MO), Average Daily Traffic (ADT), Cars (CAR), Sport Utility Vehicles (SUV), Minibuses (MBUS), Minitrucks (MTRUCK), Buses (BUS), Heavy Traffic Volume (HTV), Two-Wheelers (TW) and Non-motorised Traffic Volume (NMTV).

Model B:

$$\mu_{ij} = \exp(Alpha + B[2]\,CCR_i + B[3]\,SCR_i + B[4]\,TS_i + B[5]\,Speed_i + B[6]\,MO_i \tag{3}$$
$$+B[7]CAR_{ij} + B[8]SUV_{ij} + B[9]MBUS_{ij} + B[10]MTRUCK_{ij} + B[11]BUS_{ij}$$
$$+B[12]HTV_{ij} + B[13]TW_{ij} + B[14]NMTV_{ij} + Offset_i)$$

3. Modelling methodology applied

3.1. Poisson–Weibull (PW) model

As the name suggests, PW distribution is a mixture of Poisson and Weibull distribution. PW model is similar to most Poisson-based distributions (e.g. Poisson-gamma and Poisson-lognormal), it is also designed to accommodate the overdispersion. Interested readers are referred to Cheng et al. (2013) for further detailed information. The number of crashes "Y_{it}" for a particular ith site and time period t when conditional on its mean μ_{it} is Poisson distributed and independent over all sites and time periods.

$$Y_{it}|\mu_{it} \sim \text{Poisson}(\mu_{it}) \text{ where}, i = 1, 2, \ldots, I \text{ and } t = 1, 2, \ldots T \tag{4}$$

The mean of the Poisson is structured as:

$$\mu_{it} = \widehat{\mu_{it}}\varepsilon_{it} = f(X;\beta) \cdot \varepsilon_{it} \tag{5}$$

$$\widehat{\mu_{it}} = f(X;\beta) = \exp\left(\beta_0 + \sum_{j=1}^{q} \beta_j X_{jt}\right) \text{ where}, j = 1, 2, 3 \ldots q \text{ and } t = 1, 2, 3 \tag{6}$$

$$\varepsilon_{it} \sim \text{Weibull}(v, \lambda) \tag{7}$$

where $f(\cdot)$ is a function of the covariates (X); β is a vector of unknown regression coefficients; and ε_{it} is the model error independent of all the covariates.

Table 5. Statistical summaries of NH-58 crash data used in crash severity model					
Variable	Mean	Std. Dev.	Minimum	Maximum	Total
Direct impact collision (Code: 1)	–	–	–	–	1,361
Rear end collision (Code: 2)	–	–	–	–	1,383
Sideswipe collision (Code: 3)	–	–	–	–	5,78
Rollover collision (Code: 4)	–	–	–	–	5,17
Skidding collision (Code: 5)	–	–	–	–	195
Crash type levels	2.207	1.177	1	5	8,904
Segment length (m)	210.9	34.18495	162	350	850,621
CCR (gon/m)	0.02442	0.036	0	0.318	–
SCR (gon/m)	0.0154	0.215	−1.589	1.543	–
TS (%/m)	−2.14	2.232	−7	7	–
Speed (kmph)	84.55	5.289	75.9	95	–
MO (no./segment)	0.1326	0.339	0	1	535
ADT (veh/day)	23,383	6,158.987	13,563	40,936	–
CAR (veh/day)	10,346	3,055.177	5,057	19,263	–
SUV (veh/day)	3,089	870.2998	1,409	6,836	–
MBUS (veh/day)	233.39	101.3184	68	568.29	–
MTRUCK (veh/day)	542.2	176.0962	230.9	1,145.1	–
BUS (veh/day)	745.7	260.21	356.0	1,792	–
HTV (veh/day)	1,443.5	985.5002	693.5	6,694	–
TW (veh/day)	6,329	2,212.998	2,664	12,209	–
NMTV (veh/day)	654.41	374.6066	35	1,523	–
Day (Type coded 1–7)	3.994	2.03946	1	7	–
Time (Coded 1–24)	12.46	6.517811	1	24	–
AHT (°C)	32.13	5.937391	21.00	41.00	–
ALT (°C)	18.5	7.45330	7.0	28.0	–
PRCPTN (mm)	2.204	8.512072	0.00	123.00	–

Notes: Segment Length (Offset, meters), Curvature Change Rate (CCR), Slope Change Rate (SCR), Transverse Slope (TS), Operating Speed of Traffic (Speed), Median opening (MO), Average Daily Traffic (ADT), Cars (CAR), Sport Utility Vehicles (SUV), Minibuses (MBUS), Minitrucks (MTRUCK), Buses (BUS), Heavy Traffic Volume (HTV), Two-Wheelers (TW) and Non-motorised Traffic Volume (NMTV), Day-type (Day), Daily average high temperature (AHT), Daily average low temperature (ALT) and Precipitation (PRCPTN).

In PW model, it is assumed that ε_{it} is independent and Weibull distributed. The Weibull probability density function (p.d.f) is given as follows:

$$f(x) = \frac{v}{\lambda}\left(\frac{x}{\lambda}\right)^{v-1}\exp\left[-\left(\frac{x}{\lambda}\right)^v\right]; \quad x > 0, \quad v > 0, \quad \lambda > 0 \tag{8}$$

where λ and v are scale and shape parameters, respectively. The p.d.f. of the Weibull distribution can fit to various shapes similar to that of the gamma, gamma-like, exponential or approximate normal distributions depending on the v values. This characteristic of PW model provides a lot of flexibility to fit different kinds of data.

The mean and variance of the Weibull distribution are:

$$E(\varepsilon) = \lambda\left(1 + \frac{1}{v}\right) \tag{9}$$

$$\text{Var}(\varepsilon) = \lambda^2\Gamma\left(1 + \frac{2}{v}\right) - \left[\lambda\Gamma\left(1 + \frac{1}{v}\right)\right]^2 \tag{10}$$

The PW distribution is defined as the mixture of Poisson and Weibull distributions such that

$$P(Y = y; \mu, \lambda, v) = \int \text{Poisson}(y; \hat{\mu}\varepsilon) \text{ Weibull}(\varepsilon; \lambda, v) d\varepsilon \tag{11}$$

The mean or expected value of the PW distribution is given as:

$$E(Y) = \hat{\mu}E(\varepsilon) = \hat{\mu} \times \left[\lambda\Gamma\left(1 + \frac{1}{v}\right)\right] \tag{12}$$

and the variance is given by:

$$\text{Var}(Y) = \hat{\mu} \times \left[\lambda\Gamma\left(1 + \frac{1}{v}\right)\right] + \hat{\mu}^2 \times \lambda^2\Gamma\left(1 + \frac{2}{v}\right) - \hat{\mu}^2 \times \left[\lambda\Gamma\left(1 + \frac{1}{v}\right)\right]^2 \tag{13}$$

3.2. Bayesian-ordered logit model

The ordered logit model is commonly implemented to analyse ordered categorical data (Greene, 2007; Xie, Zhang, & Liang, 2009; Ye & Lord, 2014). The ordered logit model uses a latent variable y^*, as shown below to determine the different crash-type outcomes.

$$y_i^* = \beta X_i + \varepsilon_i \tag{14}$$

where X is a vector of independent variables for individual crashes; β is a vector of the unknown coefficients for these variables; and ε is a random error term assumed to follow standard normal distribution across observations.

Using Equation 10, the value of the crash-type variable y_i is estimated by:

$$y_i = \begin{cases} 1, & \text{if } y_i^* \leq \gamma_1 \\ k, & \text{if } \gamma_{k-1} < y_i^* \leq \gamma_k \\ C, & \text{if } \gamma_{C-1} < y_i^* \end{cases} \tag{15}$$

where $\gamma = \{\gamma_1, \ldots, \gamma_k, \ldots \gamma_{C-1}\}$ are the threshold values for all crash severity levels coded as integers in order; $k = 1, \ldots, C$ ($C = 5$ in the paper), the five crash types under consideration are: 1 = direct impact collision (DI), 2 = rear-end collision (REC), 3 = sideswipe collision (SSC), 4 = rollover collision (ROC) and 5 = skidding collision (SKC); C is the highest ordered crash-type level.

Given the value of X_i, the probability of a crash category for an individual ith crash belonging to each category is

$$\left. \begin{array}{l} P(y_i = 1) = \frac{\exp(\beta X_i - \gamma_1)}{1 + [\exp(\beta X_i - \gamma_1)]} \\ P(y_i = k) = \frac{\exp(\beta X_i - \gamma_{k-1})}{1 + [\exp(\beta X_i - \gamma_{k-1})]} - \frac{\exp(\beta X_i - \gamma_k)}{1 + [\exp(\beta X_i - \gamma_k)]} \\ P(y_i = C) = 1 - \frac{\exp(\beta X_i - \gamma_{C-1})}{1 + [\exp(\beta X_i - \gamma_{C-1})]} \end{array} \right\} \tag{16}$$

3.3. Goodness-of-fit statistics

There are many measures that can be used for estimating how well the model fits the data. There are statistics for indicating the likelihood level of a model, that is, how well the model maximises the likelihood function. Among these statistics are:

3.3.1. Pearson chi-square

Another useful likelihood statistic is the Pearson chi-square and is defined as:

$$\text{Pearson-}\chi^2 = \sum_{i=1}^{N} \frac{(y_i - \hat{\mu}_i)^2}{\text{Var}(y_i)} \tag{17}$$

3.3.2. Deviance information criterion

The deviance information criterion (DIC) (Congdon, 2006) calculation in WinBUGS was used as the measure for comparing the different Bayesian hierarchical models; DIC assigns a penalty for the complexity of the model.

The DIC for the *j*th model is given by:

$$DIC_j = D(\bar{\theta}_j) + 2pD_j = \bar{D} + pD_j \tag{18}$$

where $D(\bar{\theta}_j)$ is the deviance $D(\bar{\theta}_j | y)$ at the posterior mean $\bar{\theta}_j$ of the parameters for model j, called *Dhat* in WinBUGS, $\bar{D}$ is the expected deviance $\bar{D} = E(D | y, \theta)$, given by the mean $\bar{D}$ of the sampled deviances $D^{(t)}$ from Markov Chain Monte Carlo (MCMC) simulations, also called *Dbar* in WinBUGS, and pD_j is the effective number of parameters in the model, computed as the difference between $\bar{D}_j$ and $D(\bar{\theta}_j)$, that is, $pD_j = \bar{D}_j - D(\bar{\theta}_j)$.

While comparing between two models, a difference in DIC value greater than 10 will rule out the model giving higher value of DIC (Spiegelhalter, Thomas, Best, & Lunn, 2003). Where the difference is less than 10, the models are reasonably similar. Smaller the DIC value indicates a better model fitting.

3.4. Model error estimates

There are statistics for estimating how well the model fit the data and the converse, how much error was in the model. Two error statistics are particularly useful.

3.4.1. Mean absolute deviation

This criterion has been proposed by Oh, Lyon, Washington, Persaud, and Bared (2003) to evaluate the fit of models. The mean absolute deviance (MAD) calculates the absolute difference between the estimated and observed values.

$$MAD = \frac{1}{n} \sum_{i=1}^{n} |\hat{\mu}_i - y_i| \tag{19}$$

The model closer to zero value is considered to be best among all the available models.

3.4.2. Mean squared prediction error

The mean squared prediction error (MSPE) is a traditional indicator of error and calculates the difference between the estimated and observed values squared.

$$MSPE = \frac{1}{n} \sum_{i=1}^{n} (\hat{\mu}_i - y_i)^2 \tag{20}$$

A value closer to 1 means the model fits the data better.

3.4.3. Sum of model deviances

The sum of model deviances (G^2) is equal to zero if the model perfectly fits the complete data-set. This is a theoretical lower bound value as the observed values y_i are integers and the estimated values $\hat{\mu}_i$ are continuous (Washington, Karlaftis, & Mannering, 2011).

$$G^2 = 2 \sum_{i=1}^{n} y_i LN \left(\frac{y_i}{\hat{\mu}_i} \right) \tag{21}$$

A model with the lowest G^2 value is superior to other models fitting to the data-set.

3.4.4. Equivalent measure to R^2

Coefficient of determination, R^2 cannot be adopted for Poisson regression models due to their non-linearity of the conditional mean in the data and heteroscedastic characteristic i.e. data variables depict sub-populations with different variabilities from others. Hence, an equivalent measure based on standardised residuals can be adopted. It is the ratio of sum of square errors to total sum of squares subtracted by one (Washington et al., 2011).

$$R_p^2 = 1 - \frac{\sum_{i=1}^{n} \left[\widehat{\mu}_i - y_i / \sqrt{\widehat{\mu}_i} \right]^2}{\sum_{i=1}^{n} \left[y_i - \bar{y} / \sqrt{\bar{y}} \right]^2} \tag{22}$$

The value ranges from 0 to 1 and a value closer to 1 indicates the fitted model explains all variability in the data.

4. Data analysis and results

Bayesian framework was implemented for modelling and inference (Gelman et al., 2013). Bayesian hierarchical framework method considers the coefficients for the covariates as random variables rather than fixed values as in classical statistical inference. Hence, the model output will be a sampled posterior distribution for each of the estimated parameter. The parameter estimation and related sampling from the joint posterior probability distribution of multiple variables can be obtained by means of MCMC process using Gibbs sampler as in WinBUGS. As the Bayesian formulation requires priors for all unknown parameters, non-informative normal priors for β's and Weibull priors for error terms were adopted. For each model, three Markov chains were used in the coefficient estimation process with 20,000 iterations, and 10,000 iterations were used in burn-in process and were discarded.

Convergences of the models were inspected by monitoring the plots in WinBUGS and Gelman Rubin (G–R) diagnostics for the model parameters. If all values were within a zone without strong periodicities or tendencies, the model was considered convergent.

Output analysis and diagnostics for MCMC simulations were carried out on coda files from WinBUGS using coda package in R (R Core Team, 2013). The G–R convergence statistic is generally used to verify that the simulation runs converged properly. For model comparison, it was suggested that convergence was achieved when the G–R statistic was less than 1.2 (Mitra & Washington, 2007).

4.1. Crash frequency prediction models

4.1.1. General output interpretation for model form 'A'

This model depicts the effect of important individual independent variable, ADT on crash prediction. As revealed in Table 6, coefficient of ADT has significantly positive effect on crash occurrence. Both the techniques estimate nearly same coefficient value of ADT. The coefficient sign is comparable with past researchers (Dinu & Veeraragavan, 2011). The output of the models can be best evaluated by their goodness-of-fit measures as in Table 6, PW model reveals χ^2 of 2,136, MAD of 4.67, MSPE of 13.07, DIC of 7,166.75 and G^2 as 110.38. Equivalent measure toR_p^2 is 0.559 for major segment crash predictions.

4.1.2. General output interpretation for model form "B"

The estimated coefficient values for the model form B is as shown in Table 7. The results indicate that based on the estimates of covariate effects, the CCR, median opening, sport utility vehicles, light commercial vehicles, buses and two-wheelers are the most positively significant variables in explaining crash risk. SCR has minor positive impact on the expectancy of crash. TS, speed, car, minibuses, heavy commercial vehicles and non-motorised traffic have inverse effect on the probability of crash. Parameter coefficient of explanatory variables for both the models is comparable with each other.

Table 6. Parameter estimates and goodness-of-fit for model form A

Parameters	Value	Std. Dev.	MC error	BCI 2.5%	BCI 97.5%
Intercept (ALPHA)	−6.732	0.863	0.012	−8.433	−5.028
ADT (B[2])	0.215	0.086	0.001	0.046	0.384
v	1.413	0.052	0.001	1.313	1.517
w*	1.000	0.033	0.001	0.934	1.066
Pearson chi-square	2,136.434				
MAD	4.666				
MSPE	13.069				
DIC	7,166.75				
G^2	110.378				
R_p^2	0.559				

Table 7. Parameter estimates and goodness-of-fit for model form B

Parameters	Value	Std. Dev.	MC error	BCI 2.5%	BCI 97.5%
Intercept (ALPHA)	6.626	2.499	0.056	1.808	11.500
CCR (B[2])	0.509	0.646	0.012	−0.756	1.768
SCR (B[3])	0.018	0.101	0.002	−0.178	0.217
TS (B[4])	−0.017	0.010	1.43E-04	−0.036	0.002
Speed (B[5])	−2.670	0.444	0.008	−3.544	−1.813
MO (B[6])	0.553	0.068	0.001	0.422	0.689
CAR (B[7])	−1.255	0.192	0.007	−1.639	−0.878
SUV (B[8])	0.975	0.184	0.006	0.613	1.338
MBUS (B[9])	−0.401	0.078	0.002	−0.551	−0.244
MTRUCK (B[10])	0.217	0.113	0.003	0.002	0.442
BUS (B[11])	0.468	0.097	0.002	0.279	0.659
HTV (B[12])	−0.156	0.091	0.003	−0.336	0.020
TW (B[13])	0.414	0.146	0.006	0.132	0.703
NMTV (B[14])	−0.067	0.059	0.002	−0.182	0.048
v	1.639	0.072	0.002	1.506	1.785
w*	0.999	0.033	0.001	0.936	1.064
Pearson chi-square	2,250.546				
MAD	4.780				
MSPE	13.767				
DIC	7,104.12				
G^2	−146.055				
R_p^2	0.571				

CCR has positive effect on crash risk supporting the past crash studies (Lamm et al., 2001). Higher steep grades on the highway have minor positive effect on probability of crash occurrence. As the TS has negative effect on crash, it depicts negative TS sites are more prone to occurrence of crashes. Speed is having the highest inverse effect on crash revealing crashes occur due to lower operating speeds of traffic. Operating speeds of traffic might be varying due to congestion, bad weather and improper geometric design consistency on the highway stretch. Median openings have direct impact on the probability of crash as the manoeuvring traffic conflict with the opposite traffic stream. Cars have indirect impact on crash frequency as they constitute for the highest share in total traffic volume and are driven cautiously and majority occupants are family members. Sport utility vehicles significantly increase the probability of crash due to their higher speeds and rash driving behaviour as observed practically in the field. Minibuses have indirect impact on crash frequency as they are driven cautiously and majority occupants are tourists. Minitrucks are the light commercial vehicles which have positive effect leading to increase the probability of crash occurrence. Buses also are driven rashly at higher speeds and two-wheelers, the most vulnerable users due to their haphazardous and unpredictable movement in traffic tend to increase the crash frequency. Heavy traffic volume has negative effect on the occurrence of crash. NMTV has slight negative impact on crash risk.

The goodness-of-fit measures are as revealed in Table 7, χ^2 of 2,250, MAD of 4.78 and MSPE of 13.77. DIC of 7,104.12 and G^2 as −146.06 are lower than *model form A* test results revealing a better fit. Equivalent measure to R_p^2 is 0.571, for major segment crash predictions and relatively greater as compared to *model form A* supporting the fact as the input information is increased, prediction improves.

4.2. Crash severity prediction model

Occurrence of different crash type depends on different parameters like driver behaviour, vehicular type and their characteristics, traffic parameters, geometric condition, weather and pavement conditions. Based on the 4,034 crash records and relevant explanatory parameters as shown in Table 4, the ordered logit model was fitted using WinBUGS software package. Normal prior distributions were chosen for all parameters (explanatory variables and the thresholds) to be estimated. Mildly informed priors were chosen as the thresholds need to be in order as $\gamma_{C-1} > \gamma_{C-2}$. Since y values are 1 through C, and priors are set to match this scale, i.e. thresholds should be approximately 1.5, 2.5, ... (C−0.5). Hence, a normal prior on each threshold with a standard deviation of about 1 unit. Using MCMC simulation, samples from the posterior distribution of each parameter can be obtained. Out of these samples, an approximate density function can be drawn for each parameter and the posterior mean with standard deviation values is determined.

The parameter estimates for each independent variable, the intercept and the threshold values of the Bayesian-ordered logit model are listed in Table 8. Model coefficient values and its sign reveal the effect of each independent variable on crash types. A positive sign of an independent variable reflects, an increase in unit value of the variable will increase the probability of occurrence of higher crash category and decrease the probability of least crash category (Washington et al., 2011).

Direct impact collision includes head-on collisions, direct impact to pedestrians, animals and objects. Other collision types considered are rear-end collision, sideswipe collision, rollover collision and skid-related collisions which occur mainly in collisions involving two-wheelers. Crash-type prediction model considered additional parameters like day-type, hourly time period, daily average high temperature, daily average low temperature and daily precipitation values in addition to crash frequency parameters. Each day was coded numerically starting with Sunday as 1 to Saturday as 7. Hourly time period in 24 h format was also coded numerically with 1:00 am as 1 to midnight 12:00 am as 24.

CCR, TS, median opening, sport utility vehicles, minibuses, and minitrucks, two-wheelers in the traffic stream, time period, average daily higher temperature and precipitation have positive impact revealing their effect to lie in the higher portion of the crash-type scale under consideration. The estimated parameters for SCR, speed, presence of cars, buses, heavy vehicles, non-motorised vehicles in the traffic stream, type of day and average daily lower temperature have significant negative impact on type of crash owing to lower crash-type categories.

Sports utility vehicles, CCR and two-wheelers are the most positive significant parameters revealing the higher probability of crash types lying in the top order like direct impact, rear-end and sideswipe collision. Sports utility vehicles have comparatively higher engine power and come with recent vehicle technologies which make the driver drive rashly as practically observed on the study stretch.

Table 8. Parameter estimates and goodness-of-fit for Bayesian ordered logit crash-type model					
Parameters	Value	Std. Dev.	MC error	2.50%	97.50%
CCR (B[1])	0.792	0.879	1.03E-02	−0.915	2.535
SCR (B[2])	−0.236	0.132	1.12E-03	−0.499	0.023
TS (B[3])	0.008	0.013	1.24E-04	−0.018	0.034
Speed (B[4])	−0.681	0.605	8.23E-03	−1.862	0.490
MO (B[5])	0.156	0.089	7.86E-04	−0.019	0.331
CAR (B[6])	−1.458	0.266	7.53E-03	−1.977	−0.937
SUV (B[7])	1.018	0.266	7.47E-03	0.498	1.542
MBS (B[8])	0.118	0.105	1.56E-03	−0.086	0.323
BUS (B[9])	−0.045	0.132	2.05E-03	−0.304	0.213
MTR (B[10])	0.112	0.146	2.58E-03	−0.177	0.390
HTV (B[11])	−0.191	0.121	2.50E-03	−0.429	0.047
TW (B[12])	0.449	0.201	5.88E-03	0.061	0.837
NMTV (B[13])	−0.166	0.081	1.87E-03	−0.325	−0.007
Day (B[14])	−0.007	0.014	1.26E-04	−0.035	0.021
Time (B[15])	0.008	0.004	3.72E-05	0.000	0.017
AHT (B[16])	0.015	0.014	4.55E-04	−0.012	0.044
ALT (B[17])	−0.016	0.012	3.69E-04	−0.038	0.007
PRCPTN (B[18])	0.005	0.004	4.26E-05	−0.002	0.012
Off (B[19])	0.306	0.221	2.00E-03	−0.125	0.740
Deviance	11450	6.687	8.39E-02	11440	11460
$\tau[1]$	−0.680	0.034	3.97E-04	−0.748	−0.615
$\tau[2]$	0.772	0.034	5.16E-04	0.705	0.839
$\tau[3]$	1.564	0.042	6.75E-04	1.483	1.645
$\tau[4]$	3.012	0.074	9.28E-04	2.867	3.158
DIC	11,472.800				

Higher change in curvature without any caution to the driver makes him vulnerable to collide with the oncoming object at higher speeds. As the driver cannot judge and react within the limited time. Two-wheelers in the traffic stream have major impact on higher crash levels due to their haphazardous driving behaviour and as the rider loses control of the vehicle as compared to other four-wheelers.

TS has minimal direct effect on crash type revealing higher slope values are prone to lie in the middle of the crash-type scale. The number of median openings per segment is also significantly related to higher crash-type categories of the ordinal scale. This indicates the manoeuvring traffic opposes the oncoming traffic affecting their speed and resulting in mid-to-higher crash levels. Minibuses and minitrucks, mainly tourist vehicles, are having less direct impact on crash type leading from mid-to-higher category crash types. Time period has lower positive effect on crash type depicting mid-category crash type occurs the most during noon period. Higher crash categories have higher probability during evening and until midnight. Lower crash categories occur the most during start of the day. Average higher temperature also have minimal direct effect on crash type revealing their effect to lie in the middle to higher crashes of the ordinal scale. Precipitation too has lower effect illustrating the probability of crash type tends from mid-to-higher level.

SCR has indirect relationship with crash type depicting the category tending from sideswipe to skidding collision. Cars are having significantly negative impact on crash type as they are mostly driven at higher speeds leading to rollover and skidding collision. Buses have minimal indirect effect

Table 9. Measures of fit for Bayesian ordered logit crash severity model		
Crash severities/measures of fit	**Actual shares**	**BOL predictions**
Direct impact collision	1,361	1,364
Rear end collision	1,383	1,386
Sideswipe collision	578	576
Rollover collision	517	515
Skidding collision	195	194
RMSE	–	**2.32**
MPAE	–	**3.37E-03**

Table 10. Cross-tabulation of outcomes and predicted probabilities						
$y(i, j)$	**1**	**2**	**3**	**4**	**5**	**Total**
1	474	466	190	168	63	1,361
2	462	476	199	179	67	1,383
3	191	199	84	76	29	578
4	172	178	75	67	25	517
5	65	67	28	25	10	195
Total	1,364	1,386	576	515	194	4,034

Notes: Row = actual, Column = Prediction, Model = Ordered Logistic Value (j, m) = Sum$(i = 1, N)y(i, j)*p(i, m)$. Column totals may not match cell sums because of rounding error.

on crash type revealing crash category to lie in the middle of the crash-type scale. Heavy vehicles and non-motorised vehicles too have minimal inverse effect on crash type revealing their effect to lie in the middle of the severity scale. Daily average lower temperature has negative effect on the crash type leading to lower crash categories on the ordinal scale.

Root mean square error (RMSE) of 2.32 and mean absolute percentage error (MAPE) of 0.0033 were computed by comparing the predicted and observed shares of each crash severity level. Computed goodness-of-fit results and predicted percentage shares of each crash severity categories are presented in Table 9. Cross-tabulation of the expected outcomes and the predicted probabilities from the ordered logit model is as shown in Table 10.

5. Conclusions

This study identifies the contributing factors effecting crash frequency and different crash types on a divided four-lane national highway in India using Bayesian statistical models. Models were developed for three-year crash records on 143.5 km (both directional) of a divided four-lane highway (NH-58). The results of this study can help the policy-makers, decision-makers and road safety stakeholders in optimising allocated funds and planning effective safety countermeasures.

This paper presents two approaches viz. Poisson Weibull technique to analyse road traffic crash frequency and ordered logit model to predict crash type on national highways in India. Model parameters considered for crash analysis were CCR, SCR, TS and traffic characteristics as ADT, light vehicle traffic, light commercial vehicle traffic, heavy traffic volume (HTV), TW and NMTV, operating speed of traffic stream, day-type, hourly time period, daily average high and low temperatures and precipitation. Models were developed using WinBUGS statistical software which facilitate the computation of posterior distributions along with a measure, DIC for model comparisons. Two model forms were analysed as shown in Equations 21 and 22. Hazardous location ranking, development of crash modification factors using the crash prediction models assist the highway professionals to better understand the effect of crash contributing factors and to mitigate the same cost effectively.

Results from the crash frequency analysis accompanied by detailed examination of the road crash model, following variables significantly affect crash frequency.

(1) Model results suggest that increase in traffic volume lead to higher probability of crash risk.

(2) Model outputs strongly suggest that traffic safety is significantly distorted by higher CCR values.

(3) Operating speed of traffic stream has indirect impact on the occurrence of crashes.

(4) Cars, minibus and HTV in traffic stream have indirect impact on crashes.

(5) MO, sport utility vehicles, minitruck, bus and two wheeler share are affecting significantly higher on crash occurrence.

Following are the parameters significantly affecting different crash type:

(1) CCR, sport utility vehicles and two-wheelers have higher impact on crash type suggesting probability of higher level category crashes to occur is more.

(2) Operating speed of traffic stream has inverse effect on crash type revealing occurrence of lower collision types of ordinal scale.

(3) Cars, bus, heavy and non-motorised vehicles in traffic stream and average daily lower temperature have indirect impact on crash type.

(4) MO, minibus and minitruck have lower direct impact on crash severity revealing their effect to lie in the middle of the ordinal scale.

(5) Type of day, average daily higher temperature and precipitation have minor negative impact on crash type.

From the present study, following countermeasures/safety measures emerge from the outputs of safety performance functions in terms of enforcement and engineering terms are to improve safety:

(1) Segregation of traffic by providing dedicated lanes to reduce crashes by segregating different vehicle categories in the traffic stream.

(2) To improve curvature change rate by enhancing the deficient curve locations.

(3) Redesigning the critical transverse slope areas.

(4) Closure of illegal median openings and redesigning the unsafe median openings by providing storage lanes and enhancing by proper sign boards and pavement markings.

(5) Increasing police patrolling on the highway to enforce drivers to abide by traffic rules and regulations.

(6) To install electronic sign boards visible during night and bad weather conditions.

Acknowledgements
The inputs received from MORTH Chair Professor sponsored by Ministry of Road Transport & Highways, Government of India is thankfully acknowledged. Naveen Kumar ChikkaKrishna also wishes to express his gratitude to NHAI and WUP Toll Ltd for providing the data used for analysis.

Funding
The authors received no direct funding for this research.

[1] Department of Civil Engineering, Transportation Engineering Group, Indian Institute of Technology Roorkee, Roorkee 247667, Uttarakhand, India.

Author details
Naveen Kumar ChikkaKrishna[1]
E-mail: cnaveenkh@gmail.com
ORCID ID: http://orcid.org/0000-0001-9110-6043
Manoranjan Parida[1]
E-mail: mparida@gmail.com
Sukhvir Singh Jain[1]
E-mail: profssjain@gmail.com

References
American Association of State Highway and Transportation Officials. (2004). *A policy on geometric design of highways and streets* (5th ed.). Washington, DC: Author.
American Association of State Highway and Transportation Officials. (2010). *Highway safety manual* (1st ed.). Washington, DC: Author.

Cheng, L., Geedipally, S. R., & Lord, D. (2013). The Poisson–Weibull generalized linear model for analyzing motor vehicle crash data. *Safety Science, 54*, 38–42. http://dx.doi.org/10.1016/j.ssci.2012.11.002

Congdon, P. (2006). *Bayesian statistical modelling.* Chichester: Wiley. http://dx.doi.org/10.1002/9780470035948

Dinu, R. R., & Veeraragavan, A. (2011). Random parameter models for accident prediction on two-lane undivided highways in India. *Journal of Safety Research, 42*, 39–42. doi:10.1016/j.jsr.2010.11.007

Geedipally, S. R., & Lord, D. (2008). Effects of varying dispersion parameter of Poisson–gamma models on estimation of confidence intervals of crash prediction models. *Transportation Research Record: Journal of the Transportation Research Board, 2061*, 46–54. http://dx.doi.org/10.3141/2061-06

Geedipally, S. R., & Lord, D. (2010). Investigating the effect of modeling single-vehicle and multi-vehicle crashes separately on confidence intervals of Poisson–gamma models. *Accident Analysis & Prevention, 42*, 1273–1282. doi:10.1016/j.aap.2010.02.004

Gelman, A., Carlin, J. B., Stern, H. S., Dunson, D. B., Vehtari, A., & Rubin, D. B. (2013). *Bayesian data analysis* (3rd ed.). Boca Raton, FL: CRC Press, Taylor & Francis Group.

Greene, W. H. (2007). *Econometric analysis* (7th ed.). Boston, MA: Prentice Hall.

Hauer, E. (1997). *Observational before/after studies in road safety estimating the effect of highway and traffic engineering measures on road safety.* Oxford: Emerald.

Indian Roads Congress. (2000). *Geometric design standards for rural (non-urban) highways (IRC: 73-1980).* New Delhi: Author.

Jacob, A., & Anjaneyulu, M. V. L. R. (2013). Operating speed of different classes of vehicles at horizontal curves on two-lane rural highways. *Journal of Transportation Engineering, 139*, 287–294. doi:10.1061/(ASCE)TE.1943-5436.0000503

Krishnan, M. J., Anjana, S., & Anjaneyulu, M. V. L. R. (2013). Development of hierarchical safety performance functions for urban mid-blocks. *Procedia - Social and Behavioral Sciences, 104*, 1078–1087. doi:10.1016/j.sbspro.2013.11.203

Lamm, R., Wolhuter, K. M., Beck, A., & Rusher, T. (2001). Introduction of a new approach to geometric design and road safety. In *SATC 2001.* Pretoria: CSIR International Convention Centre.

Landge, V. S. (2006). *Development of methodology for road safety for a national highway* (PhD Thesis). Indian Institute of Technology Roorkee, Roorkee.

Lord, D., Geedipally, S. R., Persaud, B. N., Washington, S. P., van Schalkwyk, I., Ivan, J. N., ... Jonsson, T. (2008).

Methodology to predict the safety performance of rural multilane highways (NCHRP Web-Only Document 126). Washington, DC: National Cooperation Highway Research Program.

Miaou, S.-P. (1994). The relationship between truck accidents and geometric design of road sections: Poisson versus negative binomial regressions. *Accident Analysis & Prevention, 26*, 471–482.

Miaou, S. P., & Lord, D. (2003). Modeling traffic crash-flow relationships for intersections: Dispersion parameter, functional form, and Bayes versus empirical Bayes methods. *Transportation Research Record, 1840*, 31–40. http://dx.doi.org/10.3141/1840-04

Miaou, S. P., & Lum, H. (1993). Modeling vehicle accidents and highway geometric design relationships. *Accident Analysis & Prevention, 25*, 689–709.

Mitra, S., & Washington, S. (2007). On the nature of over-dispersion in motor vehicle crash prediction models. *Accident Analysis & Prevention, 39*, 459–468. doi:10.1016/j.aap.2006.08.002

Oh, J., Lyon, C., Washington, S., Persaud, B., & Bared, J. (2003). Validation of FHWA crash models for rural intersections: Lessons learned. *Transportation Research Record, 1840*, 41–49. http://dx.doi.org/10.3141/1840-05

Persaud, B., Lord, D., & Palmisano, J. (2002). Calibration and transferability of accident prediction models for urban intersections. *Transportation Research Record, 1784*, 57–64. http://dx.doi.org/10.3141/1784-08

R Core Team. (2013). *R: A language and environment for statistical computing.* Vienna: R Foundation for Statistical Computing.

Road Accidents in India. (2014). *Government of India.* New Delhi: Ministry of Road Transport & Highways, Transport Research Wing.

Spiegelhalter, D., Thomas, A., Best, N., & Lunn, D. (2003). *WinBUGS user manual.* Cambridge: MRC Biostatistics Unit.

Washington, S., Karlaftis, M. G., & Mannering, F. L. (2011). *Statistical and econometric methods for transportation data analysis* (2nd ed.). Boca Raton, FL: Chapman & Hall, CRC.

Xie, Y., Zhang, Y., & Liang, F. (2009). Crash injury severity analysis using Bayesian ordered probit models. *Journal of Transportation Engineering, 135*, 18–25. http://dx.doi.org/10.1061/(ASCE)0733-947X(2009)135:1(18)

Ye, F., & Lord, D. (2014). Comparing three commonly used crash severity models on sample size requirements: Multinomial logit, ordered probit and mixed logit models. *Analytic Methods in Accident Research, 1*, 72–85. doi:10.1016/j.amar.2013.03.001

7

Simulation and prediction for energy dissipaters and stilling basins design using artificial intelligence technique

Mostafa Ahmed Moawad Abdeen[1]*, Alaa El-Din Abdin[2] and W. Abbas[3]

*Corresponding author: Mostafa Ahmed Moawad Abdeen, Department of Engineering Mathematics and Physics, Faculty of Engineering, Cairo University, Giza 12211, Egypt

E-mail: mostafa_a_m_abdeen@ hotmail.com

Reviewing editor: Sanjay Shukla, Edith Cowan University, Australia

Abstract: Water with large velocities can cause considerable damage to channels whose beds are composed of natural earth materials. Several stilling basins and energy dissipating devices have been designed in conjunction with spillways and outlet works to avoid damages in canals' structures. In addition, lots of experimental and traditional mathematical numerical works have been performed to profoundly investigate the accurate design of these stilling basins and energy dissipaters. The current study is aimed toward introducing the artificial intelligence technique as new modeling tool in the prediction of the accurate design of stilling basins. Specifically, artificial neural networks (ANNs) are utilized in the current study in conjunction with experimental data to predict the length of the hydraulic jumps occurred in spillways and consequently the stilling basin dimensions can be designed for adequate energy dissipation. The current study showed, in a detailed fashion, the development process of different ANN models to accurately predict the hydraulic jump lengths acquired from different experimental studies. The results obtained from implementing these models showed that ANN technique was very successful in simulating the hydraulic jump characteristics occurred in stilling basins. Therefore, it can be safely

ABOUT THE AUTHOR

Mostafa Ahmed Moawad Abdeen graduated with honors in Civil Engineering Department from Cairo University, EGYPT; Master of Science in Engineering Mechanics at Cairo University, EGYPT; Doctor of Engineering in Engineering Mechanics at Yokohama National University, JAPAN. He was an associate professor of Engineering Mechanics 2004 at Dept. of Eng. Mathematics and Physics, Faculty of Engineering, Cairo University, professor of Engineering Mechanics 2012 at Dept. of Eng. Mathematics and Physics, Faculty of Engineering, Cairo University, and head of Engineering Mechanics Group 2013 at Dept. of Eng. Mathematics and Physics, Faculty of Engineering, Cairo University. He was involved in research field with the group: Engineering Mechanics, Computational Mechanics, Solid Mechanics, Fluid Mechanics, Heat and Mass Transfer, Artificial Intelligence Techniques. This paper is one of the important papers to design accurately water structures to avoid any damages caused by water jumps.

PUBLIC INTEREST STATEMENT

Water with large velocities can cause considerable damage to channels whose beds are composed of natural earth materials. Several stilling basins and energy dissipating devices have been designed in conjunction with spillways and outlet works to avoid damages in canals' structures. The dimensions of these structures have to be designed very accurately and their length mainly depends on the length of the hydraulic jump occurred in them. The current study utilized the artificial intelligence to predict these hydraulic structures' lengths. The authors could successfully model the jump lengths using the artificial neural networks in six types of different hydraulic structures. In addition, they provided an easy tool for the design engineers to safely design their energy dissipaters' structures.

utilized in the design of these basins as ANN involves minimum computational and financial efforts and requirements compared with experimental work and traditional numerical techniques such as finite difference or finite elements.

Subjects: Applied Mathematics; Applied Physics; Environmental Management

Keywords: energy dissipaters; stilling basins design; hydraulic jump; artificial neural networks

1. Introduction

Stilling basins and energy dissipating devices in conjunction with spillways, outlet works, and canal structures have been extensively investigated experimentally throughout the literature. Specifically, hydraulic jumps phenomenon associated with these structures have received lots of research attention throughout experimental works and traditional numerical modeling to accurately identify the different characteristics associated with the hydraulic jumps. Experimental researchers have agreed about the high cost and limited results for the applications associated with the identification of the hydraulic jumps characteristics and consequently provide safe design for stilling basins and energy dissipating devices. On the other hand, the traditional modeling research group has agreed on the complexity of their numerical modeling techniques toward obtaining accurate identification of the hydraulic jumps dimensions.

New technological modeling techniques such as artificial intelligence have proven its capability in simulating and predicting the behavior of different physical phenomena in most of the engineering fields. Artificial neural network (ANN) is one of the artificial intelligence techniques that has been utilized in civil engineering in general and in the water field area specifically.

Regarding water engineering field, several researchers have incorporated ANN technique in hydrology, groundwater, hydraulics, and reservoir operations to simulate their problems. In (Abdeen, 2006), a study for the development of ANN models to simulate flow behavior in open channel infested by submerged aquatic weeds has been presented. Guven, Gunal, and Cevik (2006) presented a study for the prediction of pressure fluctuations on sloping stilling basins using neural networks (NNs). Specifically, the authors developed an ANN model to predict the pressure fluctuations pattern beneath hydraulic jump occurring on sloping stilling basins. A study for estimating the scour characteristics downstream of a ski-jump bucket using NNs had been presented by Azmathullah, Deo, and Deolalikar (2005). Specifically, the authors developed NN structure as well as its connection weights and functions to predict the depth, location of maximum scour, and the width of scour hole. Water distribution systems using ANN had been investigated in Mansoor and Vairavamoorthy (2005). Specifically, the authors presented a modified network analysis program where nodal outflows were developed as function of pressure and secondary network characteristics. On the other hand, Abdin and Abdeen (2005) presented a study for predicting the impact of subsurface heterogeneous hydraulic conductivity on the stochastic behavior of well draw down in a confined aquifer using ANNs. Several ANN models were developed in this study to predict the unsteady two-dimensional well draw down and its stochastic characteristics in a confined aquifer.

Regarding the applicability of ANN in other engineering fields, several researchers have incorporated ANN to simulate the internal properties of different engineering materials. The internal properties of light weight aggregate concrete were investigated in Abdeen and Hodhod (2010). Abdeen and Abbas (2010) predicted the dynamic response of the seated human body using ANNs. The acoustic properties of some tellurite glasses using ANN technique were simulated and predicted in Gaafar, Abdeen, and Marzouk (2011). Moreover, the effect of natural and steel fibers on the performance of concrete using ANNs was studied in Hodhod and Abdeen (2011). These studies showed the wide applicability of ANN, not only in water research field, but also in other engineering research areas. Goh, Kulhawy, and Chua (2005) presented a comprehensive Bayesian neural network algorithm to model the relationship between the soil undrained shear strength, the effective overburden stress, and the undrained side resistance alpha factor for drilled shafts. Hossein Alavi, Hossein Gandomi, Mollahassani, Akbar Heshmati, and Rashed (2010) presented a study for the utilization of ANNs to predict the maximum dry density and optimum moisture content of soil-stabilizer mix. Alavi and Gandomi (2011) derived new models to predict the peak time domain characteristics of strong ground motions utilizing a novel hybrid method coupling ANN and simulated annealing (SA), called ANN/SA. Mollahasani, Alavi, Gandomi, and Rashed (2011) derived a new model to estimate undrained cohesion intercept (c) of soil using multilayer perceptron of ANNs.

It is quite clear from the previously presented literature that ANN technique showed its applicability in simulating and predicting the behavior of different engineering hydraulics as well as material problems. However, its utilization for the identification of hydraulic jump's characteristics is still limited and requires more applicable research. Therefore, the presented study is aimed toward enriching this area of research and consequently helping field engineers to adopt ANN hydraulics modeling in their stilling basins and energy dissipating designs by predicting the hydraulic jumps' characteristics occurring on stilling basins.

2. Problem description

The current paper investigates the applicability of utilizing the ANN technique in predicting the hydraulic jumps' characteristics, specifically its length, occurred in different experimental works performed by Bureau of Reclamation within the United States Department of Interior and published in The Engineering Monograph No. 25, Hydraulic Design of Stilling Basins and Energy Dissipaters (1984).

The experimental program carried out by this organization included six test flumes to obtain the experimental data for the hydraulic jumps. Throughout the research presented in the current manuscript, ANN technique was applied to each of the six experimental flumes' data to obtain one specific ANN model for each flume that is capable of predicting its hydraulic jump's length with an acceptable and high accuracy. These models can thereafter be utilized in serving the field hydraulic engineers in their design of the stilling basins where hydraulic jumps occur.

2.1. Experimental work

As mentioned previously, six test flumes were experimentally investigated to obtain accurate data for the different characteristics of the hydraulic jumps occurred in the stilling basins. Flumes A–E, as shown in Figures 1–6, contained overflow sections so that the jet entered the stilling basin at an angle to the horizontal. The degree of the angle varied in each test flume. In Flume F, the entering jet was horizontal, since it emerged from under a vertical slide gate. Each flume served a useful purpose either in verifying the similarity of flow patterns of different physical size or in extending the range of the experiments started in one flume and completed in others. The different flume sizes and arrangements also made it possible to determine the effect of flume width and angle of entry of the flow. Each flume contained a head gage, a tail gage, a scale for measuring the length of the jump, a point gage for measuring the average depth of flow entering the jump, and a means of regulating the tail water depth.

The discharge in all cases was measured through the laboratory venturi meters or portable venturi orifice meters. The tail water depth was measured by a point gage operating in a stilling well. The tail

Figure 1. Tests Flume A.
Notes: Width of basin 1.5 m, drop 0.9 m, and discharge 0.17 m³/s.

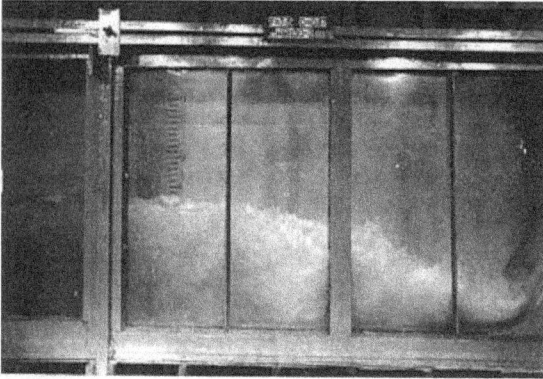

Figure 2. Tests Flume B.

Notes: Width 0.6 m, drop 1.7 m, and discharge 0.34 m³/s.

Figure 3. Tests Flume C.

Notes: Width 0.45 m, drop 3 m, discharge 1.5 m³/s, and slope 2:1.

Figure 4. Tests Flume D.

Notes: Width 1.2 m, drop 3.65 m, discharge 0.8 m³/s, and slope 0.8:1.

water depth was regulated by an adjustable weir at the end of each flume. The reader can refer to The Engineering Monograph No. 25, Hydraulic Design of Stilling Basins and Energy Dissipaters, 1984 for further details description about the entire experimental program.

Figure 5. Test Flume E.

Notes: Width 1.2 m, drop 0.15–0.45 m, and discharge 3 m³/s.

Figure 6. Test Flume F.

Adjustable tilting type, maximum slope 12°, width 0.3 m, and discharge 1.5 m³/s.

2.2. Data categories utilized for the ANN

As reported by the Bureau of Reclamation, 1984, observation of the hydraulic jump throughout its entire range required tests in all the previously described six test flumes. Specifically, this involved about 125 tests for discharges of 0.3–8.5 m³/s.

The number of flumes used enhanced the value of the results and made it possible to observe the degree of similitude obtained for the various sizes of jumps. Greatest reliance was placed on the results from the larger flumes, since the action in small jumps is too rapid for the eye to follow and, also, friction and viscosity become a measurable factor. This was demonstrated by the fact that the length of jump obtained from the two smaller flumes, A and F, was consistently shorter than that observed for the larger flumes. These jump lengths' realizations are the main outputs' type for the several developed ANN models within the current presented study.

3. Numerical models

ANN is a numerical model depends on a certain number of neurons in different layers. Every neuron acts very closely to the real neuron of the human brain. Each layer has a different function than the others. The input layer with its neurons gets the information from the external world (given data), while the hidden layers are working as detectors of these data. The output layer is the final layer of the network and it produces the required results as described in a very detailed fashion in Abdeen (2001) and Kheireldin (1998). Neuralyst software (Shin, 1996) is used to design the ANN models in the present work.

Table 1. Test flumes (cases) characteristics				
Test flume (case)	**Width characteristics**	**Width value (m)**	**Flow discharge characteristics**	**Flow discharge range (m³/s)**
Test Flume A	Maximum	1.5	Relatively small	0.9–1.5
Test Flume B	Intermediate	0.61	Intermediate	0.9–2.4
Test Flume C	Second minimum	1.5	Small	0.3–1.35
Test Flume D	Second large	0.46	Variable up to maximum	0.9–8.65
Test Flume E	Second large	1.21	Intermediate	0.74–3.05
Test Flume F	Minimum	0.3	Minimum range	0.21–1.05

4. Simulation cases

To investigate and model the hydraulic jump length using ANN technique, the experimental work performed by the Bureau of Reclamation in USA and published in 1984 was utilized in the current study. As mentioned previously, the experimental work included six flumes to capture different possible impacts of flow discharges and physical flumes' dimensions on the hydraulic jumps' length. Consequently, the current study adopts six simulation cases and develops six ANN models, one for predicting the hydraulic jump length in each experimental flume considering the other five flumes' data for training the ANN model. Table 1 summarizes all test flumes (cases) characteristics regarding their physical dimension (width) and flow discharge ranges. It is quite clear that the adopted cases for ANN development and investigation include wide range of physical dimension for the flumes and flow discharges that are expected to produce robust models that can be applied in the field for the design of stilling basins of similar characteristics.

5. Numerical models design

To develop an NN model toward simulating any physical phenomenon such as the impact of different flow discharges or physical dimensions on the hydraulic jumps length within the experimental flume mentioned previously, first, input and output variables have to be determined. Input variables are chosen according to the nature of the problem and the type of data that would be collected in the field if this was a real-field experiment. To clearly specify the key input variables for each NN simulation model and its associated outputs, Table 2 is designed to summarize all NN key input variables and outputs for all six simulation cases.

On the other hand, if the developed ANN models were to be applied to a field application, not laboratory experiment, the type of input data needs to be collected would be the same as they are listed in Table 2. Similarly, the set of output variables required for the training of the ANN models would also need to be collected and reported as they were measured in the field corresponding to their input variables conditions.

Several NN architectures are designed and tested for each of the six simulated cases investigated in the current study to finally determine the best network model to simulate and predict, very accurately, the hydraulic jump length based on minimizing the root mean square error. Table 3 shows the final NN models for each simulation case and their associated number of neurons.

Table 2. Key input and outputs variables for all six NN simulation cases					
Simulation case	**Input variables**				**Output variable**
All cases (Test Flumes A–F)	Flow discharge (m³/s)	Flume width (m)	Jump upstream depth (m)	Jump downstream depth (m)	Hydraulic jump length (m)

Table 3. The design of the developed NN models for all the simulated cases

Simulation case	Number of layers	Number of training epochs	Number of neurons in each layer				
			Input	1st hidden	2nd hidden	3rd hidden	Output
Test Flume A	4	39223	4	3	3		1
Test Flume B	5	567608	4	4	3	2	1
Test Flume C	4	82418	4	4	3		1
Test Flume D	3	161472	4	3			1
Test Flume E	4	62010	4	4	4		1
Test Flume F	5	108963	4	3	2	4	1

Table 4. Training parameters for all the ANN models

Training parameter	Value used for all ANN models
Learning rate (LR)	0.5
Momentum (M)	0.7
Training tolerance (TRT)	0.005
Testing tolerance (TST)	0.01
Input noise (IN)	0
Function gain (FG)	1
Scaling margin (SM)	0.1
Learning algorithm	Back propagation with float calculation method
Epochs per update	1.0
Epoch limit	0.0
Activation function	Hyperbolic activation function

The input and output layers represent the key input and output variables described previously in Table 2 for each simulation case. Regarding the adopted activation function within the current developed ANN models, it is important to mention here that the developed models for test Flumes A, B, C, and D incorporated the hyperbolic activation function while ANN models for test Flumes E and F utilized the sigmoid and linear activation functions, respectively.

Table 4 presents the different parameters' values for all network models developed in the current study for all the simulation cases according to their tasks.

The definitions of each parameter can be found clearly in any NN text book (Alavi & Gandomi, 2011).

6. Results and discussion

This section is mainly devoted to present each of ANN model's prediction results as well as their accuracy for each of the simulation cases (test flumes) described in the previous sections. The accuracy of each model's prediction was evaluated based on the percentage relative error computed for each single data value according to Equation 1 as follows:

$$PRE = \left(\text{Absolute Value} \left(ANN\_PR - AMV \right) / AMV \right) \times 100 \qquad (1)$$

where PRE is the percentage relative error, ANN_PR is the prediction results using the developed ANN model, and AMV is the actual measured value

6.1. Test Flume A

Test Flume A is characterized by the maximum width value among all tested flumes with relatively small water flow discharges as mentioned in Table 1. The ANN model designed to predict the

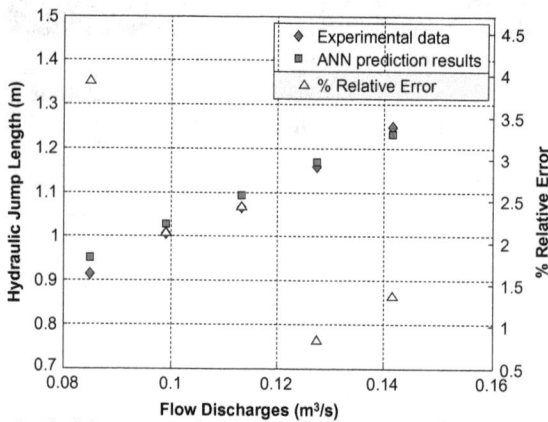

Figure 7. Comparison between experimental data and ANN model prediction results as well as percentage relative error for test Flume A.

hydraulic jump length within this flume was trained using the other five flumes data and the model's network design is described in Table 3. Figure 7 shows comparison between the experimental data for this flume and the ANN model predicted results as well as the percentage relative error between these two series. The results presented in this figure show that the developed ANN model was very successful in predicting the hydraulic jump length for this flume with maximum percentage relative error less than 4%.

6.2. Test Flume B
Test Flume B is characterized by intermediate width value among all tested flumes with intermediate water flow discharges as mentioned in Table 1. The ANN model designed to predict the hydraulic jump length within this flume was trained using the other five flumes data and the model's network design is described in Table 3. Figure 8 shows comparison between the experimental data for this flume and the ANN model predicted results as well as the percentage relative error between these two series. The results presented in this figure shows that the developed ANN model was very successful in predicting the hydraulic jump length for this flume with maximum percentage relative error less than 7%.

6.3 Test Flume C
Test Flume C is characterized by second minimum width value among all tested flumes with small water flow discharges as mentioned in Table 1. The ANN model designed to predict the hydraulic jump length within this flume was trained using the other five flumes data and the model's network

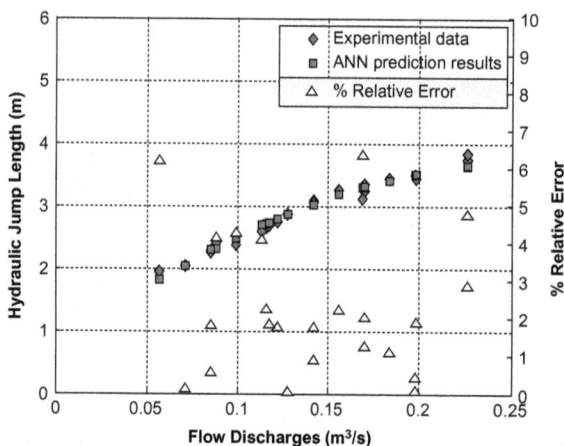

Figure 8. Comparison between experimental data and ANN model prediction results as well as percentage relative error for test Flume B.

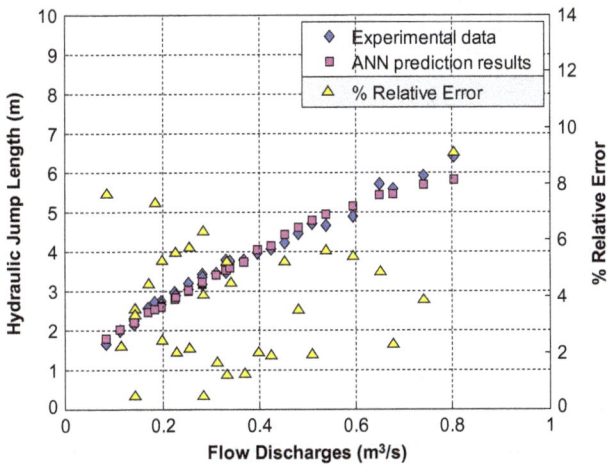

Figure 9. Comparison between experimental data and ANN model prediction results as well as percentage relative error for test Flume C.

design is described in Table 3. Figure 9 shows comparison between the experimental data for this flume and the ANN model predicted results as well as the percentage relative error between these two series. The results presented in this figure show that the developed ANN model was very successful in predicting the hydraulic jump length for this flume with maximum percentage relative error less than 5%.

6.4. Test Flume D

Test Flume D is characterized by second large width value among all tested flumes with variables water flow discharges that vary from intermediate values (3.0 cfs) up to maximum value (28.37 cfs) as mentioned in Table 1. The ANN model designed to predict the hydraulic jump length within this flume was trained using the other five flumes data and the model's network design is described in Table 3. Figure 10 shows comparison between the experimental data for this flume and the ANN model predicted results as well as the percentage relative error between these two series. The results presented in this figure show that the developed ANN model was very successful in predicting the hydraulic jump length for this flume with maximum percentage relative error equals 9%.

6.5. Test Flume E

Test Flume E is characterized by second large width value among all tested flumes with intermediate water flow discharges between 2.4 and 10.0 cfs as mentioned in Table 1. The ANN model designed to predict the hydraulic jump length within this flume was trained using the other five flumes data and the model's network design is described in Table 3. Figure 11 shows comparison between the

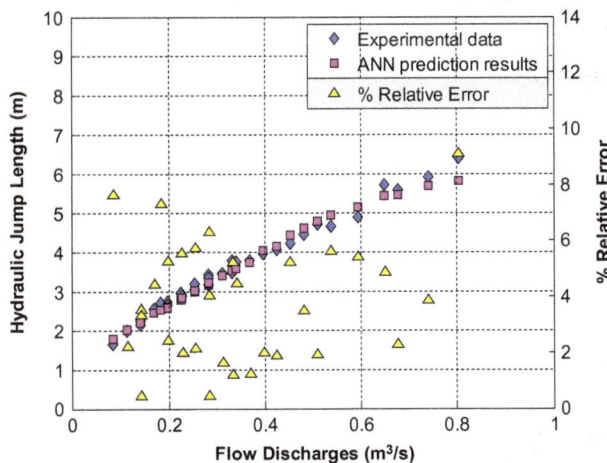

Figure 10. Comparison between experimental data and ANN model prediction results as well as percentage relative error for test Flume D.

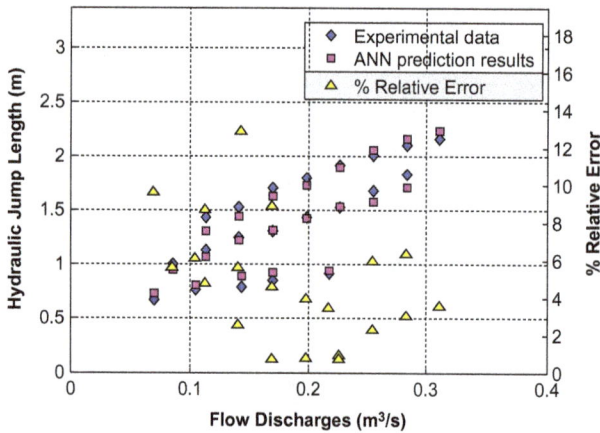

Figure 11. Comparison between experimental data and ANN model prediction results as well as percentage relative error for test Flume E.

experimental data for this flume and the ANN model predicted results as well as the percentage relative error between these two series. The results presented in this figure show that the developed ANN model was very successful in predicting the hydraulic jump length for this flume with maximum percentage relative error less than 13%.

6.6. Test Flume F

Test Flume F is characterized by minimum width value among all tested flumes with minimum water flow discharges that range between 0.68 and 3.46 cfs as mentioned in Table 1. The ANN model designed to predict the hydraulic jump length within this flume was trained using the other five flumes data and the model's network design is described in Table 3. Figure 12 shows comparison between the experimental data for this flume and the ANN model predicted results as well as the percentage relative error between these two series. The results presented in this figure show that the developed ANN model was very successful in predicting the hydraulic jump length for this flume with maximum percentage relative error less than 14%.

It can be easily seen from Figures 7 to 12 that all ANN models, designed and developed for the different simulation cases, could predict the hydraulic jump lengths in all these simulation cases separately when they are trained with different set of data. On the other hand, Table 5 summarizes the maximum percentage relative error results produced from the different models and shows that this percentage was 13.55% in simulating case F and 12.88% for case E while the other four simulation cases recorded less than 10%. In addition, Table 5 shows the correlation coefficient (R) for the

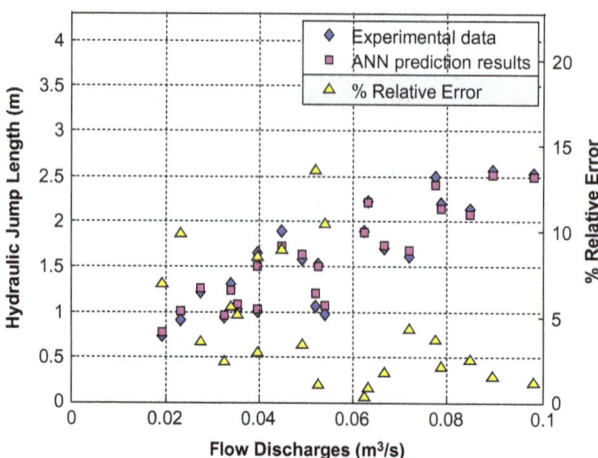

Figure 12. Comparison between experimental data and ANN model prediction results as well as percentage relative error for test Flume F.

Table 5. Summary of percentage relative errors and correlation coefficient for each simulation case

Test flume (case)	Width characteristics	Width value (m)	Flow discharge characteristics	Flow discharge range (m³/s)	Maximum percentage relative error (%)	(R) test	(R) train
A	Maximum	1.5	Relatively small	0.9–1.5	3.93	0.998	0.998
B	Intermediate	0.61	Intermediate	0.9–2.4	6.34	0.988	0.999
C	Second minimum	1.5	Small	0.3–1.35	4.76	0.996	0.998
D	Second large	0.46	Variable up to maximum	0.9–8.65	9.08	0.99	0.998
E	Second large	1.21	Intermediate	0.74–3.05	12.88	0.99	0.998
F	Minimum	0.3	Minimum range	0.21–1.05	13.55	0.993	0.998

Table 6. Summary of testing and training data numbers used for the development of each model

Test flume (case)	Number of training data points	Number of testing data points
A	115	5
B	96	24
C	105	15
D	87	33
E	99	21
F	98	22

testing and training cases for all five experiments. All these results give an indication that ANN can successfully, with high acceptable accuracy, simulate hydraulic jump phenomenon with much less computational efforts compared with the other traditional numerical approaches.

For the sake of full describing the developed model, Table 6 shows the number of testing and training data used in developing each of the models for all cases.

7. Summary and conclusion

Stilling basins and energy dissipating devices are essential in protecting open channels from the damages that can be caused by large water velocities. Lots of experimental and traditional mathematical numerical works have been performed to profoundly investigate the accurate design of these stilling basins and energy dissipaters. The most important element in this design is the identification of the hydraulic jump lengths occurred when spillways and outlet works are constructed to dissipate the energy associated with high water velocities. Traditional numerical approaches for the determination of hydraulic jumps' lengths involve lots of computational efforts and they are time consuming.

The current research introduced the utilization of one of artificial intelligence techniques to, accurately, determine the hydraulic jumps' lengths with minimum numerical efforts. Specifically, ANN was adopted in the current manuscript for this determination in six experimental flumes. Several ANN models were tested for each of the six simulation cases to finally design one model for each case. The designed ANN model for each case was trained, first, utilizing the data from the other five simulation cases. Thereafter, this case-specific model is tested to predict the hydraulic jump's length for its own case data. As it was presented in the results and discussion section, all ANN designed models were very successful in predicting their cases hydraulic jumps' lengths with high accuracy. The maximum percentage relative error encountered in all six simulation flumes was 13.55% in the minimum characteristics flume. In addition, it was observed from the current study that the numerical computations involved in the utilization of the ANN were much less than those required for the adaptations of the traditional numerical techniques.

Therefore, it can be concluded that ANN technique can be safely adopted as a supporting tool in the design of various stilling basins and energy dissipaters that are associated with hydraulic jumps occurrence. Due to its simple implementation procedures, field and design engineers can easily be encouraged to adopt this ANN technique in their design and field works that involve stilling basins and energy dissipaters design and construction. On the other hand, future research could address the robustness of the developed ANN models using sensitivity and parametric analyses.

8. Guidelines for applying the ANN to a field scale channel

As mentioned previously, if the ANN models were to be applied to a field application, not laboratory experiment, the type of input data needs to be collected from the field would be the same as they are listed in Table 2. Similarly, the set of output variables required for the training of the ANN would also need to be collected and reported as they were measured in the field corresponding to their input variables conditions. However, in most of real-field earth open channels, the natural cross-section might not be rectangular. Therefore, full cross-sections' dimensions should be considered among the input variables and not only the channel width as it is presented in Table 2.

Funding
The authors received no direct funding for this research.

Author details
Mostafa Ahmed Moawad Abdeen[1]
E-mail: mostafa_a_m_abdeen@hotmail.com
Alaa El-Din Abdin[2]
E-mail: alaa_ea_abdin@yahoo.com
W. Abbas[3]
E-mail: wael_abass@hotmail.com
[1] Department of Engineering Mathematics and Physics, Faculty of Engineering, Cairo University, Giza 12211, Egypt.
[2] National Water Research Center, Ministry of Water Resources and Irrigation, Cairo, Egypt.
[3] Basic and Applied Science Department, College of Engineering and Technology, Arab Academy for Science, Technology, and Maritime Transport (Cairo Branch), Cairo, Egypt.

References
Abdeen, M. A. M. (2001). Neural network model for predicting flow characteristics in irregular open channels. *Scientific Journal, 40*, 539–546.

Abdeen, M. A. M. (2006). Development of artificial neural network model for simulating the flow behavior in open channel infested by submerged aquatic weeds. *Journal of Mechanical Science and Technology, KSME International Journal, 20*, 1576–1589.

Abdeen, M. A. M. & Abbas, W. (2010). Prediction the biodynamic response of the seated human body using artificial intelligence technique. *Computer Science Journals (IJE), 4*, 491–506.

Abdeen, M. A. M., & Hodhod, H. (2010). Experimental investigation and development of artificial neural network model for the properties of locally produced light weight aggregate concrete. *Engineering, 2*, 408–419.

Abdin, A. E., & Abdeen, M. A. M. (2005). Predicting the impact of subsurface heterogeneous hydraulic conductivity on the stochastic behavior of well draw down in a confined aquifer using artificial neural networks. *Journal of Mechanical Science and Technology, KSME International Journal, 19*, 1582–1596.

Alavi, A. H., & Gandomi, A. H. (2011). Prediction of principal ground-motion parameters using a hybrid method coupling artificial neural networks and simulated annealing. *Computers & Structures, 89*, 2176–2194. http://dx.doi.org/10.1016/j.compstruc.2011.08.019

Azmathullah, H. Md., Deo, M. C., & Deolalikar, P. B. (2005). Neural networks for estimation of scour downstream of a ski-jump bucket. *Journal of Hydraulic Engineering, 131*, 898–908. http://dx.doi.org/10.1061/(ASCE)0733-9429(2005)131:10(898)

Gaafar, M. S., Abdeen, M. A. M., & Marzouk, S. Y. (2011). Structural investigation and simulation of acoustic properties of some tellurite glasses using artificial intelligence technique. *Journal of Alloys and Compounds, 509*, 3566–3575. http://dx.doi.org/10.1016/j.jallcom.2010.12.064

Goh, A. T. C., Kulhawy, F. H., & Chua, C. G. (2005). Bayesian neural network analysis of undrained side resistance of drilled shafts. *Journal of Geotechnical and Geoenvironmental Engineering, 131*, 84–93. http://dx.doi.org/10.1061/(ASCE)1090-0241(2005)131: 1(84)

Guven, A., Gunal, M., & Cevik, A. (2006). Prediction of pressure fluctuations on sloping stilling basins using neural networks. *Canadian Journal of Civil Engineering, 33*, 1379–1388.

Hodhod, H., & Abdeen, M. A. M. (2011). Simulation and prediction for the effect of natural and steel fibers on the performance of concrete using experimental analyses and artificial neural networks numerical modeling. *KSCE Journal of Civil Engineering, 15*, 1373–1380. http://dx.doi.org/10.1007/s12205-011-1053-8

Hossein Alavi, A., Hossein Gandomi, A., Mollahassani, A., Akbar Heshmati, A., & Rashed, A. (2010). Modeling of maximum dry density and optimum moisture content of stabilized soil using artificial neural networks. *Journal of Plant Nutrition and Soil Science, 173*, 368–379. http://dx.doi.org/10.1002/jpln.200800233

Kheireldin, K. A. (1998, August). Neural network application for modeling hydraulic characteristics of severe contraction. In *Proceeding of the Third International Conference, Hydroinformatics* (pp. 24–26). Copenhagen, Denmark.

Mansoor, M. A. M., & Vairavamoorthy, K. (2005). Application of neural networks for estimating nodal outflows as a function of pressure in water distribution systems. In *Proceeding of the ASCE International Conference on Computing in Civil Engineering*. Cancum, Mexico.

Mollahasani, A., Alavi, A. H., Gandomi, A. H., & Rashed, A. (2011). Nonlinear neural-based modeling of soil cohesion intercept. *KSCE Journal of Civil Engineering, 15*, 831–840. http://dx.doi.org/10.1007/s12205-011-1154-4

Shin, Y. (1996). *NeuralystTM user's guide* [Neural Network Technology for Microsoft Excel]. Monrovia, CA: Cheshire Engineering Corporation.

United States Department of Interior Report. (1984). *Hydraulic design of stilling basins and energy dissipaters* (The Engineering Monograph No. 25).

8

Heavy metal contamination of groundwater due to fly ash disposal of coal-fired thermal power plant, Parichha

Chanchal Verma[1], Sangeeta Madan[1]* and Athar Hussain[2]

*Corresponding author: Sangeeta Madan, Department of Environmental Sciences, Gurukul Kangri University, Haridwar, 249407, Uttarakhand, India

E-mail: snmadan21@gmail.com

Reviewing editor: Shashi Dubey, Hindustan College of Engineering, India

Abstract: The present study focused on the groundwater contamination due to fly ash disposal of coal-fired thermal power plant into a non liner ash pond. Six villages were selected as study site around ash pond of Parichha thermal power plant, Jhansi. Groundwater samples were collected on seasonal basis; winter season (January 2015), pre monsoon season (May 2015), and monsoon season (August 2015) using composite sampling method. Five heavy metals (Pb, Ni, Cr, Mn, and Fe) were detected in coal, fly ash, and groundwater samples. Heavy metal concentration in coal and fly ash was assessed by Energy Dispersive X-ray Fluorescence, while AAS was used for groundwater assessment. Heavy metal concentration in groundwater was ranged as Pb (0.170–0.581 ppm), Ni (0.024–0.087 ppm), Fe (0.186–11.98 ppm), Cr (0.036–0.061 ppm), and Mn (0.013–0.178 ppm). The observed results revealed the exceeding value of heavy metals prescribed by WHO for groundwater.

Subjects: Environmental Health; Pollution; Water Science

Keywords: groundwater; fly ash; thermal power plant; ash pond; heavy metal; environmental sustainability engineering

1. Introduction

Heavy metal contamination in air, soil and water is a global problem that is a growing threat to human beings. There are hundreds of sources of heavy metal pollution, including coal combustion in thermal power plants (Khan, Din, Ihsanullah, & Ahmad, 2011). Fly ash produced during the combustion of coal contains several toxic heavy metals like lead (Pb), nickel (Ni), zinc (Zn), manganese (Mn), etc. The extent of heavy metals in fly ash depends on both the mineralogy and particle size distribution of the raw material being burnt and combustion temperature. Fly ash is found to be more enriched with several toxic elements as compared to its parent coal (Baba, Gurdal, Sengunalp, & Ozay, 2008) because many of trace elements present in parent coal is vaporized during the combustion.

Coal-based thermal power plants contributing to 61% of total installed capacity, are the major source of electricity generation in India (Tiwari, Bajpai, Dewangan, & Tamrakar, 2015). Thermal

ABOUT THE AUTHORS

Chanchal Verma is a research scholar and Sangeeta Madan is an assistant professor, affiliated to the Department of Environmental Sciences, Gurukul Kangri University, Haridwar, India. Athar Hussain is working as an associate professor, Gautam Buddha University, Greater Noida, India. Our research work is focused on water quality assessment.

PUBLIC INTEREST STATEMENT

The present work has highlighted the groundwater contamination with heavy metal due to uncontrolled fly ash disposal from thermal power plant and possible health impacts on the population residing in the vicinity of power plant, those are consuming groundwater for drinking purpose.

power plants generate a large amount of fly ash as solid waste material from coal combustion. Currently, the generation of fly ash from coal-based thermal power plants in India is about 131 million tons per annum (Singh, Gupta, & Guha, 2014). Disposal of such a huge quantity of fly ash is a major environmental concern. Much of the ash is disposed off in the settling ponds with effluent outlets that enter into local water ways. If the ash pond is unlined, then a significant volume of the ash leachate percolates underground to the underlying water table (Praharaj, Powell, Hart, & Tripathy, 2002).

Water quality around ash pond is altered due to the leaching of soluble ions present in fly ash. The leaching potential of heavy metals from an open system (ash pond) is expected to be greater due to diurnal and seasonal variation in temperature, moisture, and other parameters. Movement of water through materials containing soluble components significantly influences the surrounding soil, groundwater, and surface water. Due to leaching characteristics of fly ash, the soluble heavy metals gradually percolate downward to contaminate nearby groundwater aquifers. This raises the potential threat of percolation of hazardous elements from fly ash to groundwater and subsoil degradation from the ash pond (Singh et al., 2014).

Parichha thermal power plant is a coal-fired power plant in Jhansi, India. Fly ash produced as a byproduct in power plant is being disposed in form of ash slurry into a non liner ash pond. Ash slurry water contains toxic heavy metals leached from fly ash and percolates downward through soil strata and may contaminate groundwater. The objective of present study is to assess the heavy metal concentration in coal, fly ash, and groundwater in surrounding of ash pond.

2. Materials and methods

2.1. Sample collection
Coal and fly ash samples were collected from Parichha Thermal power plant, Jhansi, India. Groundwater samples were collected from villages around ash pond of Parichha thermal power plant using composite sampling method. Sampling of groundwater was carried out in winter season (January 2015), pre monsoon season (May 2015), and monsoon (August 2015). Groundwater samples were acidified with nitric acid. Three ml of nitric acid was added in one liter of sample to stabilize heavy metals (Figure 1).

2.2. Sample preparation
Fine powder of coal and fly ash samples was mixed with boric acid to make separate pellets. The mixture of sample and boric acid is filled in a ring-shaped cup and pressed with a press machine. Boric acid was used as a binding material. The process is known as pelletization. Compressed pellet, shown in Figure 2, is used for Energy Dispersive X-ray Fluorescence (EDXRF) analysis.

Figure 1. Study area map with sampling sites.

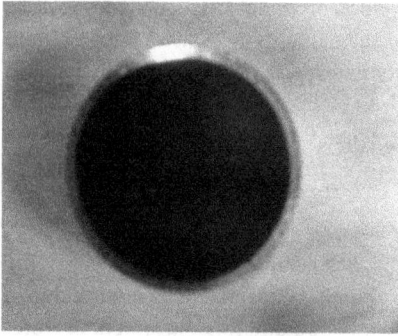

Figure 2. Pellet used in EDXRF.

Table 1. Different wavelengths used for heavy metal detection on AAS	
Element	**Wavelength (nm)**
Pb	217
Ni	232
Fe	248.3
Cr	357.9
Mn	279.5

Groundwater samples were digested with nitric acid. Five ml nitric acid was added to 100 ml of water sample. Sample was heated at a very low temperature until 10–20 ml was left. The sample was cooled and filtered with Whatmann 42 filter paper and the final volume was made up to 100 ml (American Public Health Association, 2005).

2.3. Sample analysis

Heavy metals like lead (Pb), nickel (Ni), iron (Fe), chromium (Cr), and manganese (Mn) were assessed in coal, fly ash and groundwater samples. EDXRF technique was used for coal and fly ash samples and atomic absorption spectrophotometer was used for groundwater samples. The detection wavelengths of heavy metals on atomic absorption spectrophotometer are given in Table 1.

3. Results and discussion

Fly ash is normally considered hazardous to living organisms. Heavy metals in coal can be emitted into the atmosphere with flue gas during combustion process (Yuan, 2009). Heavy metal concentration obtained in coal, fly ash and groundwater samples are presented in Tables 2 and 3 respectively.

3.1. Heavy metal concentration in coal and fly ash

Coal is carbon-rich combustible material containing organically bound mineral matter. This organic material is released during coal combustion to form an ash residue (Zandi & Russell, 2007). Various factors that control the concentrations of trace elements in the coal and ashes include element

Table 2. Heavy metal concentration in coal and fly ash samples		
Heavy metal	**Coal (ppm)**	**Fly ash (ppm)**
Pb	4.117	34.809
Ni	18.695	51.609
Fe	1,929.48	2,635
Cr	7.976	64.772
Mn	35.084	286.205

Table 3. Heavy metal concentration in groundwater samples									
Heavy metals	Season	S1	S2	S3	S4	S5	S6	Average	WHO Permissible limit (ppm)
Pb	Pre Monsoon	0.411	0.507	0.438	0.509	0.435	0.477	0.463	0.01
	Monsoon	0.534	0.581	0.501	0.511	0.533	0.501	0.527	
	Winter	0.17	0.263	0.307	0.273	0.316	0.301	0.272	
Ni	Pre Monsoon	0.061	0.073	0.074	0.079	0.04	0.024	0.059	0.02
	Monsoon	0.064	0.081	0.083	0.087	0.058	0.028	0.067	
	Winter	0.056	0.046	0.052	0.077	0.029	0.028	0.048	
Fe	Pre Monsoon	0.291	0.352	9.8	0.374	0.63	0.205	1.942	0.3
	Monsoon	0.297	0.67	11.98	0.539	0.845	0.292	2.437	
	Winter	0.25	0.348	4.59	0.35	0.426	0.186	1.025	
Cr	Pre Monsoon	0.048	0.053	0.054	0.048	0.046	0.039	0.048	0.05
	Monsoon	0.049	0.061	0.059	0.05	0.05	0.042	0.052	
	Winter	0.048	0.051	0.053	0.043	0.041	0.036	0.045	
Mn	Pre Monsoon	0.016	0.145	0.175	0.121	0.089	0.027	0.096	0.1
	Monsoon	0.027	0.162	0.178	0.142	0.091	0.034	0.106	
	Winter	0.013	0.067	0.111	0.092	0.047	0.017	0.058	

sources, modes of element occurrence, combustion conditions, volatilization–condensation mechanism, and particle size of the ash (Ram et al., 2015). Concentration of heavy metals in coal was observed as Pb (4.117 ppm), Ni (18.695 ppm), Fe (1929.48 ppm), Cr (7.976 ppm), and Mn (35.084 ppm). Concentration of heavy metals in fly ash sample was found as Pb (34.809 ppm), Ni (51.609 ppm), Fe (2635 ppm), Cr (64.772 ppm), and Mn (286.205 ppm).

Heavy metal concentration in fly ash was found to be higher than feed coal. It is due to loss of organic components during volatilization and enrichment of inorganic heavy metals. Cao et al. (2008) has reported that fly ash is a fine particle having more surface area for condensation which shows high affinity with trace elements.

3.2. Heavy metal contamination in groundwater
Amount of heavy metals released from fly ash in groundwater from stock piles of fly ash depends largely on the pH, bonding between the element and fly ash, its chemical form, and physicochemical properties of water (Fulekar & Dave, 1991; Pandey, 2014). The heavy metal concentration assessed in groundwater samples near ash pond are as follows.

3.2.1. Lead
Lead is a dangerous element; it is harmful even in small amount and enters the human body in many ways. High concentration of lead in the body can cause death or permanent damage to the central nervous system, the brain, and kidneys (Jennings, Sneed, & Clair, 1996; Mandour, 2012). The permissible limit of lead in drinking water prescribed by WHO is 0.01 ppm although in present study concentration ranged from 0.17 ppm to 0.581 ppm. Lead concentration in all groundwater samples around ash pond was observed exceeding the prescribed limit of WHO (Figure 3).

3.2.2. Nickel
Nickel is regarded as essential trace metal but toxic in large amount to the health. Concentration of Ni in groundwater was observed from 0.024 ppm to 0.087 and it was above the prescribed permissible limit (0.02 ppm) by WHO. Drinking contaminated water causes hair loss and can be related to dermatoxicity in hypersensitive human (Hanaa, Eweida, & Farag, 2000) (Figure 4).

Lead (ppm)

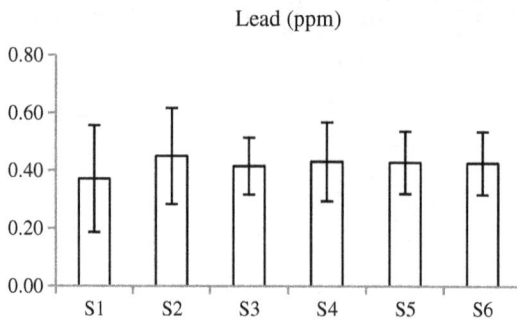

Figure 3. Average concentration of lead (Pb) in groundwater samples at different sites.

Nickel (ppm)

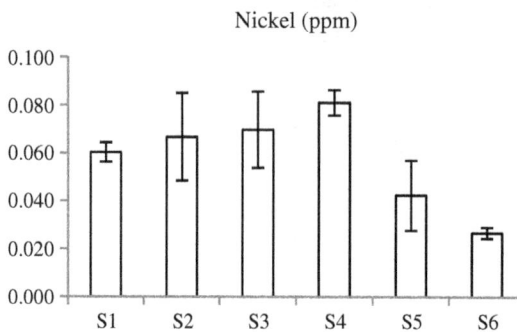

Figure 4. Average concentration of Nickel (Ni) in groundwater samples at different sites.

Iron (ppm)

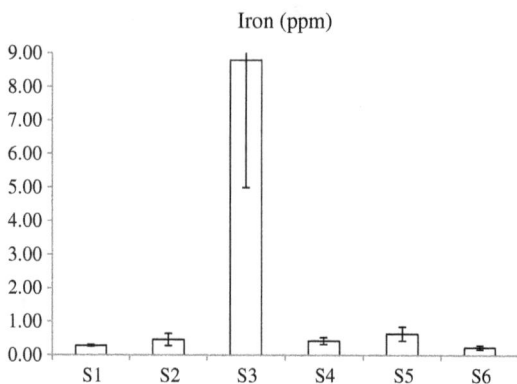

Figure 5. Average concentration of Iron (Fe) in groundwater samples at different sites.

3.2.3. Iron

Iron is readily found in soil and water. Coal and pond ash are also a rich source of iron into groundwater. Concentration of iron in groundwater samples was observed from 0.186 ppm to 11.98 ppm. Prescribed limit of WHO for iron in drinking water is 0.3 ppm while in most of the sample of groundwater it is exceeding the limit. Liver cirrhosis is found to be related to drinking water contaminated with iron (Mandour, 2012) (Figure 5).

3.2.4. Chromium

Maximum permissible limit of WHO in drinking water for chromium is 0.05 ppm but the concentration estimated in groundwater samples near ash pond ranged from 0.036 ppm to 0.061 ppm. This is showing higher concentration than prescribed value in more than 50% of samples. Chromium in excess amount can be toxic especially the hexavalent form. Long-term exposure of chromium can

Chromium (ppm)

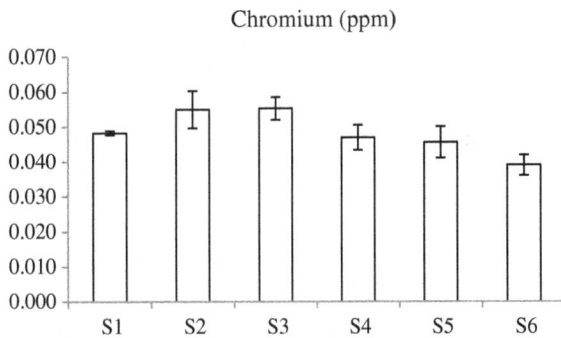

Figure 6. Average concentration of Chromium (Cr) in groundwater samples at different sites.

Manganese (ppm)

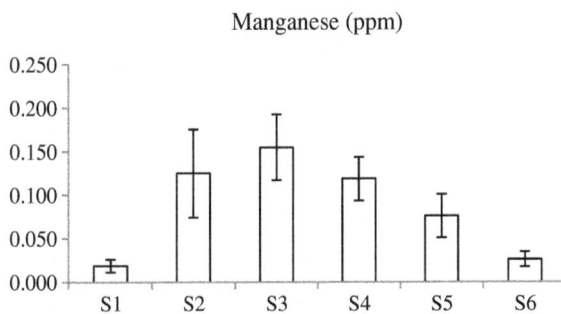

Figure 7. Average concentration of Manganese (Mn) in groundwater samples at different sites.

cause kidney and liver damage and can damage too circulatory and nerve tissue (Hanaa et al., 2000) (Figure 6).

3.2.5. Manganese

Manganese concentration in groundwater samples ranged from 0.013 ppm to 0.178 ppm while the permissible limit is 0.1 ppm (WHO). Only one-third of the samples were exceeding the prescribed limit and rests of the samples were within the limit. Higher concentration of manganese in drinking water is reported to cause neurological impairment and manganism (Figure 7).

The results obtained above have showed high concentration of heavy metals in groundwater surrounding the ash pond of thermal power plant. Groundwater is the only source of drinking water in the region of thermal power plant which is being contaminated due to fly ash disposal. The presence of heavy metals in drinking water sources is a serious matter of concern due to health damaging effects.

4. Conclusions

Fly as is observed to be enriched with various heavy metals. Fly ash in slurry form may be a major source of groundwater contamination into an unlined pond. As hydraulic water of ash slurry infiltrates through soil, heavy metals dissolve and percolate out of soil and reach the groundwater. Heavy metal concentration assessed in groundwater around ash pond was found to be exceeding the prescribed permissible values of WHO in almost all samples. This may cause significant health effects on population depending on groundwater only for drinking purpose. The heavy metal concentration was observed maximum in rainy season into groundwater. The heavy metal leaching from ash pond mainly depends on pH not and on the amount present in fly ash. Heavy metal release from ash slurry increases with decreasing pH. Lowering of pH in ash lagoon is caused due to SO_2 which may be adsorbed on the surface of fly ash or reach through atmosphere. Rain water is already found to have acidic pH up to 5.5 and dissolved atmospheric SO_2 in the surrounding of thermal power plant further lowers pH. This makes more acidic conditions into ash pond and heavy metal release from ash slurry into groundwater increases.

Acknowledgments
The authors of this paper are thankful to Gautam Buddha University for using facilities of Environmental Engineering Laboratory of Civil Engineering Department at School of Engineering.

Funding
The authors received no direct funding for this research.

Author details
Chanchal Verma[1]
E-mail: chanchalverma008@gmail.com
ORCID ID: http://orcid.org/0000-0002-2611-9929
Sangeeta Madan[1]
E-mail: snmadan21@gmail.com
ORCID ID: http://orcid.org/0000-0002-2611-9929
Athar Hussain[2]
E-mail: athariitr@gmail.com

[1] Department of Environmental Sciences, Gurukul Kangri University, Haridwar 249407, Uttarakhand, India.

[2] Civil Engineering Department, School of Engineering, Gautam Buddha University, Greater Noida 201310, Uttar Pradesh, India.

References
American Public Health Association. (2005). *Standard methods for examination of water and wastewater* (21st ed.). Washington, DC: Author.
Baba, A., Gurdal, G., Sengunalp, F., & Ozay, O. (2008). Effects of leachant temperature and pH on leachability of metals from fly ash. A case study: Can thermal power plant, province of Canakkale, Turkey. *Environmental Monitoring and Assessment, 139,* 287–298. http://dx.doi.org/10.1007/s10661-007-9834-8
Cao, Y., Cheng, C., Chen, C., Liu, M., Wang, C., & Pan, W. (2008). Abatement of mercury emissions in the coal combustion process equipped with a fabric filter baghouse. *Fuel, 87,* 332–3330.
Fulekar, M. H., & Dave, J. M. (1991). Release and behaviour of Cr, Mn, Ni and Pb in a fly-ash/soil/water environment: Column experiment. *International Journal of Environmental Studies, 38,* 281–296. http://dx.doi.org/10.1080/00207239108710673
Hanaa, M. S., Eweida, A. E., & Farag, A. (2000). *Heavy metals in drinking water and their environmental impact on human health* (pp. 542–556). Cairo: ICEHM, Cairo University.
Jennings, G. D., Sneed, R. E., & Clair, M. B. (1996). *St. Metals in drinking water* (Publication No: AG-473-1. Electronic version 3/1996). North Carolina Cooperative Extension Service.
Khan, S. A., Din, Z. U., Ihsanullah, Z. A., & Zubair, A. (2011). Levels of selected heavy metals in drinking water of Peshawar city. *International Journal of Science and Nature, 2,* 648–652.
Mandour, R. A. (2012). Human health impacts of drinking water (surface and ground) pollution Dakahlyia Governorate, Egypt. *Applied Water Science, 2,* 157–163. http://dx.doi.org/10.1007/s13201-012-0041-6
Pandey, S. K. (2014). Coal fly ash: Some aspects of characterization and environmental impacts. *Journal of Environmental Science, Computer Science and Engineering & Technology, 3,* 921–937.
Praharaj, T., Powell, M. A., Hart, B. R., & Tripathy, S. (2002). Leachability of elements from sub-bituminous coal fly ash from India. *Environment International, 27,* 609–615. http://dx.doi.org/10.1016/S0160-4120(01)00118-0
Ram, L. C., Masto, R. E., Srivastava, N. K., George, J., Selvi, V. A., Das, T. B., Pal, S. K., Maity, S., & Mohanty, D. (2015). Potentially toxic elements in lignite and its combustion residues from a power plant. *Environmental Monitoring Assessment, 187,* 4148. http://dx.doi.org/10.1007/s10661-014-4148-0
Singh, R. K., Gupta, N. C., & Guha, B. K. (2014). pH dependence leaching characteristics of selected metals from coal fly ash and its impact on groundwater quality. *International Journal of Chemical and Environmental Engineering, 5,* 218–222.
Tiwari, M. K., Bajpai, S., Dewangan, U. K., & Tamrakar, R. K. (2015). Assessment of heavy metal concentrations in surface water sources in an industrial region of central India. *Karbala International Journal of Modern Science, 1,* 9–14. http://dx.doi.org/10.1016/j.kijoms.2015.08.001
Yuan, C. G. (2009). Leaching characteristics of metals in fly ash from coal-fired power plant by sequential extraction procedure. *Microchimica Acta, 165,* 91–96. http://dx.doi.org/10.1007/s00604-008-0103-5
Zandi, M., & Russell, N. V. (2007). Design of a leaching test framework for coal fly ash accounting for environmental conditions. *Environmental Monitoring Assessment, 131,* 509–526. http://dx.doi.org/10.1007/s10661-006-9496-y

9

Theoretical model for the mechanical behavior of prestressed beams under torsion

Sérgio M.R. Lopes[1] and Luís F.A. Bernardo[2*]

*Corresponding author: Luís F.A. Bernardo, Department of Civil Engineering and Architecture, University of Beira Interior, Edificio II das Engenharias, Calçada Fonte do Lameiro, 6201-001 Covilhã, Portugal
E-mail: lfb@ubi.pt

Reviewing editor: Amir H. Alavi, Michigan State University, USA

Abstract: In this article, a global theoretical model previously developed and validated by the authors for reinforced concrete beams under torsion is reviewed and corrected in order to predict the global behavior of beams under torsion with uniform longitudinal prestress. These corrections are based on the introduction of prestress factors and on the modification of the equilibrium equations in order to incorporate the contribution of the prestressing reinforcement. The theoretical results obtained with the new model are compared with some available results of prestressed concrete (PC) beams under torsion found in the literature. The results obtained in this study validate the proposed computing procedure to predict the overall behavior of PC beams under torsion.

Subjects: Civil, Environmental and Geotechnical Engineering, Concrete & Cement, Structural Engineering

Keywords: prestress concrete, beams, torsion, theoretical model

1. Introduction

Since the original space-truss analogy (STA) proposed by Rausch (1929), several developments have been proposed to predict the behavior of reinforced concrete (RC) beams under torsion. Several studies from many authors up to 1980 can be found in the literature, for instance: Andersen, Cowan,

ABOUT THE AUTHORS

Sérgio M.R. Lopes is a full professor in the Department of Civil Engineering of the University of Coimbra, Portugal. He is a member of the Research Centre CEMUC—Centre for Mechanical Engineering, Portugal. His research interests include behavior of Structural Concrete, Bridges, High Strength Concrete Structures, Alkali Activated Materials and Structures, Non-Metallic Reinforcement, and Composite Timber-Concrete Structures.

Luís F.A. Bernardo is an assistant professor at the Department of Civil Engineering and Architecture of the University of Beira Interior, Portugal. He is a member of the Research Centre C-made: Centre of Materials and Building Technologies, Portugal. His research interests include the mechanical behavior of structural concrete and the development of new structural materials and building systems.

PUBLIC INTEREST STATEMENT

This article presents a theoretical study on the mechanical behavior of structural concrete, in particular of structural beams under torsion. A previous theoretical model from the authors is extended in order to be able to predict the global and realistic behavior of prestressed concrete beams under torsion. Such structural elements are widely used in special buildings and bridges. From this theoretical model, a new computational software was edited, which can be made available for researchers and structural engineers for research and structural design practice. The results of this study are also useful for future revisions of Codes of Practice for structural engineers.

Walsh, Lampert, Thurlimann, Elfgren, Müller, Collins, and Mitchell. The STA as a high historical value, it constitutes the base for torsion design of the European Model Code since 1978 and also the American Code since 1995.

Among the analytical models based on the STA, the variable angle truss-model (VATM) proposed by Hsu and Mo (1985) is probably the most used theoretical truss model to predict the ultimate behavior of beams under torsion. This model firstly aimed to unify the torsion design of small and large sections and also of RC and prestressed concrete (PC) beams. When compared to the STA, the VATM incorporates a compressive stress (σ)–strain (ε) relationship for the concrete in the struts which allows to account for the softening effect, instead of a conventional σ–ε relationship for uni-axial compression. Several authors proposed further versions of the VATM, for instance: Rahal and Collins (1996), Bhatti and Almughrabi (1996) and Wang and Hsu (1997). Generally, these models are only able to compute the ultimate torsional strength of the beams and are mainly focused on RC beams. VATM is able to predict the global behavior of RC and PC beams under torsion throughout the entire loading history, although very good results are observed only for high loading levels (Bernardo, Andrade, & Lopes, 2012a; Hsu & Mo, 1985). This is because VATM assumes that the beam is exten-sively cracked in all levels of loading, which is true only for high level of loading. Furthermore, for low level of loading, the concrete core of solid sections (neglected in the VATM) also influences the tor-sional stiffness of the beams.

Recently, Jeng and Hsu (2009), and Jeng, Chiu, and Chen (2010), extended the softened mem-brane model, previously developed to treat theoretically RC membrane elements under shear (Hsu & Zhu, 2002), to RC and PC beams under torsion. The new model was called softened membrane model for torsion (SMMT). Bernardo, Andrade, and Nunes (in press) extended the VATM to RC beams under torsion. The new model was called generalized softened variable angle truss-model for rein-forced concrete beams under torsion (GSVATM). These analytical methods (SMMT and GSVATM) in-corporate the influence of the concrete in tension in the perpendicular direction of the concrete struts and are able to predict the entire Torque (T)–Twist (θ) curve for all the loading levels. When compared with experimental data available in the literature (namely cracking and resistance torque, as well as the corresponding twists), the authors found that the theoretical predictions from the SMMT and the GSVATM are generally good. However, it was also found that the theoretical models don't predict very well the torsional stiffness of the post-cracked ascending branch of the T–θ curves, mainly for beams with lower reinforcement ratio. In general, the post-cracked stiffness is overesti-mated. This is because in the theoretical T–θ curves a sudden "fall" of the torques is observed after the concrete cracking (Bernardo et al., in press; Jeng et al., 2010; Jeng & Hsu, 2009). This behavior is not experimentally observed in RC and PC beams under torsion. This observation constitutes a draw-back of the referred theoretical models because it is very important to predict well the post-cracked branch of the T–θ curves, for instance to allow to check the internal deformation and stress states of the beams for the serviceability limit states.

Bernardo and Lopes (2008, 2011) developed a simple calculation procedure in order to also predict the overall theoretical behavior of RC beams under torsion. The theoretical approach was firstly performed by studying different behavioral states, each of one identified with the different states observed experimentally. These behavioral states were characterized individually by the authors by using different theories, namely:

- Linear elastic analysis in non-cracked state (State I): state characterized with Theory of Elasticity, Skew-Bending Theory, and Bredt's Thin Tube Theory;

- Linear elastic analysis in cracked state (State II): state characterized with a simplest version of the STA, assuming an angle of 45° for the concrete struts and assuming a linear σ–ε relationship for the materials; and

- Non-linear analysis: state characterized with VATM, assuming non-linear σ–ε relationships for the materials and also the softening effect for the concrete in compression.

The transition between the different theoretical states was performed by adopting semi-empirical criteria. This general approach proposed by Bernardo and Lopes (2008, 2011) has been already validated by the authors. From the comparative analysis between the theoretical predictions from the analytical model and the experimental results, the authors showed that the procedure was adequate to predict the global behavior of normal-strength (NS) and high-strength (HS) RC beams under torsion, including the post-cracked ascending branch of the $T–\theta$ curves.

In this article, the previously proposed theoretical model from Bernardo and Lopes (2008, 2011) is revised and corrected in order to incorporate PC beams. The theoretical predictions obtained from the new computing procedure, namely through the calculation of the overall $T–\theta$ curves, are compared with the results of some experimental tests on PC beams under pure torsion which were found in literature. In this study, only beams with uniform longitudinal prestress are analyzed because this situation is the simplest solution for the problem of pure torsion.

2. Research significance
When applied rationally, prestress increases resistance to cracking of a RC beam under bending. This is due to the fact that prestress induces a compressive stress which opposes tensile stress caused by the bending moment (Alnuaimi & Bhatt, 2006). Cracking can only occur when tensile stress, due to bending, exceeds compressive stress due to prestress.

Prestress also increases resistance to cracking of a RC beam under shear or under torsion. In these cases, prestress induces a compressive stress which, combined with tangential stress induced by shear or torsion, results in a state of biaxial stress (shear+compression). This state of biaxial stress delays the cracking of concrete.

In order to justify the advantages of including longitudinal prestress in beams under high levels of torsional forces, some relevant aspects should be highlighted.

Tensile strength of concrete is not incremented in direct proportion to its compressive strength. For this reason, concrete's full potential cannot be reached in structures in which resistance to cracking or ultimate member resistance is governed by tensile stresses often induced by shear or torsion. Thus, an appropriate prestress increases global resistance capacity to shear or torsion and enables a higher portion of concrete in the transversal section to be effective.

When compared with normal-strength concrete (NSC), prestress is particularly important for high-strength concrete (HSC) members because such members are expected to have less stiffness due to smaller transversal sections. Prestress can increase the stiffness of structural elements with small transversal sections and allows for a better control of deformation and cracking. Thus, prestressed HSC combines the best characteristics of HSC with the advantages of prestress technique. This is also true for structural members under shear or torsion. The study of the mechanical behavior of prestressed beams under torsion is, therefore, important.

It should be referred that the incorporation of longitudinal prestress for solving practical problems of pure torsion is a not common case. However, the incorporation of longitudinal prestress in structural elements under interaction forces, with high level torsional forces, is common cases. The torsional effects in RC beams under interaction forces still need to be studied (Bairan Garcia & Mari Bernat, 2006a, 2006b; Kim & Yoo, 2006; Koechlin, Andrieux, Millard, & Potapov, 2008).

In a hypothetical situation with pure torsion, the existence of a longitudinal compression state due to longitudinal prestress corresponds to a situation of interaction forces. Thus, a re-evaluation and correction of the theoretical models for RC beams under torsion must be carried out. So far, consistent theoretical models only exist to predict the global behavior of RC beams under torsion (Bernardo et al., 2012a; Bernardo, Andrade, & Lopes, 2012b; Bernardo & Lopes, 2008, 2011).

3. Theoretical idealization of the T–θ curve

From tests of RC beams under pure torsion until failure, a typical experimental T–θ curve can be drawn (Figure 1). Three zones corresponding to different behavioral states can be differentiated, namely: Zone 1, Zone 2, and Zone 3 (Figure 1). In Figure 1, the following parameters were used to characterize the typical experimental T–θ curve (see Figure 1):

T_{cr} cracking torque

θ_{cr}^{I} twist corresponding to T_{cr} for the final part of Zone 1 (linear elastic analysis in non-cracked state)

θ_{cr}^{II} twist corresponding to T_{cr} for the initial part of Zone 2.b (linear elastic analysis in cracked state)

T_{ly}; T_{ty} torque corresponding to yielding of torsional reinforcement (longitudinal and transversal, respectively)

θ_{ly}; θ_{ty} twist corresponding to T_{ly} and T_{ty}, respectively

T_r maximum torque

θ_{Tr} twist corresponding to T_r

$(GJ)^{I}$ torsion stiffness in Zone 1 (linear elastic analysis in non-cracked state)

$(GJ)^{II}$ torsion stiffness in Zone 2.b (linear elastic analysis in cracked state)

In Figure 1, Zone 1 corresponds to the non-cracking behavior of the beam (State I). This stage can be characterized by using a linear elastic analysis in non-cracked state. In the present study, the theoretical models adopted to characterize Zone 1 are the following: Theory of Elasticity, Skew-Bending Theory and Bredt's Thin Tube Theory.

Zone 2 corresponds to the cracked behavior beam (State II). The upper limit of Zone 2 is attributed to the point in which elastic behavior of materials is no longer valid. Zone 2.a corresponds to a sudden increase of twist for a constant torque after T_{cr} is reached. This zone is not easily observed in hollow beams (Bernardo & Lopes, 2009). The beam in State II (mainly in Zone 2.b) can be characterized with a linear elastic analysis in cracked state. In this study, the model adopted was based on the STA with concrete struts at an angle of 45° and assuming linear behavior for the materials.

State II is considered no longer valid from the point when the torsional reinforcement (longitudinal or transversal) starts to yield or when the compression concrete struts start to behave non-linearly due to the high stress level. From this point, Zone 3 begins (see Figure 1). To characterize the behavior of the beam in Zone 3, the VATM is used considering the non-linear behavior of the materials and the softening effect for the concrete in compression in the struts.

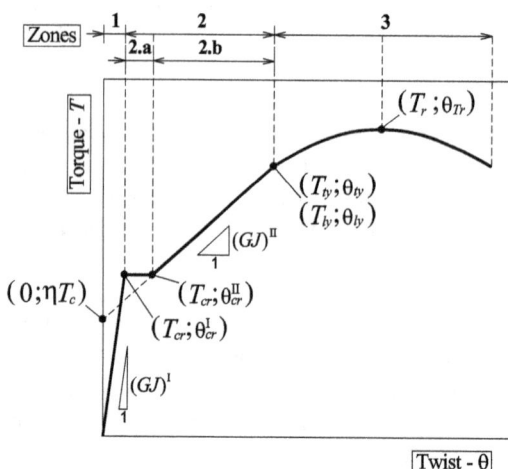

Figure 1. Typical T–θ curve for a RC beam under torsion (Bernardo & Lopes, 2011).

In previous articles from the authors (Bernardo & Lopes, 2008, 2011), Zones 1, 2, and 3 identified in the $T-\theta$ curve of Figure 1 were separately studied. To accomplish the transition within the three zones, semi-empirical criteria were established so that the theoretical $T-\theta$ curve could be entirely drawn. This procedure was accomplished for NSC RC beams (Bernardo & Lopes, 2008) and also for HSC RC beams (Bernardo & Lopes, 2011), with plain and hollow cross sections. In the referred articles, a computing procedure was also proposed to compute the global behavior of RC beams under torsion, namely by drawing the theoretical $T-\theta$ curve. The theoretical results were compared with experimental results of tested RC beams under pure torsion. Such comparative analysis enabled to validate the proposed theoretical model and also the computing procedure.

In the following section, the theoretical models previously presented by the authors are reviewed and modified in order to cover PC beams under torsion. Only longitudinal uniform prestress is considered.

4. Modification of the theoretical models

4.1. Criterion of concrete failure under a biaxial stress state
In order to evaluate the efficiency of prestress in increasing the resistance torque for a RC beam, it is necessary to adopt a concrete failure criterion for biaxial stress state. From that failure criterion, it is possible to compute a simple prestress factor, which is defined as the ratio between the resistance of a PC beam with respect to the resistance of a non-prestressed beam (Hsu, 1984).

A rectangular beam under a torque, T, and under a longitudinal prestress stress, σ, is illustrated in Figure 2. The element A, half way up the side surface of the beam, is under biaxial stress state, with a tangential stress τ due to torsion in each of the four sides and a compressive normal stress σ due to prestress in each of the vertical sides.

Failure of element A occurs when biaxial stresses reach a critical value. The failure criterion more widely accepted for concrete is Mohr's failure theory (Hsu, 1984), which establishes that failure occurs due to the sliding of a plan inside the material. At failure, the tangential and normal stress in that plan (τ and σ, respectively) is related by the following general equation:

$$\tau = F(\sigma) \tag{1}$$

The relationship established by Equation 1, which constitutes a characteristic of the material, is illustrated in Figure 3(a). This relationship, which represents the failure envelope, may be obtained through experimental tests on concrete samples until rupture for several stress conditions, namely: uniaxial compression, uniaxial tension, and pure shear (Mohr's Circles C_1, C_2 and C_3 in Figure 3(a), respectively). In Figure 3(a), the dashed circle represents an arbitrary biaxial stress condition.

The Mohr's failure envelope was simplified by Cowan (1952), through two straight lines BD and DE (Figure 3(b)). From this simplified criterion, Cowan derived the following two equations which express, adimensionally, the shear stress due to torsion, τ, at failure as function of the normal stress σ:

$$\frac{\tau}{f_c'} = \sqrt{0.0396 + 0.120\frac{\sigma}{f_c'} - 0.1594\left(\frac{\sigma}{f_c'}\right)^2} \tag{2}$$

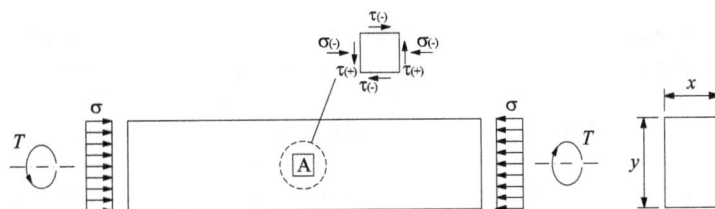

Figure 2. State of stress for a beam under torsion and longitudinal prestress.

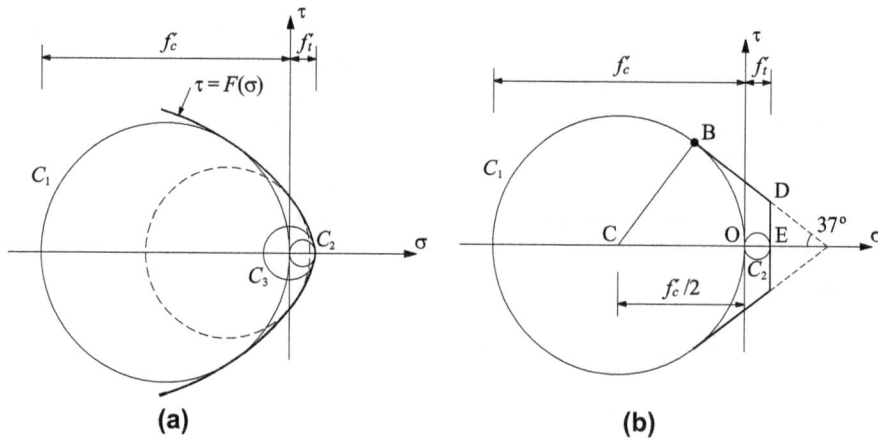

Figure 3. Failure envelops: (a) by Mohr and (b) by Cowan.

$$\frac{\tau}{f_c'} = \frac{1}{\left(\frac{f_c'}{f_t'}\right)} \sqrt{1 + \left(\frac{f_c'}{f_t'}\right) \frac{\sigma_c}{f_c'}} \tag{3}$$

Equation 2 is applicable when failure firstly occurs by compression, while Equation 3 is applicable when failure firstly occurs by tension. In the previous equations, f_c' and f_t' are the concrete compressive and tensile stresses, respectively.

4.2. Linear elastic behavior in non-cracked state (State I)

For the non-cracked state, classical theories still continue to be used to characterize the torsional effect in beams (Romano, Barretta, & Barretta, 2012). The Theory of Elasticity, Skew-Bending Theory, and Bredt's Thin-Tube Theory, to characterize the linear elastic behavior in non-cracked state of RC beams under pure torsion as previously used by the authors (Bernardo & Lopes, 2008, 2011), must be corrected in order to be applied to PC beams. The original formulations of the earlier theories used by the authors for RC beams are summarized in Table 1.

By using Cowen's failure criterion (Cowan, 1952), and according to Hsu (1984), the effect of prestress may be included through a simple prestress factor assuming that, at this behavior stage, the effect of the torsional reinforcement may be neglected. For torsion problems, the prestress factor must be based on concrete uniaxial tensile stress, f_t' (Hsu, 1984). Assuming that f_t' may be related to f_c' as $f_c'/f_t' = 10$, the prestress factor is defined as follow:

$$\gamma = \sqrt{1 + 10 \frac{\sigma}{f_c'}} \tag{4}$$

In order to calculate the cracking torque for a prestress beam, T_{cr}^P, the original equations to compute T_{cr} (RC beam) from each theory (see Table 1) must be multiplied by the prestress factor, γ:

$$T_{cr}^P = T_{cr} \gamma = T_{cr} \sqrt{1 + 10 \frac{\sigma}{f_c'}} \tag{5}$$

The general calculation, as previously presented by the authors for RC beams (Bernardo & Lopes, 2008, 2011), in order to predict the behavior in State I (namely to compute T_{cr}, θ_{cr}^I and $(GJ)^I$), remains unchanged (see Table 1). In order to consider the influence of reinforcements to compute the effective cracking torque, the total reinforcement ratio, ρ_{tot}, may or may not account for the longitudinal prestress reinforcement. If the prestress reinforcement is bonded to the concrete and if it is located in the external region of the section, then it can be considered effective to compute the cracking torque and should be included in $\rho_{l,tot}$ according to Equation 6:

Table 1. Equations for linear elastic behavior in non-cracked state (State I) (Bernardo & Lopes, 2008, 2011)	
Skew-Bending Theory	**Bredt's Thin-Tube Theory**
NSC beams	NSC beams
$T_{cr} = 6y(x^2+10)\sqrt[3]{f_c'}$ (plain section)	$T_{cr} = 2A_c(1.2A_c/p_c)\left(2.5\sqrt{f_c'}\right)$ (plain section)
$T_{cr} = 6(x^2+10)y\sqrt[3]{f_c'}(4h/x)$ for $h \le x/4$ (hollow section) If $h > x/4$ then $h = x/4$	$T_{cr} = 2A_c t\left(2.5\sqrt{f_c'}\right)$ (hollow section)
HSC beams	HSC beams
$T_{cr} = 5.1y(x^2+10)\sqrt[3]{f_c'}$ (plain section)	$T_{cr} = 1.7A_c(1.2A_c/p_c)\left(2.5\sqrt{f_c'}\right)$ (plain section)
$T_{cr} = 5.1(x^2+10)y\sqrt[3]{f_c'}(4h/x)$ (hollow section)	$T_{cr} = 1.7A_c t\left(2.5\sqrt{f_c'}\right)$ (hollow section)

Theory of Elasticity

$T_{cr} = W_T f_{ctm}$

$W_T = \alpha x^2 y$ (plain section)

$W_T = 2Ah$ (hollow section)

General equations

$\theta_{cr}^I = \dfrac{T_{cr,ef}}{K(GJ)^I}$ $J = \beta x^3 y$ (plain section) $J = 4A^2 h/u$ (hollow section)

$T_{cr,ef} = (1+4\rho_{tot})T_{cr}$ $\rho_{tot} = \rho_l + \rho_t$ $\rho_l = \dfrac{A_l}{A_c}$ $\rho_t = \dfrac{A_t u_t}{A_c s}$

Symbology

f_{ctm}	= average concrete strength in tension
f_c'	= concrete strength in compression
h, t	= wall thickness of the hollow section
p_c	= perimeter of the outer line of the section
s	= longitudinal stirrups' spacing
u	= perimeter of area A: $u = 2x_1 + 2y_1$
u_t	= perimeter of the center line of the stirrups
x, y	= minor and major external dimension of the plain section, respectively
x_1, y_1	= minor and major dimension of the rectangular area defined by the center lines of walls of the hollow section, respectively
A	= area limited by the center line of the wall of the hollow section ($A = x_1 y_1$)
A_c	= area of the section limited by the outer perimeter (includes the hollow area)
A_l	= total area of the longitudinal reinforcement
A_t	= area of one leg of the transversal reinforcement
G	= shear modulus ($G = E_c/[2(1+v)]$, E_c is the Young's modulus of concrete and v is the Poisson's coefficient)
J	= stiffness factor
K	= correcting factor ($K \approx .7{-}1$)
$T_{cr}; T_{cr,ef}$	= cracking torque (nominal value and effective value, respectively)
W_T	= elastic modulus of torsion
α, β	= St. Venant's coefficients
ρ_l	= longitudinal reinforcement ratio
ρ_t	= transversal reinforcement ratio
ρ_{tot}	= total reinforcement ratio
θ_{cr}^I	= twist corresponding to T_{cr}

$$\rho_{l,tot} = \frac{A_{sl} + nA_p}{A_c} \tag{6}$$

where A_p is the area of longitudinal prestress reinforcement and $n = E_p/E_s$, being E_s and E_p the Young's modulus for the ordinary and prestress reinforcement, respectively.

The other parameters in Equation 6 are defined in Table 1.

4.3. Linear elastic behavior in cracked state (State II)

After the decompression, a PC beam under torsion behaves similarly to a RC beam under torsion. Thus, prestress only affects the equations for the beam's longitudinal equibrium. The unique change in the equations for this behavior stage (see Table 2) is the introduction of the influence of the longitudinal prestress reinforcement. Such influence is introduced in the same way as explained in Section 4.2 for State I, by computing the total longitudinal reinforcement ratio $\rho_{l,tot}$ (Equation 6). This ratio substitutes ρ_l in the equations for State II (see Table 2).

For the longitudinal equilibrium equations, it is necessary to calculate the stress in the prestress reinforcement, f_{ps}, previously knowing the strain, ε_{ps}. By using the concept of concrete decompression, the strain in the prestress reinforcement is calculated from:

$$\varepsilon_{ps} = \varepsilon_{dec} + \varepsilon_l \tag{7}$$

where ε_{dec} is the strain in the prestress reinforcement at decompression and ε_l is the strain in the ordinary longitudinal reinforcement ($\varepsilon_l = \sigma_l/E_s$).

When prestress is applied, an initial tensile strain in the longitudinal prestress reinforcement and an initial compressive strain in the longitudinal ordinary reinforcement, ε_{li}, are imposed. Such strains are calculated as follows:

$$\varepsilon_{pi} = \frac{f_{pi}}{E_p} \tag{8}$$

$$\varepsilon_{li} = \frac{A_{ps}f_{pi}}{A_l(E_s - E_c) + (A_c - A_h - A_{ps})E_c} \tag{9}$$

where f_{pi} is the initial strain in the prestress reinforcement, E_c is the Concrete Young's modulus, A_c is the area limited by the outer perimeter of the transversal section of concrete, and A_h is the area of the hollow section part (for plain sections $A_h = 0$).

Table 2. Equations for linear elastic behavior in cracked state (State II) (Bernardo & Lopes, 2008, 2011)

General equations
$(GJ)^{II} = \dfrac{E_s x_1^2 y_1^2 xy}{(x_1+y_1)^2 \left[\frac{2nxy}{(x_1+y_1)h_e} + \frac{1}{\rho_l} + \frac{1}{\rho_t}\right]}$ $\quad \rho_l = \frac{A_l}{A_c} ; \rho_t = \frac{A_t u_t}{A_c s}$
$h_e = 1.4(\rho_l + \rho_t)x$
$\eta = 0.57 + 2.86h/x ; \eta = 2$ (plain section)
NSC beams: $T_c = \frac{x^2 y}{3} 2.4\sqrt{f_c'}$ HSC beams: $T_c = \frac{x^2 y}{3} 2.04\sqrt{f_c'}$

Symbology	
h_e	= effective wall thickness
n	$= E_s/E_c$ (E_s is the Young's modulus of the reinforcement)
x_1, y_1	= minor and major dimensions of the rectangular area defined by the center lines of the stirrups
T_c	= concrete torsional strength

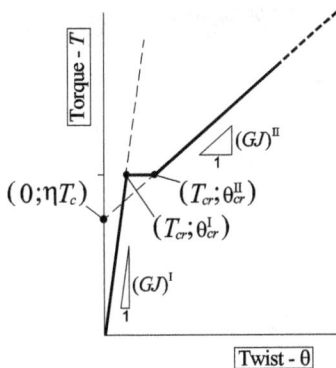

Figure 4. Transition from non-cracked and cracked states.

When the torque increases in a PC beam, the compressive strain in the longitudinal ordinary reinforcement decreases until it becomes null (at decompression). The strain in the longitudinal prestress reinforcement at decompression is:

$$\varepsilon_{dec} = \varepsilon_{pi} + \varepsilon_{li} \tag{10}$$

For torque levels higher than the torque corresponding to ε_{dec}, the PC beam behaves similarly to a RC beam.

Since the stresses immediately after the decompression stage are relatively low, a linear σ–ε relationship may be assumed for the prestress reinforcement. Thus, the stress in the prestress reinforcement, f_{ps}, may be calculated from:

$$f_{ps} = \varepsilon_{ps} E_p \tag{11}$$

Following the same methodology as for RC beams (Bernardo & Lopes, 2008, 2011), the ordinate at the origin of the theoretical straight line in linear elastic behavior in State II is given by $\eta T'_c$, being T'_c the contribution of concrete (Figure 4). From an extensive parametric analysis, Bernardo (2003) proposed Equation 12 to calculate T'_c, where γ_1 represents a prestress factor which is function of γ.

$$T'_c = T_c \gamma_1 = T_c \left(2.20 \sqrt{1 + 10 \frac{\sigma}{f'_c}} - 1.53 \right) \tag{12}$$

4.4. Transition between linear elastic analysis in non-cracked and cracked states

The same criteria proposed by the authors and validated for RC beams under torsion (Bernardo & Lopes, 2008, 2011), for the transition between the two early states of behavior, will be also adopted for PC beams.

Figure 4 shows the typical theoretical T–θ curve for linear elastic non-cracked and cracked states (Zones 1 and 2 of Figure 1). It should be noted that the horizontal line of the transition from non-cracked to cracked states is hardly observed in tests of hollow beams ($\theta^{II}_{cr} - \theta^{I}_{cr}$ is normally very small or zero).

4.5. Non-linear analysis

The original formulation for non-linear analysis based on VATM, for RC beams under torsion, is presented in Table 3. As referred in Section 4.3, in order to account for the influence of prestress reinforcement, the force in the ordinary longitudinal reinforcement ($A_l f_l$) must be substituted by the total force including both the ordinary and the prestress reinforcement ($A_l f_l + A_{ps} f_{ps}$). Thus, the new equations to compute the angle of the concrete struts, α, and the effective thickness of the concrete struts, t_d, are:

Table 3. Equations for non-linear analysis (Bernardo & Lopes, 2008, 2011)

General equations

$$T = 2A_o t_d \sigma_d \sin\alpha\cos\alpha \quad \cos^2\alpha = \frac{A_l f_l}{p_o \sigma_d t_d} \quad t_d = \frac{A_l f_l}{p_o \sigma_d} + \frac{A_t f_t}{s\sigma_d} \quad \varepsilon_t = \left(\frac{A_o^2 \sigma_d}{p_o T \,\text{tg}\,\alpha} - \frac{1}{2}\right)\varepsilon_{ds} \quad \varepsilon_l = \left(\frac{A_o^2 \sigma_d}{p_o T \cot g\alpha} - \frac{1}{2}\right)\varepsilon_{ds} \quad \theta = \frac{\varepsilon_{ds}}{2t_d \sin\alpha\cos\alpha}$$

Softened $\sigma-\varepsilon$ relationship for NSC	Softened $\sigma-\varepsilon$ relationship for HSC
$\sigma_d = f_c'\left[2\left(\frac{\varepsilon_d}{\varepsilon_o}\right) - \lambda\left(\frac{\varepsilon_d}{\varepsilon_o}\right)^2\right]$ if $\varepsilon_d \le \varepsilon_o/\lambda$	$\sigma_d = \zeta_{\sigma_o} f_c'\left[2\left(\frac{\varepsilon_d}{\zeta_{\varepsilon_o}\varepsilon_o}\right) - \left(\frac{\varepsilon_d}{\zeta_{\varepsilon_o}\varepsilon_o}\right)^2\right]$ if $\varepsilon_d/\zeta_{\varepsilon_o}\varepsilon_o \le 1$
$\sigma_d = \frac{f_c'}{\lambda}(1-\eta^2)$ if $\varepsilon_d > \varepsilon_o/\lambda$	$\sigma_d = \zeta_{\sigma_o} f_c'\left[1 - \left(\frac{\varepsilon_d/\varepsilon_o - \zeta_{\varepsilon_o}}{2-\zeta_{\varepsilon_o}}\right)^2\right]$ if $\varepsilon_d/\zeta_{\varepsilon_o}\varepsilon_o > 1$
$\lambda = \sqrt{\frac{\gamma_m}{\varepsilon_d} - 0.3} \quad \eta = \frac{\varepsilon_d - \varepsilon_p}{2\varepsilon_o - \varepsilon_p} \quad \varepsilon_p = \varepsilon_o/\lambda$	$\zeta_{\sigma_o} = \frac{0.81}{\sqrt{1+K_f\varepsilon_{dt}}} \quad \zeta_{\varepsilon_o} = \frac{0.9}{\sqrt{1+400\varepsilon_{dt}}}$
$\gamma_m = \varepsilon_l + \varepsilon_t + 2\varepsilon_d \; ; \quad \varepsilon_d = \varepsilon_{ds}/2$	$\varepsilon_{dt} = \varepsilon_l + \varepsilon_t + \varepsilon_d \quad K_f = \frac{10f_c'}{\eta} \quad \eta = \frac{\rho_t f_t}{\rho_l f_l}$
$\sigma_d = k_1(1/\lambda)f_c'$	$\sigma_d = k_1\zeta_{\sigma_o}f_c'$
$k_1 = \lambda\frac{\varepsilon_{ds}}{\varepsilon_o} - \frac{\lambda^2}{3}\left(\frac{\varepsilon_{ds}}{\varepsilon_o}\right)^2$ if $\varepsilon_{ds} \le \varepsilon_o/\lambda$	$k_1 = \frac{\varepsilon_{ds}}{\varepsilon_p}\left(1 - \frac{\varepsilon_{ds}}{3\varepsilon_p}\right)\varepsilon_p = \zeta_{\varepsilon_o}\varepsilon_o$ if $\varepsilon_{ds} \le \zeta_{\varepsilon_o}\varepsilon_o$
$k_1 = \left(1 - \frac{1}{(2\lambda-1)^2}\right)\left(1 - \frac{1}{3}\frac{\varepsilon_p}{\varepsilon_{ds}}\right) + \frac{1}{(2\lambda-1)^2}\frac{\varepsilon_{ds}}{\varepsilon_p}\left(1 - \frac{1}{3}\frac{\varepsilon_{ds}}{\varepsilon_p}\right)$ if $\varepsilon_{ds} > \varepsilon_o/\lambda$	$k_1 = \left[1 - \frac{1}{(2/\zeta_{\varepsilon_o}-1)^2}\right]\left(1 - \frac{1}{3}\frac{\varepsilon_p}{\varepsilon_{ds}}\right) + \frac{1}{(2/\zeta_{\varepsilon_o}-1)^2}\frac{\varepsilon_{ds}}{\varepsilon_p}\left(1 - \frac{1}{3}\frac{\varepsilon_{ds}}{\varepsilon_p}\right)$ $\varepsilon_p = \zeta_{\varepsilon_o}\varepsilon_o$ if $\varepsilon_{ds} > \zeta_{\varepsilon_o}\varepsilon_o$

Symbology

f_l, f_t	= longitudinal an transversal reinforcement stress, respectively
t_d	= effective thickness of the concrete strut
p_o	= perimeter of the area A_o
A_o	= area limited by the center line of the flow of shear stresses
T	= torque
α	= angle of the concrete struts
ε_d	= average strain in the concrete strut
ε_{ds}	= strain at the surface of the concrete strut
$\varepsilon_l, \varepsilon_t$	= strains in the longitudinal and transversal reinforcement, respectively
ε_{dt}	= tension strain in the perpendicular direction to diagonal strut
ε_o	= compressive strain corresponding to the maximum stress in the concrete
θ	= twist
γ_m	= maximum distortion
σ_d	= average stress in the diagonal concrete strut
$\zeta_{\sigma o}, \zeta_{\varepsilon o}, \lambda$	= reducing factors to account for the softening effect

$$\cos^2\alpha = \frac{A_l f_l + A_{ps} f_{ps}}{p_o \sigma_d t_d} \tag{13}$$

$$t_d = \frac{A_l f_l + A_{ps} f_{ps}}{p_o \sigma_d} + \frac{A_t f_t}{s\sigma_d} \tag{14}$$

All the remaining equations of VATM for RC beams under torsion (equilibrium and compatibility equations) remains unchanged for PC beams (see Table 3).

A classical bilinear $\sigma-\varepsilon$ relationship (with horizontal landing after yielding) is assumed for the ordinary reinforcement. For prestress reinforcement, the Ramberg–Osgood equation (Hsu & Mo, 1985) will be used to describe the $\sigma-\varepsilon$ relationship (Figure 5):

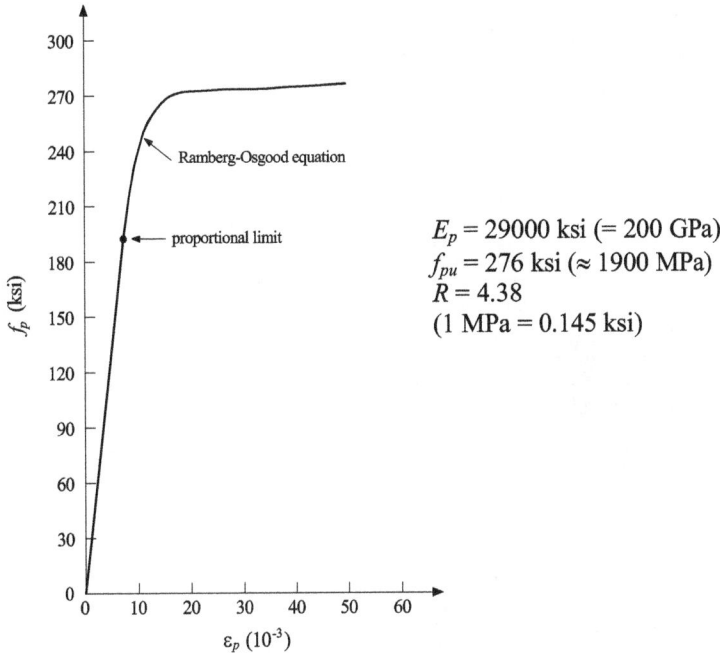

Figure 5. σ–ε relationship based on Ramberg–Osgood equation.

$$f_{ps} = E_p \varepsilon_{ps} \left[1 + \left(\frac{E_p \varepsilon_{ps}}{f_{pu}} \right)^R \right]^{1/R} \tag{15}$$

where f_{pu} is the failure tensile stress of prestress reinforcement and R is the coefficient determined from experimental tests and equal to 4.38 for common prestress steel (Hsu & Mo, 1985).

Since Equation 15 is not exactly linear for low stress levels, it can only be used above the proportional conventional limit stress $f_{p0.1\%}$ (corresponding to a strain limit of .1%). Below this limit, a linear relationship can be used (Equation 11), by adjusting E_p so that Equations 11 and 15 coincide at the limit point corresponding to $f_{p0.1\%}$.

For PC beams, parameter η, which represents the ratio between the transversal and longitudinal reinforcement tensile forces (Table 3), must be corrected in order to account for the influence of the longitudinal prestress reinforcement:

$$\eta = \frac{\rho_t f_t}{\rho_l f_l + \rho_{ps} f_{ps}} = \frac{A_t u f_t}{s A_l f_l + s n A_{ps} f_{ps}} \tag{16}$$

where $n = f_{p0.1\%}/f_{ly}$.

The original flowchart for the algorithm previously proposed by the authors (Bernardo & Lopes, 2008, 2011) to compute the T–θ curve from VATM for RC beams under torsion can also be used, with some corrections, for PC beams under torsion. The new algorithm flowcharts are illustrated in Figure 6.

4.6. Transition between linear elastic analysis in cracked state and non-linear analysis
Te same criteria proposed and validated by the authors (Bernardo & Lopes, 2008, 2011) for RC beams under torsion and for the transition between the two early states of behavior will also be adopted for PC beams.

Figures 7 and 8 show the T–θ curve for the linear cracked state and for the non-linear state (Zones 2.b and 3 of Figure 1) for the two possible failure modes: ductile and fragile failure, respectively.

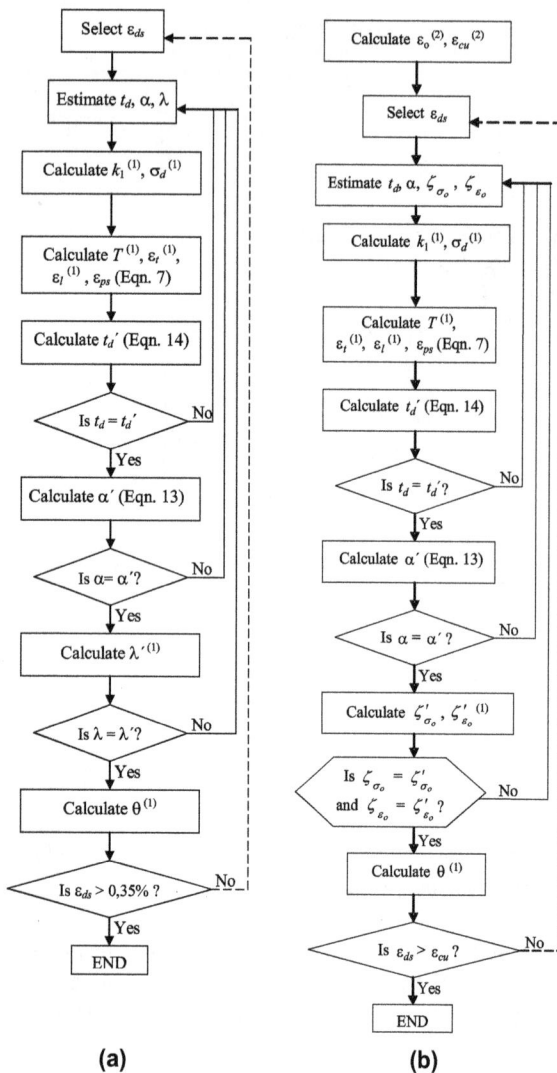

Figure 6. Flowcharts for the calculation of $T-\theta$ curve: (a) NSC prestressed beams and (b) HSC prestressed beams.

Notes: [1] See Table 3 and [2] Eurocode 2 (2010).

For the ductile failure (Figure 7), it is assumed that the linear cracked state is valid until the torsional reinforcement (transversal and/or longitudinal) yields. Once the transition point is defined, $\Delta\theta$ is calculated from $\Delta\theta = \theta_y - \theta_y^{II}$ (see Figure 7). Then, the non-linear zone of the $T-\theta$ curve is horizontally shifted to the left by $\Delta\theta$ to reset the continuity of the whole curve.

Figure 8 shows the procedure to perform the transition when the concrete in the struts reaches its ultimate strain before the yielding of the torsional reinforcement. The transition point is defined from the slopes of the curves from both states: linear cracked and non-linear sates (there's a point from which the slope of non-linear state curve equals the slope of the linear state curve). The torque level of this point is identified in Figure 8 by T_A. The corresponding twists are: θ_A^{II} and θ_A. $\Delta\theta$ is the value of the horizontal shift of the non-linear branch of the curve to the left to make the whole curve continuous.

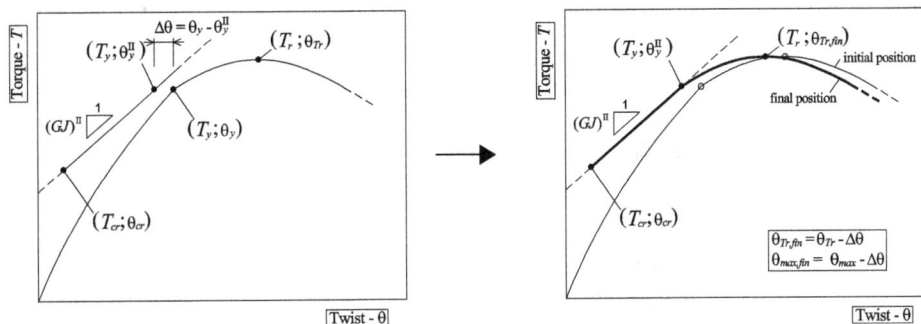

Figure 7. Transition from linear cracked state to non-linear state—Ductile failure (Bernardo & Lopes, 2011).

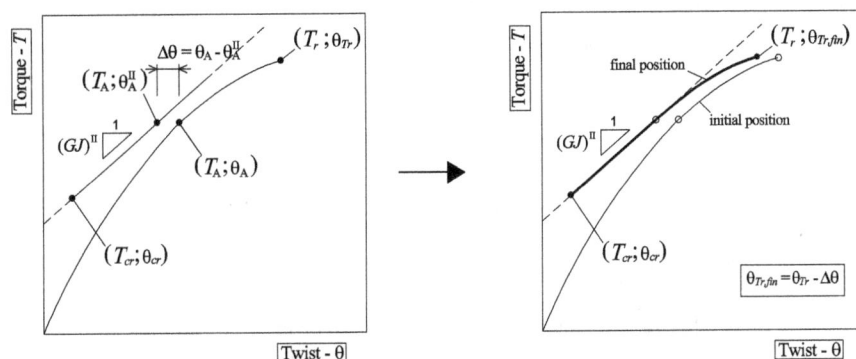

Figure 8. Transition from linear cracked state to non-linear state—Fragile failure (Bernardo & Lopes, 2011).

5. Comparative analysis with experimental results

Based on the theoretical models and criteria described in Section 4, the computing procedure previously developed and validated by the authors for RC beams under torsion (Bernardo & Lopes, 2008, 2011) is extended in order to compute the T–θ curve for NS and HS PC beams under pure torsion, with rectangular plain and hollow cross-section. In order to validate the computing procedure, the predictions will be compared to some experimental results available in the literature.

The amount of experimental studies focused on PC beams under torsion is limited. Furthermore, some of the tests reported in literature cannot be used in this study for some reasons. For instance, some older studies simply do not meet the basic design recommendations found in current codes of practice (minimum reinforcement, maximum spacing, minimum wall thickness of hollow sections, etc.). In these cases, beams show atypical behaviors. In some other experimental studies, the authors present an average twist for all the beam length, and not the twist of the critical section. Therefore, concentrated twists on the critical section cannot be derived from tests. In this case, the experimental twists cannot be directly compared with the theoretical predictions because these latter represent the twists in the critical section. This aspect is particularly important in slender beams because failure zone is located in a small region of the beam.

As a consequence, experimental tests related with only five prestressed beams under torsion found in literature will be used for comparative analysis. In these tests, the rotations were recorded in the failure zone of the beams, which enabled comparative analysis with theoretical predictions. The beams are the following ones: Beams P2 and P3 (Mitchell & Collins, 1974), Beam P8 from (Hsu, 1984) and Beams D1 and D2 (Bernardo, 2003). The experimental results of these beams enable to plot the full T–θ curves in the critical zone. Such curves are illustrated in Figures 9–13, where f_{cp} represents the initial concrete stress due to prestress.

Table 4 summarizes the comparative analysis of the linear elastic stage in State I. Thus, the experimental values for the cracking torque ($T_{cr, exp}$) and the torsional stiffness in State I ($(GJ)^{I}_{exp}$) are

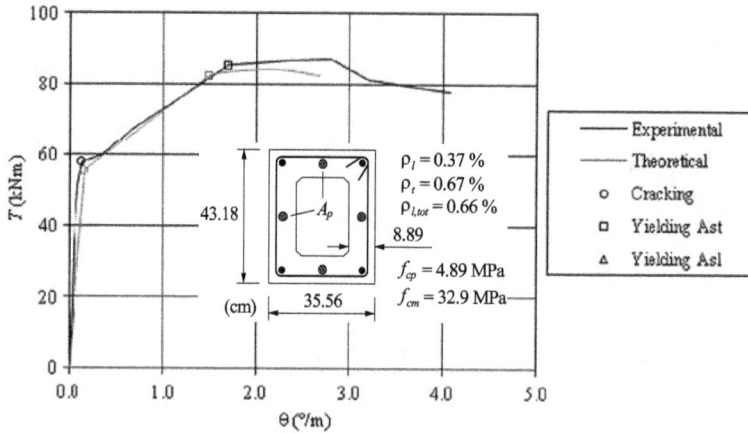

Figure 9. *T–θ* curve for Beam P2 (Mitchell & Collins, 1974),

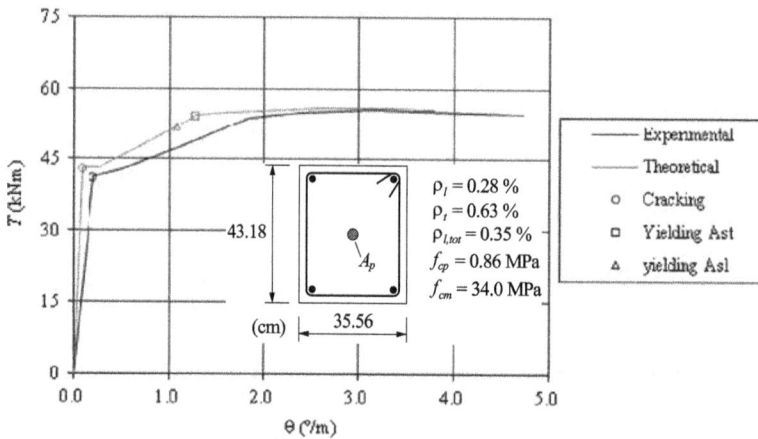

Figure 10. *T–θ* curve for Beam P3 (Mitchell & Collins, 1985).

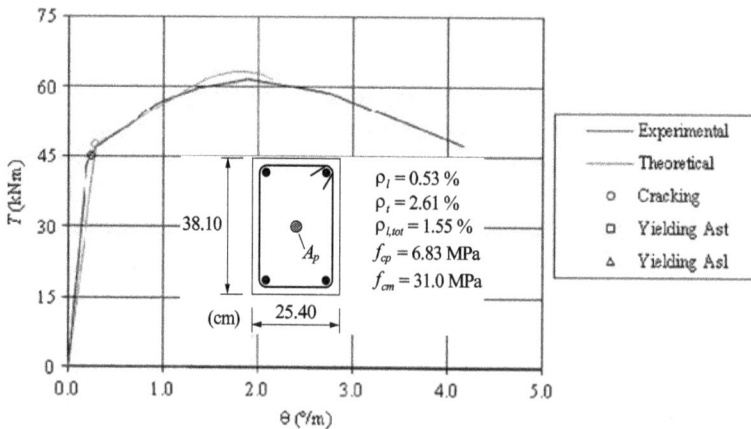

Figure 11. *T–θ* curve for Beam P8 (Hsu, 1984).

indicated. The theoretical cracking torque was calculated from the following theories: Theory of Elasticity ($T_{cr,calc}^{TE}$), Skew-Bending Theory ($T_{cr,calc}^{SBT}$), and Bredt's Thin-Tube Theory ($T_{cr,calc}^{BTT}$). The theoretical values for the torsional stiffness in State I ($(GJ)_{calc}^{I}$) was calculated for the two limit values of the minorative factor $K = .7$ and $K = 1$ (Bernardo & Lopes, 2008). The quotients between the experimental and theoretical values of the several parameters are also presented.

Table 1 shows that for beams with plain section, it is not very clear which theory generally leads to the best predictions for the cracking torque.

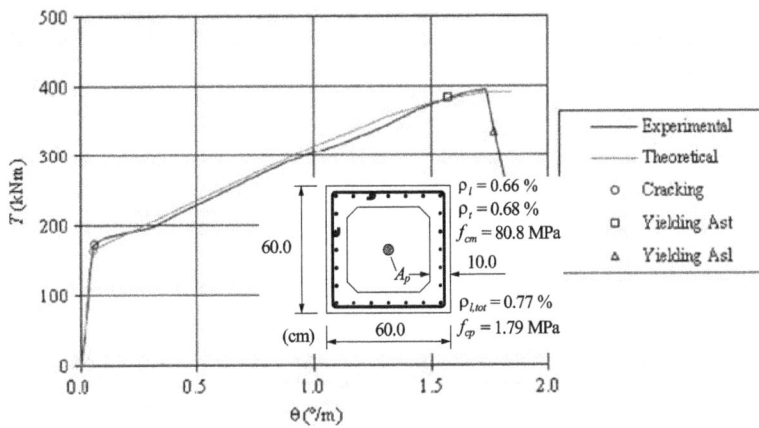

Figure 12. T–θ curve for Beam D1 (Bernardo, 2003).

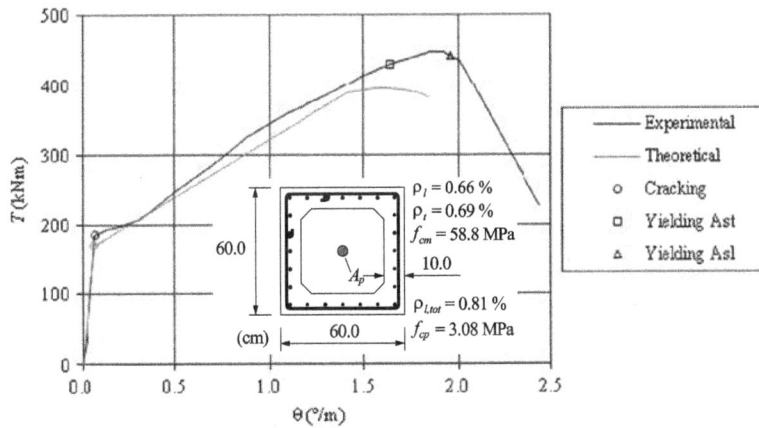

Figure 13. T–θ curve for Beam D2 (Bernardo, 2003).

For beams with hollow sections, the best prediction of the cracking torque is obtained with Bredt's Thin-Tube Theory. Despite the limited number of tested beams, it is possible to state that the trend is similar to the one observed in previous articles (Bernardo & Lopes, 2008, 2011) related with RC beams with hollow section.

Table 4 shows considerable deviations for all the beams between the theoretical and experimental values of the torsional stiffness in State I $((GJ)^I)$. It is possible that the difficulty associated to the experimental measurements of very low twists at this stage led to an inconclusive analysis.

Table 5 summarizes the comparative analysis for the linear elastic stage in State II. The experimental values of the ordinate at the origin $((\eta T_c)_{exp})$ and the torsional stiffness in State II $((GJ)^{II}_{exp})$

Table 4. Comparative analysis for elastic cracked stage (State I)

Beam	Section type	$T_{cr,exp}$ (kNm)	$T^{TE}_{cr,calc}$ (kNm)	$\frac{T_{cr,exp}}{T^{TE}_{cr,calc}}$	$T^{SBT}_{cr,calc}$ (kNm)	$\frac{T_{cr,exp}}{T^{SBT}_{cr,calc}}$	$T^{BTT}_{cr,calc}$ (kNm)	$\frac{T_{cr,exp}}{T^{BTT}_{cr,calc}}$	$(GJ)^I_{exp}$ (kNm²)	$(GJ)^I_{calc}$ (kNm²)	$\frac{(GJ)^I_{exp}}{(GJ)^I_{calc}}$	$(GJ)^I_{calc}$ (kNm²)	$\frac{(GJ)^I_{exp}}{(GJ)^I_{calc}}$
										$K = .7$		$K = 1.0$	
P2	Hollow	58.0	88.6	.65	68.3	.85	55.5	1.04	24013	19338	1.24	27617	.87
P3	Plain	41.2	43.1	.96	44.8	.92	48.2	.86	11294	26212	.43	37433	.30
P8	Plain	45.2	40.8	1.11	42.9	1.05	47.6	.95	13126	9399	1.40	13422	.98
D1	Hollow	172.9	251.2	.69	166.4	1.04	165.4	1.05	152188	157283	.97	224612	.68
D2	Hollow	184.7	208.8	.89	174.2	1.06	170.5	1.08	143187	143376	1.00	204752	.70

Table 5. Comparative analysis for elastic cracked stage (State II)

Beam	$(\eta T_c)_{exp}$ (kNm)	$(\eta T_c)_{calc}$ (kNm)	$\dfrac{(\eta T_c)_{exp}}{(\eta T_c)_{calc}}$	$(GJ)^{II}_{exp}$ (kNm²)	$(GJ)^{II}_{calc}$ (kNm²)	$\dfrac{(GJ)^{II}_{exp}}{(GJ)^{II}_{calc}}$
P2	54.1	52.2	1.04	1061.6	1173.4	.91
P3	39.1	40.0	.98	451.2	645.4	.70
P8	43.2	43.8	.99	757.0	736.8	1.03
D1	156.9	155.9	1.01	9261.2	9018.5	1.03
D2	166.0	159.5	1.04	9529.6	9219.1	1.03

Table 6. Comparative analysis for non-linear stage

Beam	$T_{r,exp}$ (kNm)	$T_{r,cal}$ (kNm)	$\dfrac{T_{r,exp}}{T_{r,calc}}$	$\theta_{Tr,exp}$ (°/m)	$\theta_{Tr,cal}$ (°/m)	$\dfrac{\theta_{Tr,exp}}{\theta_{Tr,calc}}$
P2	87.1	84.1	1.04	2.80	2.03	1.38
P3	55.8	56.0	1.00	3.14	3.38	.93
P8	61.8	63.3	.98	1.89	1.88	1.01
D1	396.0	390.8	1.01	1.73	1.84	.94
D2	447.7	395.0	1.13	1.93	1.64	1.18

are indicated. Table 5 also includes the theoretical values calculated for the same parameters ($(\eta T_c)_{calc}$ and $(GJ)^{II}_{calc}$). Torsional stiffness can be calculated with or without the influence of prestress reinforcement. For Beams P3 and P8 (centered prestressed reinforcement) and for Beams D1 and D2 (external prestress), the calculation of $(GJ)^{II}$ should not include the influence of prestress reinforcement. For Beam P2 (prestressed reinforcement located in the peripheral shell area of the section) the calculation of $(GJ)^{II}$ should include the influence of prestress reinforcement.

Table 5 shows excellent predictions for the ordinate at the origin, ηT_c, of the straight line of linear elastic behavior in State II. Except for Beam P3, the predictions for the torsional stiffness in State II, $(GC)^{II}$ are good.

Table 6 summarizes the comparative analysis for the non-linear stage. This table presents the experimental and theoretical values of the strength or maximum torque ($T_{r,exp}$ and $T_{r,calc}$, respectively) as well as the corresponding twist values ($\theta_{Tr,exp}$ and $\theta_{Tr,calc}$, respectively).

For the non-linear state, Table 3 generally shows good predictions, namely for the maximum torque.

Despite the limited number of experimental results, it seems to be possible to state that the corrections introduced in the equations of the theoretical models described in Section 4, in view to predict the theoretical behavior of PC beams under torsion, are valid.

Figures 9–13 illustrate graphically the results transcribed in Tables 4–6, through the $T–\theta$ Curves. For the theoretical point related to the cracking torque and for the torsional stiffness, the best predictions from Table 1 were used. Figures 9–13 are in line with the general conclusions previously established from the analysis of Tables 4–6.

6. Conclusions

In this article, some corrections were introduced in a global theoretical model previously proposed by the authors for RC beams under torsion in order to predict the global behavior of beams under torsion with uniform longitudinal prestress. These corrections were based on the introduction of prestress factors and on the modification of the longitudinal equilibrium equations for the non-linear stage.

From the comparative analyzes performed in this study, the following conclusions can be drawn:

- For PC beams with hollow sections, the best prediction of the cracking torque is obtained with Bredt's Thin-Tube Theory. For PC beams with plain section, it is not very clear which theory generally leads to the best predictions for the cracking torque.

- Considerable deviations for all the test beams were observed between the theoretical and experimental values of the torsional stiffness in State I. This is probably associated with the very small magnitude of the twists at this stage, which can lead to high dispersions of the experimental measurements.

- In general, good predictions for the ascending branch of the $T-\theta$ curves in the cracked stage (State II) were obtained.

- For the non-linear state (ultimate stage), good predictions were obtained, namely for the maximum torque.

Despite the limited number of experimental results, it can be stated that the correction of the global theoretical model previously developed and validated by the authors for RC beams under torsion (Bernardo & Lopes, 2008, 2011) led to generally good predictions when compared to some available experimental results of PC beams under torsion found in the literature. However, given the limited number of analyzed beams, it is nevertheless important to carry out more analyzes based on further experimental results. Therefore, new projects aiming to testing PC beams under torsion would be very useful.

Funding
The authors received no direct funding for this research.

Author details
Sérgio M.R. Lopes[1]
E-mail: sergio@dec.uc.pt
Luís F.A. Bernardo[2]
E-mail: lfb@ubi.pt
[1] FCTUC, University of Coimbra, Portugal.
[2] Department of Civil Engineering and Architecture, University of Beira Interior, Edifício II das Engenharias, Calçada Fonte do Lameiro, 6201-001 Covilhã, Portugal.

References
Alnuaimi, A. S., & Bhatt, P. (2006). Direct design of partially prestressed concrete hollow beams. *Advances in Structural Engineering, 9*, 459–476. http://dx.doi.org/10.1260/136943306778812741

Bairan Garcia, J. M., & Mari Bernat, A. R. (2006a). Coupled model for the non-linear analysis of anisotropic sections subjected to general 3D loading. Part 1: Theoretical formulation. *Computers and Structures, 84*, 2254–2263. http://dx.doi.org/10.1016/j.compstruc.2006.08.036

Bairan Garcia, J. M., & Mari Bernat, A. R. (2006b). Coupled model for the nonlinear analysis of sections made of anisotropic materials, subjected to general 3D loading. Part 2: Implementation and validation. *Computers and Structures, 84*, 2264–2276. http://dx.doi.org/10.1016/j.compstruc.2006.08.035

Bernardo, L. F. A. (2003). *Torção em Vigas em Caixão de Betão de Alta Resistência* [Torsion in reinforced high-strength concrete hollow beams] (PhD thesis). University of Coimbra, Portugal, 692 pp. (in Portuguese).

Bernardo, L. F. A., Andrade, J. M., & Lopes, S. M. R. (2012a). Softened truss model for reinforced NSC and HSC beams under torsion: A comparative study. *Engineering Structures, 42*, 278–296. http://dx.doi.org/10.1016/j.engstruct.2012.04.036

Bernardo, L. F. A., Andrade, J. M., & Lopes, S. M. R. (2012b). Modified variable angle truss-model for torsion in reinforced concrete beams. *Materials and Structures, 45*, 1877–1902. http://dx.doi.org/10.1617/s11527-012-9876-4

Bernardo, L. F. A., Andrade, J. M. A., & Nunes, N. C. G. (in press). Generalized softened variable angle truss-model for reinforced concrete beams under torsion. *Materials and Structures*. doi:10.1617/s11527-014-0301-z

Bernardo, L. F. A., & Lopes, S. M. R. (2008). Behaviour of concrete beams under torsion: NSC plain and hollow beams. *Materials and Structures, 41*, 1143–1167. http://dx.doi.org/10.1617/s11527-007-9315-0

Bernardo, L. F. A., & Lopes, S. M. R. (2009). Torsion in HSC hollow beams: Strength and ductility analysis. *ACI Structural Journal, 106*, 39–48.

Bernardo, L. F. A., & Lopes, S. M. R. (2011). Theoretical behavior of HSC sections under torsion. *Engineering Structures, 33*, 3702–3714. http://dx.doi.org/10.1016/j.engstruct.2011.08.007

Bhatti, M. A., & Almughrabi, A. (1996). Refined model to estimate torsional strength of reinforced concrete beams. *Journal of the American Concrete Institute, 93*, 614–622.

Cowan, H. J. (1952). Strength of reinforced concrete under the action of combined stresses, and the representation of the criterion of failure by a space model. *Nature, 169*, 663 pp. http://dx.doi.org/10.1038/169663a0

Hsu, T. T. C. (1984). *Torsion of reinforced concrete*. New York, NY: Van Nostrand Reinhold.

Hsu, T. T. C., & Mo, Y. L. (1985). Softening of concrete in torsional members—Prestressed concrete. *Journal of the American Concrete Institute, 82*, 603–615.

Hsu, T. T. C., & Zhu, R. R. H. (2002). Softened membrane model for reinforced concrete elements in shear. *ACI Structural Journal, 99*, 460–469.

Jeng, C. H., Chiu, H. J., & Chen, C. S. (2010). Modelling the initial stresses in prestressed concrete members under torsion. *ASCE Conference Proceedings, 369*, 1773–1781.

Jeng, C. H., & Hsu, T. T. C. (2009). A softened membrane model for torsion in reinforced concrete members. *Engineering*

Structures, 31, 1944–1954.
http://dx.doi.org/10.1016/j.engstruct.2009.02.038

Kim, K., & Yoo, C. H. (2006). Ultimate strength interaction of bending and torsion of steel/concrete composite trapezoidal box girders in positive bending. *Advances in Structural Engineering, 9*, 707–718. http://dx.doi.org/10.1260/136943306778827529

Koechlin, P., Andrieux, S., Millard, A., & Potapov, S. (2008). Failure criterion for reinforced concrete beams and plates subjected to membrane force, bending and shear. *European Journal of Mechanics A/Solids, 27*, 1161–1183. http://dx.doi.org/10.1016/j.euromechsol.2007.12.009

Mitchell, D., & Collins, M. P. (1974, March). *The behavior of structural concrete beams in pure torsion.* Civil Engineering Publication No.74-06. Department of Civil Engineering, University of Toronto, 88 pp.

NP EN 1992-1-1. (2010, March). *Eurocode 2: Design of concrete structures—Part 1: General rules and rules for buildings.*

Rahal, K. N., & Collins, M. P. (1996). Simple model for predicting torsional strength of reinforced and prestressed concrete sections. *Journal of the American Concrete Institute, 93*, 658–666.

Rausch, E. (1929). *Design of reinforced concrete in torsion* (PhD thesis). Berlin (in German).

Romano, G., Barretta, A., & Barretta, R. (2012). On torsion and shear of Saint-Venant beams. *European Journal of Mechanics A/Solids, 35*, 47–60.
http://dx.doi.org/10.1016/j.euromechsol.2012.01.007

Wang, W., & Hsu, C. T. T. (1997). Limit analysis of reinforced concrete beams subjected to pure torsion. *Journal of Structural Engineering, 123*, 86–94. http://dx.doi.org/10.1061/(ASCE)0733-9445(1997)123:1(86)

10

Distributed GIS for automated natural hazard zonation mapping Internet-SMS warning towards sustainable society

Devanjan Bhattacharya[1*], Jayanta Kumar Ghosh[2], Jitka Komarkova[3], Santo Banerjee[4] and Hakan Kutoglu[1]

*Corresponding author: Devanjan Bhattacharya, Faculty of Engineering, Geomatics Engineering Department, Bulent Ecevit University, Zonguldak 67100, Turkey

E-mail: devanjan.bhattacharya@beun.edu.tr

Reviewing editor: Sanjay Shukla, Edith Cowan University, Australia

Abstract: Today, open systems are needed for real time analysis and warnings on geo-hazards and over time can be achieved using Open Source Geographical Information System (GIS)-based platform such as GeoNode which is being contributed to by developers around the world. To develop on an open source platform is a very vital component for better disaster information management as far as spatial data infrastructures are concerned and this would be extremely vital when huge databases are to be created and consulted regularly for city planning at different scales, particularly satellite images and maps of locations. There is a big need for spatially referenced data creation, analysis, and management. Some of the salient points that this research would be able to definitely contribute with GeoNode, being an open source platform, are facilitating the creation, sharing, and collaborative use of geospatial data. The objective is development of an automated natural hazard zonation system with Internet-short message service (SMS) warning utilizing

ABOUT THE AUTHORS

Devanjan Bhattacharya, assistant professor of Geomatics, Bulent Ecevit University, Turkey, works in GIS, remote sensing, KBES & intelligent DSS, GPS, and image processing for utilization of Geomatics towards sustainable development. Jayanta Kumar Ghosh, associate professor of Civil Engineering, Indian Institute of Technology, Roorkee, has interests in Geomatics Engineering, GPS, Remote Sensing, AI, Soft Computations, etc. Jitka Komárková, an associate professor, Institute of Systems Engg & Informatics, faculty of Economics and Administration, vice dean of Study and Teaching at the University of Pardubice, has interests in GIS, Information Systems, Systems Analysis and quality of web-based GIS. Santo Banerjee, research scientist (associate prof. grade), at INSPEM, University Putra, Malaysia, has interests in Dynamical Systems and its chaotic properties, Synchronization of chaotic signals and its applications, etc. Hakan Kutoglu is head of Department, director of Hazard Application Research Center, Bulent Ecevit University, Turkey, with interests in geodesy, satellite geodesy, and computational methods.

PUBLIC INTEREST STATEMENT

Systems are needed for real time analysis and warnings on geo-hazards through Open Source Geographical Information System (OS-GIS) platform. To develop such a system is vital for better disaster information management and spatial data infrastructure creation. This also benefits when huge databases are created and consulted regularly for city planning at different scales through satellite images and maps of locations. There is need for spatially referenced data creation, analysis, and management. The objective is development of automated natural hazard zonation system with Internet-SMS warning utilizing geomatics for sustainable societies. There has been a need to develop automated integrated system to categorize hazard and issue warning that reaches users directly. At present, no web-enabled warning system exists which can disseminate warning after hazard evaluation in real time. Research work formalizes a notion of an integrated, independent, generalized, and automated geo-hazard warning system making use of geo-spatial data under popular usage platform.

geomatics for sustainable societies. A concept of developing an internet-resident geospatial geohazard warning system has been put forward in this research, which can communicate alerts via SMS. There has been a need to develop an automated integrated system to categorize hazard and issue warning that reaches users directly. At present, no web-enabled warning system exists which can disseminate warning after hazard evaluation at one go and in real time. The objective of this research work has been to formalize a notion of an integrated, independent, generalized, and automated geo-hazard warning system making use of geo-spatial data under popular usage platform. In this paper, a model of an automated geo-spatial hazard warning system has been elaborated. The functionality is to be modular in architecture having GIS-graphical user interface (GUI), input, understanding, rainfall prediction, expert, output, and warning modules. A simplified but working prototype of the system without the GIS-GUI module has been already tested, validated, and reported. Through this paper, a significantly enhanced system integrated with web-enabled-geospatial information has been proposed, and it can be concluded that an automated hazard warning system has been conceptualized and researched. However, now the scope is to develop it further.

Subjects: Geographic Information Systems; Georisk & Hazards, GIS, Remote Sensing & Cartography; Information & Communication Technology (ICT); Information Technology

Keywords: internet-based; geo-spatial; short message service; interfacing; warning; communication; graphical interface

1. Introduction

A warning system capable of disseminating adequate and timely warning to the public is of great value. At present, hazard warning is communicated statically through telephone, television, radio, and internet but not through the short message service (SMS) (Pries, Hobfeld, & Gia, 2006). The increasing number of mobile phone subscriptions, around five billion worldwide, shows that a large number of persons can be reached via a mobile phone service through SMS (Cioca, Cioca, & Buraga, 2008). Real time analysis and warnings on geo-hazards can be issued using Open Source Geographical Information System (GIS)-based platform such as GeoNode allowing a graphical user interface (GUI) and which could be contributed to by developers around the world. To develop on an open source platform is a very vital component for better disaster information management as far as spatial data infrastructures (SDIs) are concerned and this would be extremely vital when huge databases are to be created and consulted regularly for city planning at different scales particularly satellite images and maps of locations. There is a big need for spatially referenced data creation, analysis, and management. Some of the salient points that this research would be able to definitely contribute with GeoNode being an open source platform are facilitating the creation, sharing, and collaborative use of geospatial data and benefitting masses by real-time communication (Figure 1).

The research is aimed to create a dynamic and real-time SDI solution by the way of continual sharable activity imparted by internet and GeoNode (Figures 2 and 3). At its core, the system is based on open source components GeoServer, GeoNetwork, Django, and GeoExt, elaborated in Figure 3, that provide a platform for sophisticated web browser spatial visualization and analysis. Building on this stack, the present work utilizes a map composer and viewer, tools for analysis, and reporting tools which are facilitated by GeoNode. It is designed on Web 2.0 principles to make it extremely simple to share data; easily add comments, ratings, tags connecting between GeoNode and existing GIS tools. To enhance distribution, the GeoNode enables simple installation and distribution; automatic metadata creation; and search via catalogs and search engines. And to promote data collection, the system is aimed to align incentives to create a sustainable SDI to align efforts so that

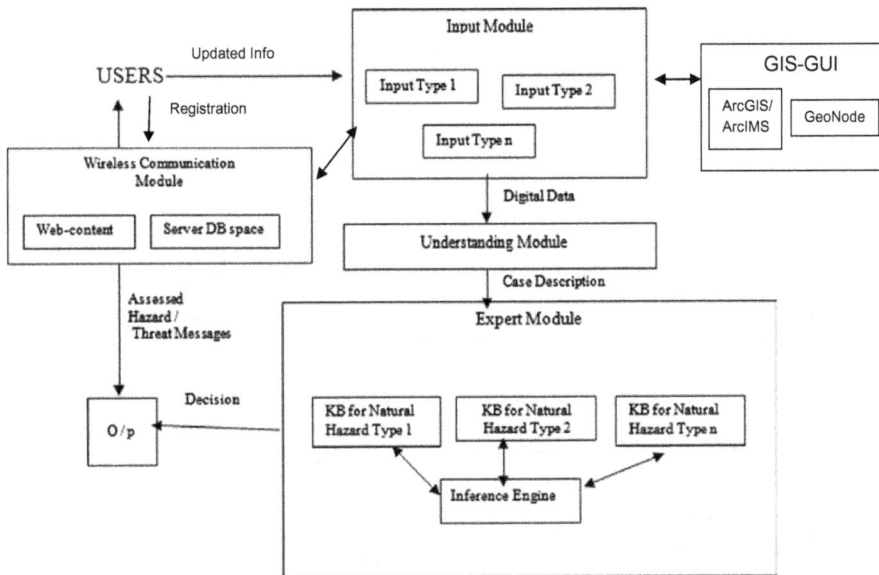

Figure 1. Modules of the integrated internet based GIS-GUI warning communication system with KBs for generalized hazard warning capability (modified on Bhattacharya, Ghosh, Boccardo, & Samadhiya, 2011; Ghosh, Bhattacharya, Boccardo, & Samadhiya, 2012).

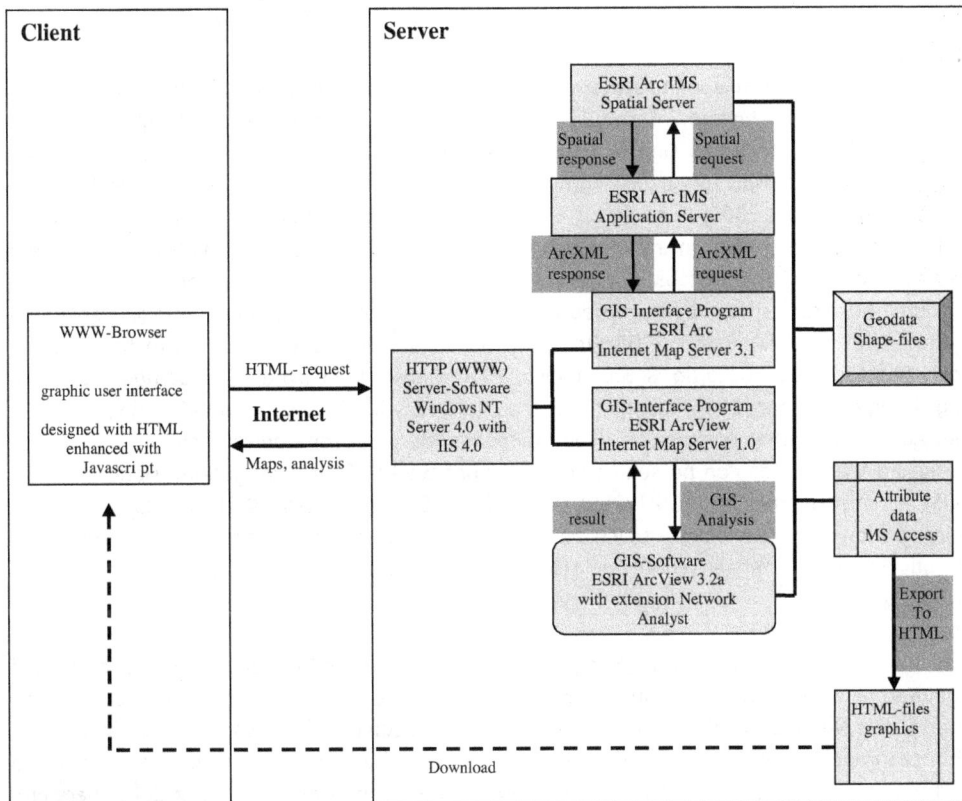

Figure 2. GIS-GUI shared internal architecture of the system.

amateur, commercial, non-governmental organizations, and governmental creators all naturally collaborate, figure-out workflows, tools, and licenses that work to assure data quality, in order to promote data, constantly evolving, convincing, and always up to date. The idea is to create a full featured platform for helping decision-makers easily compose and share developments with spatial data.

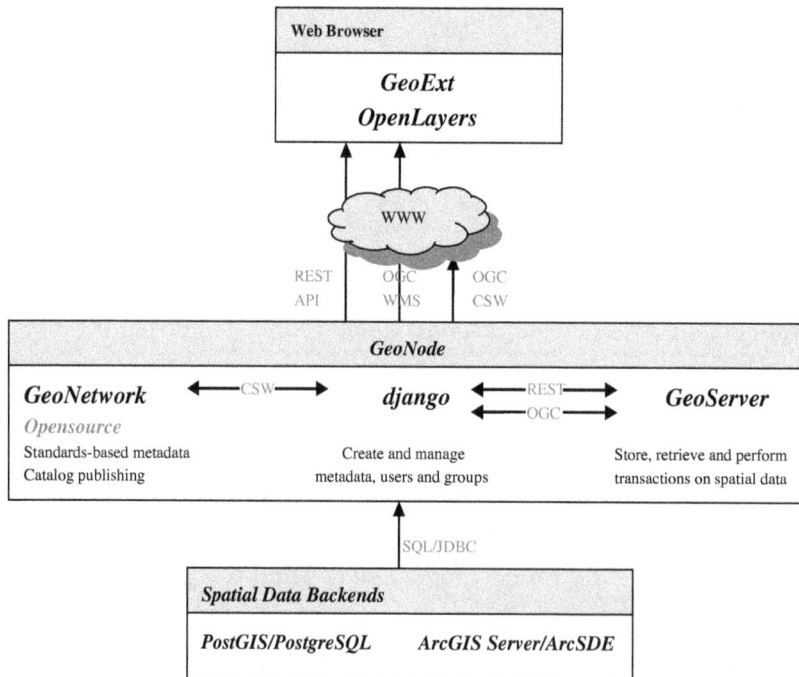

Figure 3. Conceptual schema of geospatial data manipulation for open-source sharing in GIS-GUI.

2. Background

A fully automated hazard warning system consists of hazard classification followed by progressive dissemination of the hazard information. There have been many hazard zonation systems (Albayrak, 2006; Bhattacharya, Ghosh, & Samadhiya, 2012; Ghosh & Bhattacharya, 2010; Montoya, 2003) as well as proposals for hazard warning systems (Cioca et al., 2008; Montanari, Mehrotra, & Venkatasubramanian, 2007; Pries et al., 2006; Waidyanatha, Gow, & Anderson, 2007) but an amalgamated system having integration of both is missing. A well tested fully working such system looked even more distant. A look into previous developments in the two respective fields shows that numerous studies have been carried out towards evaluation of geo-hazard occurrence and their levels of threat (Bhandari, 2008; Flax, Jackson, & Stein, 2002; Ghosh & Bhattacharya, 2010; Ghosh, Bhattacharya, Boccardo, & Samadhiya, 2010; Ghosh, Bhattacharya, & Samadhiya, 2009; Hong, Adler, & Huffman, 2007; Wang, 2002; Zhou, Lee, Li, & Xu, 2002), to name just a few. On the other hand many other researchers have independently proposed communication systems for geo-hazard warning, as can be seen from important works of Albayrak (2006), Darienzo et al. (2005), McGinley, Bennet, and Turk (2006), McLoughlin (1985), Sorensen (2000), and Roy (2009). But the integration of these two areas to yield a complete end-to-end system has not been taken place yet, aptly points out Sorensen (2000) and a United Nations Organization global report (Bhattacharya et al., 2012).

The introduction of internet and GIS together or for that matter a full-fledged WebGIS dedicated to geohazard warning and communication has been rare and the examples available are the United States National Weather Service under National Oceanic and Atmospheric Administration, in France through research associated with Meteo France, United Nations Platform for Space-based Information for Disaster Management and Emergency Response, and groups of developers and scientists like in Information Technology for Humanitarian Assistance Cooperation and Action at Politecnico di Torino, Italy. These are web enabled knowledge base (KB)-GIS which issue alerts on websites and also SMS those who are subscribed. These WebGIS are repositories of classified hazard information and are not reaching the effected masses in time (Montoya, 2003; Samarajiva, 2005; Siddiqui, 2005; Tobita, Fukuwa, & Mori, 2009; Yang, Lin, Chen, & Su, 2007). These are government/internationally funded big projects where dedicated teams monitor each module of the huge system like external data capturing, data analysis, then decision-making and ultimately publishing the forecast. What is needed is integration of all functionalities and external hardware interfaces also to directly receive causative factors parameters. A hardware design development of some proposed interfaces with real time monitoring field-based systems would make a system effective.

Thus, literature shows existing systems work either as a hazard classification or as a hazard proclamation system but not both together. Further, the focus has been primarily on the hazard classification and hazard information communication has taken a backseat. Thus, integration of both needs to be kept in perspective and learnt before embarking on development of a complete geo-hazard warning communication system.

Bhattacharya et al. (2012) report that the utilization of geo-hazard warning systems across the multitudes has to be done and the report postulates some specifications. It says economic ways of developing geo-hazard warning systems must be explored by tapping the potential of ICT-based development. The report cites the broad picture by showing the advantages in using ICT, cost of developing country-wide systems as well as limitations of systems due to demographic barriers. Bhattacharya et al. (2012) explore the interoperability of next generation wireless data networks, real time alarms, instantaneous warning generation, and propagation to masses in an area, ultimately connecting WiFi, WIMAX, mobile networks, and propagating the assessed hazard through the interconnection. These are regional extent systems, having setup of transmitters and receivers and may be GSM-based. The benefits reported were integrated framework and knowledge-based approach but the lacunae were ill-defined interfaces, and undefined modularity. These approaches have drawbacks such as: repetitive coverage of area is difficult, involve the association of experts at every step of the functioning are not adaptable with communication technologies, and the concept of interfacing between different modules is missing. A solely GIS-based hazard system only allows creating a knowledge base (KB) restricted to the options available in the package, and the development is controlled and limited by the facilities available in the GIS (Roy, 2009; Zlatanova, 2009). It has been seen that an efficient knowledge representation scheme (KRS) is essential for good performance of a KB system (Backhaus & Beuleb, 2005; Bhattacharya & Ghosh, 2008; Carrara et al., 2004; Gomarasca, 2007; Zlatanova & Dilo, 2010). The flexibility to make use of a customized KRS is missing in a GIS. The concept of integrating with an external interface like a warning technology is difficult in a GIS. Hence, an approach is needed to amalgamate open source GIS platforms with advanced architecture of proprietary packages like ArcGIS and develop a customized and integrated system addressing the pressing needs.

With this background, the design of the proposed system has been taken up and the different functionalities have been assigned to different modules as explained in its architecture. The system has been proposed to be based on integration of two broad divisions: the hazard evaluation and the warning proclamation. The system methodology includes interpretation of causative factors from their input data, addressing of expert knowledge as rules, re-classification of geomorphologic images, evaluation of hazard susceptibility intensity based on causative factors ratings, and warning proclamation by communicating through the extant cellular network infrastructure (Cioca et al., 2008; Montanari et al., 2007; Waidyanatha et al., 2007; Wattegama, 2007; Xu & Zlatanova, 2007) available in a locality.

It is expected that the proposed system would have many advantages over existing systems on many counts, such as: first, interfacing of functional modules is facilitated; second, implementation of an indigenously developed faster knowledge representation is possible; and third, the implementation of a customized framework to deal with multiple hazards inventory.

3. Methodology

The initial architecture for the shared data concept using the GIS-GUI module of the proposed system is shown in Figure 1. The input module is a highly interactive interface (Bhattacharya &

Ghosh, 2010; Bhattacharya et al., 2011; Ghosh et al., 2012) having connectivity to GIS-GUI as well as Wireless Communication for warning module. The types of inputs correspond to the various causative factors sets for different hazard types. The Understanding Module is the intelligence embedded into the system for deciphering the input and access the correct knowledge-base. The understanding consists of a matching algorithm based on Complete Matching with Exact String match approach. The algorithm is a variant of brute force algorithm that has been adapted to the needs of the KB. This leads to understanding of the digital maps to correlate the information with the next functional module, i.e. the KB housed in the Expert module.

Expert module houses the inference engine and knowledge database of the system (Bhattacharya & Ghosh, 2010; Bhattacharya et al., 2011; Ghosh et al., 2012). The Output module (O/p) is responsible for accepting the classified hazard map and location-based communication details. The Wireless Communication module is the warning functionality of the system and will be responsible for system information manipulation, processing, and dissemination; web-content handler sub-module for web-based processing; Trigger sub-module for Threat Extraction; and Communication sub-module for sending warning messages using interfacing with the GSM network. The GIS-GUI is to interface to the Input module of the system and is responsible for the features creation pertaining to geospatial datasets. This is proposed to be interactive and shareable in nature with functionalities like geodata shape files, attribute data, web-content graphics and the click and point interface. The two way communication with the input module allows the GIS-GUI to effectively create client–server computer architecture (Figure 2).

The input module in-turn communicates with the warning module to extract the mobile communication details which the user might want to display on the console via the GIS-GUI. Domain specific SDI including data models, applications, and services based on Open Geospatial Consortium (OGC) standards and their benchmarking/evaluation are the building blocks of this proposed research, being taken care by the concept of GIS-GUI module. The conceptual schema (Figure 2) provides insight about the components and the way they are used to create the final product. The main components are: The GeoSpatial Data Manager, GeoServer, GeoNetwork, and Map Composer. GeoServer provides an OGC compatible data store that can speak WMS, WFS, WCS, and others in common formats like GML, GeoJSON, KML, and GeoTiff. It can be connected to different spatial backends including PostGIS, Oracle Spatial, ArcSDE, and others.

GeoNetwork provides a standard catalog and search interface based on OGC standards. It is used via the CSW interface to create and update records when they are accessed in GeoNode. This is a Django-based project that allows the user to easily tweak the content and look and feel and to extend GeoNode to build Geospatial. It includes tools to handle user registration and accounts, avatars, and helper libraries to interact with GeoServer and GeoNetwork in a programmatic and integrated way. There is a wide range of third party apps that can be plugged into a GeoNode-based site including tools to connect to different social networks, to build content management systems and more. The Map Composer: GeoNode Client: The main map interface for GeoNode is the Map Composer/Editor. It talks to the other components via HTTP and JSON as well as standard OGC services.

The interactive GUI allows for data visualization, manipulation, and sharing (Figures 2 and 3) and it integrates with the broad functionalities of the system as in Figure 1. The overall architecture depends on the creation of KBs for natural hazards to deduce the extremity of the occurrence. The methodology is that the input module of the system implements extraction, based on legend matching, of information about causative factors from thematic maps, satellite images, and GIS layers, addresses expert knowledge rules (qualitative approach), conducts pixel-based reclassification of input (compatible to KB), results in evaluation of intensity of hazard on ratings of causative factors (deterministic method), and communication to user is achieved using existing cellular network infrastructure in a region. Proposed research should contribute to the development and application of OGC standards. The proposal brings out the benefits of the study towards these goals and the overall requirement of setting up of SDIs in the country.

The proposed system architecture is based on the concepts of interactivity between geo-spatial data management, internet and web-based processing, logical inferencing, and communication technology. Hence the development of different modules, each of which achieves a specific set of tasks related to the mentioned technologies, such as the data needed by the geo-hazard warning communication system and the structure of data maintenance adopted inside the database module.

3.1. Input data to the system

The data utilized by the system comes in many basic formats like string, numeric, alphanumeric, and arrays. The aggregated data are stored in the database as geo-referenced data, threat strings, communication numbers, and instructive messages if any. The data-sets required by the geo-hazard warning communication system are as follows:

3.2. Geo-referenced data

The information pertaining to assessed hazard and subscriber mobile data those have been registered in the system and mapped to the region (geo-referenced threat locations) where the messages are to be disseminated. The validation procedure works on landslide threat in a region evaluated a priory as a hazard map. The mobile numbers to be utilized for sending messages are the numbers lying in the region of the map. There could be many maps whose threat data are stored in the database of the warning system at any given time. To select the correct mobile numbers for that region, the hazard location as well as subscriber data both have been geo-referenced. The latitude and longitude for a given location describes the threat level in that location in one table and the same latitude–longitude describes the mobile numbers in that region. The separation of regions has been kept as .25° × .25° latitude × longitude. The latitude–longitude combination has been used as indexes for accessing the tables in the database.

3.2.1. Location data

The location data consists of spatial as well as the threat details of an area, contained in the server database. The server database holds in its table *hazard_details* threat messages in association with their geo-location (Table 1). The index column represents the pixel location of the rasterized hazard data having geo-referenced match with the ground location shown in the second column of the table. The classified hazard description constitutes the third column of the table which notifies the local area name as well. The geo-location in the second column is accessed by the client table described next.

3.2.2. Subscriber data

The subscriber data consists of the spatial details and the mobile numbers existing in that area, and populated with registered users (Table 2). The client database stores the subscribers' registered mobile numbers in association with their geo-locations in a manner which corresponds to the format that the server database stores its geo-locations. Each entry from the first column of the

Table 1. Sample table entries for location data in server database

Index	Ground location	Perceived threat at that location
000 000	30°04′00″ N and 30°19′00″ N	Moderate landslide threat in Devaprayag
	77°50′00″ E and 77°65′00″ E	
010 020	30°20′00″ N and 30°35′00″ N	Low landslide threat in Muni ki reti
	77°66′00″ E and 77°81′00″ E	
So on ...	So on ...	So on ...

Table 2. Sample table entries for subscriber data in client database		
client_location	client_mobile	hazard_string
30°04'00" N and 30°19'00" N	9634317343	Moderate
77°50'00" E and 77°65'00" E	9448734189	
30°20'00" N and 30°35'00" N	9986572110	Low
77°66'00" E and 77°81'00" E	9411149587	
So on ...	So on ...	So on ...

client database table *client_location* is searched in the server database table *hazard_details* and on successful match, the location threat details are extracted from the server database table. The perceived threat in server database is matched with the hazard_string of the client database table. If the hazard_string occurs in the perceived threat string then the hazard level is confirmed to be correct and valid for use by higher modules of the geo-hazard warning system.

3.3. System information manipulation, processing, and dissemination
One of the tasks of each of the modules in the warning system is the handling of data received from the previous modules over the interfaces. Once the external hazard is received by the warning system, the database module automatically creates data-tables to store it. It then keeps transferring the data through function calls to web-content handler module from the data-tables. Web-content handler creates packets of the data automatically and transfers it to trigger module. Trigger module utilizes the data and also creates its own data then calls a function to create the interface towards communication module. Communication module extracts the packet and calls the GSM interface system method to disseminate the message in the mobile network.

Through the *external_event* call to the geo-hazard warning communication system, it undergoes initialization with a fixed and distinct digital identification for each pixel which is a region having its distinct latitude–longitude information stored as DBMS tables. The system is coded in Java as this programming language has facilities for implementing internet-based and intranet-based applications, and software for devices that communicate over a network. Java programs consist of pieces called classes. Classes include pieces called methods that perform tasks and return information when they complete execution. Java programs take advantage of rich collection of existing classes in the Java class libraries, which are known as Java Application Programming Interfaces (APIs). The connectivity with the database has been provided by developing a Java Database Connectivity-Open Database Connectivity (JDBC-ODBC) bridge with the help of JDBC-ODBC driver provided in the JDK. This facility has been used by the system to obtain location and range of mobile network, and to store the output of the *external_event* as the input, i.e. the warning messages for the danger zones/areas.

3.4. Database module: system data management
The functions defined under the class *Database* are: *Connection Pool, Create Database, Initialize Server Database*, and *Initialize Client Database* function. The connection pool function creates connections to the data tables and maintains the list of open connections. *Create database* function utilizes a connection to latch on to the database to start creating tables as well as, initializes database to handle the read and append modes of data handling which are needed for both server and client data tables.

The various functions executed by the database module, as and when the requests come from higher modules, are: *Get Zone for Pixel* receives the pixel value as input and utilizes "Get Database Connection" sub-module to query the zone data associated with the pixel value; then invokes "Release Database Connection" and returns the retrieved zone (geo-location). Similar sequence of commands are executed for *Get Subscriber Data* to receive the geo-location (zone) as input and query all subscriber mobile numbers associated with the input geo-location and return the retrieved subscriber mobile numbers for the zone. *Get Location Threat Level* receives the geo-location (zone) as input and runs the query to access the threat level associated with the input geo-location and returns the retrieved threat message. Likewise there are other procedures for insertion, authentication and storage of server (operator) and client (subscriber) data available to the database module.

3.5. Web-content handler module: web-based processing

The web-content handler in the warning system application is based on a web platform. Data transmission security is ensured by using the HTTP-Secured, which provides a secure channel between the client station and the system residing on the server. The proposed warning communication system, being web-based, has interfaces through the external events of hazard classification, local servers, the internet and the warning system server. Web-content handler module has defined classes *Initialize GUI*, *Create New Subscriber*, and *User Login* for managing various data processing and transmission tasks. These classes have further sub-divided tasks as methods and functions. The information exchange in the warning system starts from receiving the hazard classification of an area, storing it in the local server, invoking the web-content handler which is internet-resident using HTTP connection, selecting the mobile numbers stored in the system server, utilizing the trigger conditions for checking the level of threats, and finally, calling the communication module to disseminate the message. The telecommunication commands handle the modem functionality and thus the communication with the mobiles gets established.

Web-content handler creates the GUI environment of the system using HTML (HTML, 1997) and controls the web (internet) application data transmission applying HTTP. This module receives the data from the database module and gets these encapsulated in the GSM SMS format (3GPP, 1998). Independent packet gets formed for each location consisting of, in sequence, subscriber number, and geo-location and threat message (Figure 4). The packet has a header part at the beginning and a marker at the end. Finally, it sends the encapsulated packet to the trigger module of the system.

The web-content handler controls the display related configuration in the web (internet) application creating a GUI environment for the operator. This module utilizes the data from the database module as well as the triggering conditions from the trigger module. It also encapsulates and sends the parameters for launching the SMS utility to the communication module. The web-handler sends the strings inserted into empty form-fields as per the requirements of GSM short messaging. The variables to be used in a GSM environment are declared at the top. Then the opening of a message is handled. Various headers are defined. Next would come the handling the body of the message. This segment places the appropriate size limits, and the *time to communicate* parameters are filled in the data packet sent to the next module. Firstly, the code has been instructed with the mechanism to monitor the availability of internet connectivity. This part of web-content module code does the important task of retrieving the internet parameters and storing them in variables to be used by higher up modules.

The important parameters that the web-content module deals with are *gsmIncomingSMS, gsmPower, DebugFmt, clockNow, strConcat, dintToStr, boardSerialNumber, DebugMsg,* and *gsmSendSMS*. These functions and methods are incorporated for the packet formation and defining the utilities of the parameters for higher modules to which the web-content handler module sends the packets.

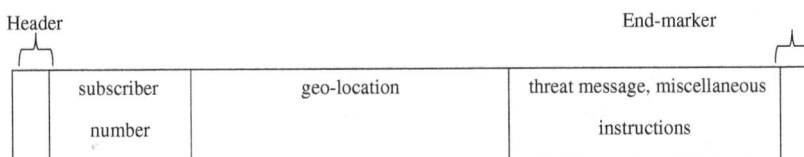

Header			End-marker	
	subscriber number	geo-location	threat message, miscellaneous instructions	

Figure 4. The format of the data packet formed by web-content handler module for transferring.

3.6. Trigger module: threat extraction

The trigger module holds the logic and the algorithm to check that the correct string is being sent as SMS. The level of hazard (at present, there are three levels of hazards, such as low, moderate, and high) is stored as a string which comes from the server database. This string is identified as high or moderate or low and accordingly is determined the trigger to be generated. Trigger module accepts the encapsulated information packet from web-content handler and extracts the appropriate message for a geo-location and passes it on as a parameter passing mechanism to the communication module. Trigger module has the declarations for setting the THREAT_TO_SMS_FREQUENCY_MAPPING variable which contains the mapping between threat level and frequency of SMS, for defining *Process Information Packet* sub-module, for *Searching and Matching* sub-module, for *Set SMS Frequency* sub-module, and defining *Trigger Hazard Warning* sub-module.

The mapping variable transfers the value assigned for message dissemination for each level of threat. The *process information packet* sub-module contains codes for executing the packet data processing tasks. The *searching and matching* sub-module performs the string operations to search for hazard level and come to a conclusion about the threat. *Set SMS Frequency* sub-module modifies the packet with new data of frequency of dissemination. Finally, the *Trigger Hazard Warning* sub-module sends the triggered packet to communication module. The various events taking place in the execution of the trigger-module are, *the external event*—an event that requires a warning alert (landslide, flooding, etc.); *the initiator*: part of code that acknowledges and identifies the external event and sends the alert request to the handler; *the main method*: is invoked in the server where the request is processed; and subsequently is executed the *GSM Method* which is a specialized software component that transfers the alert message to the communication module for propagating over mobile communication network.

3.7. Communication module: sending warning messages

The communication module is responsible for sending the warning messages and hence defines parameters such as definition of DELIVERY_MODE variable to either of INTERNET_SOFTWARE or SELF_HARDWARE, and further defining internet broadcast to mobiles. Finally, the module defines and executes *Send SMS* sub-module. *Send SMS* accepts the subscriber mobile numbers, zone, associated threat message, and SMS frequency as input parameters, validates the input parameters, and initializes a parameter called ROUND_NUMBER to 1 which is meant to execute rounds of hazard warning dissemination. The following steps are processed while ROUND_NUMBER is less than or equal to the value contained in variable *SMS frequency*:

> display the ROUND_NUMBER of the SMS being sent in the client GUI
>
>> if DELIVERY_MODE is *internet broadcast*
>>
>>> call "Send SMS INTERNET_SOFTWARE"
>>
>> else if DELIVERY_MODE is SELF_HARDWARE
>>
>>> call "Send SMS SELF_HARDWARE" sub-module
>>
>> else
>>
>>> DELIVERY_MODE is not specified, display error message to user
>
> display the status of SMS sent in client GUI
>
> increment ROUND_NUMBER variable by 1

The communication module receives packet consisting of subscriber number, threat message, and SMS frequency, in sequence. The primary objective of the communication module is to open the SMS sending utility and ensure that the communication goes through the correct gateway. The gateway is responsible for channeling data from internet to mobile network. The network communication broadcast facility will be used to freely send SMS to all users of mobiles moving into an area affected and perceived threat prone. The SMS sending utility is invoked/opened by the communication module and is entered with the initial parameters such as username, password and application identity (API ID). These initial parameters are required for authorizing the delivery of each SMS, hence the communication module inserts these parameters into the SMS program each time the program is called.

The warning system uses the location details from the server database and accesses the threat message strings corresponding to each. In its client database, the warning module has the mobile numbers in a pre-determined storage format. Hence, as soon as the mobile numbers in the region are extracted from the table, the SMS Protocol program is called and the mobile numbers filled in the program as command line parameters and the respective hazard messages are sent. The number of mobile numbers selected per region is fed in a loop and the SMS program is called for each number for sending SMS.

The SMS program connects to the SMS gateway via the internet and this gateway forwards the text message to the mobile environment. The steps correspond to the logic of sending an SMS. The first step is to verify the user. The second step consists of creating the SMS message body including gsm_number, sender_name, text_message, name_of_packet, gateway_identification, quality_of_message, and delivery_code. These ensure that the message is delivered correctly; if not then appropriate negative acknowledgment is sent. In the flow of controls, it can be seen that it is mandatory to create a connection and close it upon exit each time SMS is to be sent. After opening a connection, login can be proceeded to, and if login is positive then SMS sending is allowed. Upon detecting the final header of SMS the connection is closed.

3.8. Interfacing with the GSM network
The internal processing involving the database and web servers maintains the actual data flow controlled by the http/s and TCP/IP commands. When an http request is generated by the system after creating the data packet, server hosting web-content module starts processing the requests and accesses the database through a TCP/IP channel. Further internal processing involves the function calls in sequential manner to the trigger module and communication module. The communication module executes the server command ComX (present in *attention* (*AT*) command-set) to connect to the modem over a physical channel RS232.

The AT (attention/initiate) communication protocol command-set is a terminal protocol (Cioca et al., 2008). The protocol activates the modem and directly issues modem commands. It facilitates sending of SMS by executing the AT communication protocol command-set. The procedure consists of typing several text commands in a terminal window (Figure 5). The first command shown initializes the modem connection, then the mode is set to SMS mode, after that the commands setup the message parameters like size and headers. In order to transmit a text message (SMS) through modem specific instructions are used on the serial interface.

The AT commands follow a sequence as per the logic within the system. The logical steps of sending an SMS are as follows: the first step verifies the authenticity of the user. In the second step, appropriate SMS message body (consisting of gsm_number, sender_name, text_message, name_of_packet, gateway_identification, quality_of_message, and delivery_code) is created to ensure the message gets delivered to correct users. And if not, then negative acknowledgment gets sent. As soon as a message is ready to be sent, a connection gets opened, for permissible login SMS gets sent and on detecting the final header of SMS, the connection is closed. The cycle repeats for each SMS.

Hence, as soon as the mobile numbers in the region are extracted from the table, the SMS protocol program is called and the mobile numbers filled in the program as command line parameters and

```
AT    :                  Initiate modem connection
OK  :                    Result
AT+CMFG=1  :    Setting mode – SMS Mode
OK  :                    Result
AT+CMGW="+0742****"  :  Setting Number
>Example Text  :     Message ends with Ctrl Z
+CMGW: 1   :         Message Index
OK    :                  Result
AT+CMSS=1  :      Send message
+CMSS:20    :         Sent message index
OK    :                  Result
```

Figure 5. A snapshot of AT commands for sending warning SMSs in GSM network.

the respective hazard messages are sent. The number of mobile numbers selected per region is fed in a loop and the SMS program is called for each number for sending SMS. The SMS program connects to the SMS gateway via the internet and this gateway forwards the message to the mobile numbers.

And subsequently, the *GSM_Method* is executed which provides interfacing between the communication module and external GSM environment. It transfers the alert message from the communication module over mobile communication network. The telecommunication commands handle the modem functionality and thus the communication with the mobiles gets established and hence providing interfacing. The *GSM method* also sends the strings to be inserted into empty form-fields as per the requirements of GSM short messaging. The variables to be used in a GSM environment are declared at the initiation of the GSM program. Next, the program handles the opening of a message. Various headers are defined within the body of the program. Next in the program is the section of the code that handles the body of the message. This segment places the appropriate size limits, and the time to communicate parameters are filled in the SMS technical specifications.

The communication module is equipped with two ways of interfacing with the GSM network to send SMS messages from the warning system to mobile phones. The two methods are:

(1) Connectivity of the geo-hazard warning system to the SMS center (SMSC) or SMS gateway of a wireless carrier or SMS service provider through the internet. Subsequently the communication module sends SMS messages using a protocol/interface supported by the SMSC or SMS gateway. This is the software method of message sending.

(2) Connectivity of GSM modem to the geo-hazard warning system and execution of AT commands to instruct the GSM modem to send SMS messages. This is the hardware method.

The SMS gateway is the responsible entity to disseminate messages in an SMS messaging system. Hence, the developed system utilizes programming interfaces to SMS gateway (Figure 6) using an open source SMS gateway software package Kannel (2010), which is programmable. Through Kannel, the geo-hazard warning communication system can handle connections to SMSCs, mobile phones, and GSM modems. It has an HTTP/HTTPS interface for the sending and receiving of SMS messages.

Figure 6. SMS gateway acts as a relay between two SMSCs.

To connect to an SMS gateway, the developed system uses an SMSC protocol called Short Message Peer to Peer (SMPP). Some features of the developed SMS application are programmed using the HTTP/HTTPS interface also. HTTP/HTTPS are easier to use than SMSC protocol SMPP.

3.8.1. Customizing communication module to send messages using software interface

To configure SMS sending capability of the geo-hazard warning system (Bhattacharya et al., 2011; Ghosh et al., 2012), registration was done with a wireless carrier SMS service provider, and to start sending SMS messages a protocol/interface to the SMSC or SMS gateway was programmed. To communicate with the SMSC, an SMSC protocol SMPP which is an open-source SMSC protocol. The first step executed by the customized SMS application is to run the SMS gateway (Figure 7).

3.8.2. Running SMS gateway

SMS gateways of SMS service providers and wireless carriers very often support one or more of the following protocols/interfaces: HTTP, HTTPS (HTTP + SSL encryption), XML over HTTP/HTTPS, SMTP (email to SMS), FTP, SMPP, CIMD, etc. SMPP protocol imparts advanced capabilities to a system hence the communication module of the warning system has been coded to a customized application using SMPP protocol. The developed warning communication system's *communication module* connects to the SMSC as shown in Figure 8.

The developed warning system has been programmed with a *terminal executable* to send AT commands to a mobile phone or GSM modem is to use a terminal program. A terminal executable program's function is like this: it sends the characters typed to the mobile phone or

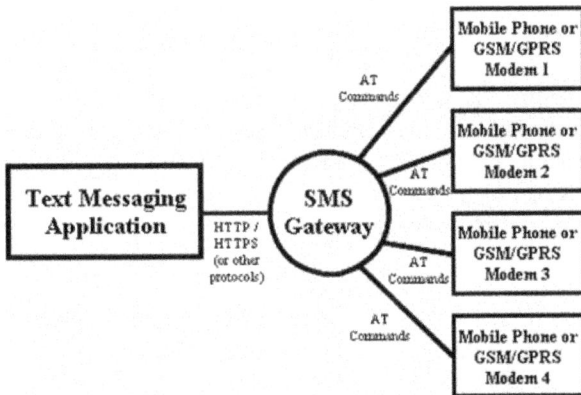

Figure 7. SMS text messaging application connects to a pool of mobile phones or GSM modems through an SMS gateway.

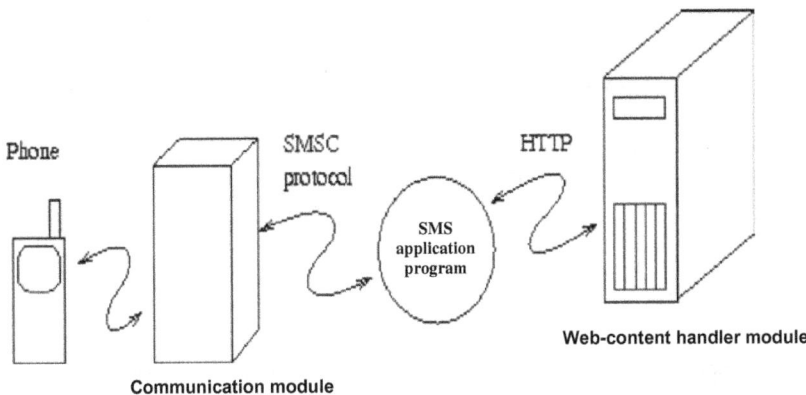

Figure 8. SMS text messaging application (such as the developed system utilizing internet) connects to SMSCs through an SMS gateway.

```
AT
OK
AT+CMGF=1
OK
AT+CMGW="+85291234567"
> Low/no landslide threat in Pratapnagar.
+CMGW: 1

OK
AT+CMSS=1
+CMSS: 20

OK
```

Figure 9. The sequence of AT commands executed by the terminal executable program (resident in communication module) of the developed system.

GSM modem. It then displays the response it receives from the mobile phone or GSM/GPRS modem on the screen. The terminal program interfaced with the developed system executes a system call on Microsoft Windows HyperTerminal. Figure 9 shows a snapshot of the commands run by the system on Windows platform that demonstrates its use of AT commands to send an SMS text message. The lines in bold type are the command lines that the system generates as commands to be entered in HyperTerminal. The other lines are responses returned from the GSM modem or mobile phone, handled by the buffer provided to the warning system.

The system uses the constructed methods, functions, and commands to send SMS messages from the SMS application for connecting to and sending AT commands to the mobile phone or GSM modem in a sequential and automated mode.

The JAVA class containing the above methods supports many of the URL parameters that are defined for the warning system communication module application, and could easily be adapted to support additional parameters. The URL parameters are supported as methods for the *sendsms* class, with *methodnames* matching the URL parameter names, except that all methods are in lower case. In addition, the URL parameter methods, the following additional methods are defined as shown below.

4. Conclusion

An interactive web-based geo-hazard communication system having geospatial capabilities has been conceptualized and proposed in which once hazard level is fed in acceptable format (which could be best done by storing the information latitude and longitude area-wise and associating threat messages and mobile numbers in succeeding columns of the table), warning could be sent through mobile communication. The system functions by searching and matching the geo-address of the area as a tag to search the associated mobile information and message string to be trans-mitted. Once the table is accessed, the SMS program is initiated, and the mobile numbers filled in for the messages to be sent. It is sensible to develop a system using open-source technologies and software to as much an extent as possible. This avoids the problem of any proprietary issues and is cost effective. Hence the proposed warning system has been developed using ArcIMS/GeoNode as the database management systems providing online accessibilities, HTTP, and HTML as the internet technologies, JAVA programming environment for system programming, and mobile programming.

The developed system is capable of utilizing the software interfaces through the internet in order to send messages to user mobiles, as well as utilizing the hardware interfaces in the case when a GSM modem or a mobile phone is directly installed in the computer where the developed system is running. The necessity of utilizing the hardware interface arises in case the availability of internet is scarce at any location. The methods, parameters, and classes all remain constant in both software and hardware interfaces procedures, only the situation demand changes. The integrated working of all these interfaces leads to successful delivery of important warning messages to threat prone areas. The techniques are also useful for natural resource optimization, agricultural yield calculations and betterment, policy planning, and long-term goal setting.

Funding
The Ministry of Education, Youth and Sports of the Czech Republic, Project [grant number CZ.1.07/2.3.00/30.0021] "Strengthening of Research and Development Teams at the University of Pardubice", financially supported this work.

Author details
Devanjan Bhattacharya[1]
E-mail: devanjan.bhattacharya@beun.edu.tr
Jayanta Kumar Ghosh[2]
E-mail: gjkumfce@iitr.ernet.in
Jitka Komarkova[3]
E-mail: Jitka.Komarkova@upce.cz
Santo Banerjee[4]
E-mail: santo@iscass.org
Hakan Kutoglu[1]
E-mail: kutogluh@hotmail.com

[1] Faculty of Engineering, Geomatics Engineering Department, Bulent Ecevit University, Zonguldak 67100, Turkey.

[2] Department of Civil Engineering, Indian Institute of Technology Roorkee, Roorkee, Uttarakhand 247667, India.

[3] Faculty of Economic & Administrative Science, Institute of System Engineering & Informatics, University of Pardubice, Pardubice, Czech Republic.

[4] Institute for Mathematical Research, University Putra Malaysia, Serdang, Malaysia.

References
Albayrak, O. (2006, July 9–13). Management and diffusion of technology for disaster management. In *Proceedings technology management for the global future* (Vol. 1, pp. 1742–1748). Istanbul: Portland International Center for Management of Engineering and Technology, PICMET.

Backhaus, R., & Beuleb, B. (2005). Efficiency evaluation of satellite data products in environmental policy. *Space Policy*, Elsevier, *21*, 173–183. Retrieved from http://linkinghub.elsevier.com/retrieve/pii/S0265964605000482; http://dx.doi.org/10.1016/j.spacepol.2005.05.008

Bhandari, R. K. (2008). *Early warning systems against landslides, landslide management—Present scenario and future directions—2008.* Roorkee: CBRI.

Bhattacharya, D., & Ghosh, J. K. (2008). Evaluation of knowledge representation schemes as a prerequisite toward development of a knowledge-based system. *Journal of Computing in Civil Engineering, 22,* 348–359. Retrieved from http://cedb.asce.org/cgi/WWWdisplay.cgi?167914; http://dx.doi.org/10.1061/(ASCE)0887-3801(2008)22:6(348)

Bhattacharya, D., Ghosh, J. K., Boccardo, P., & Samadhiya, N. K. (2011). Wireless hazard communication system. *Journal of Systems and Information Technology,* Emerald Publishing, *13*, 408–424.
http://dx.doi.org/10.1108/13287261111183997

Bhattacharya, D., Ghosh, J. K., & Samadhiya, N. K. (2012). A review of geo-hazard warning systems towards development of a popular usage geo-hazard

warning communication system. *American Society of Civil Engineers—Natural Hazards Review.* Retrieved from http://link.aip.org/link/doi/10.1061/(ASCE)NH.1527-6996.0000078

Carrara, P., Fortunati, L., Fresta, G., Gomarasca, M., Bonati, L. P., & Poggioli, D. (2004). A methodological approach to the development of applications in a SDI environment. In *7th AGILE Conference on geographic information science* (pp. 743–746). Heraklion, Greece. Retrieved from http://plone.itc.nl/agile_old/conference/greece2004/papers/P-02_Carrara.pdf

Cioca, M., Cioca, L. I., & Buraga, S. C. (2008). SMS disaster alert system programming. In *Second IEEE International Conference on digital ecosystems and technologies* (Vol. 1, pp. 260–264). Phitsanulok, Thailand.

Darienzo, M., Aya, A., Crawford, G. L., Gibbs, D., Whitmore, P. M., Wilde, T., & Yanagi, B. S. (2005). Local Tsunami warning in the Pacific coastal United States. *Natural Hazards, 35,* 111–119. http://dx.doi.org/10.1007/s11069-004-2407-z

Flax, L. K., Jackson, R. W., & Stein, D. N. (2002). Community vulnerability assessment tool methodology. *Natural Hazards Review, 3,* 163–176.
http://dx.doi.org/10.1061/(ASCE)1527-6988(2002)3:4(163)

Ghosh, J. K., & Bhattacharya, D. (2010). A knowledge based landslide susceptibility zonation system. *Journal of Computing in Civil Engineering,* American Society of Civil Engineers, *24,* 325–334. Retrieved from link.aip.org/link/doi/10.1061/(ASCE)CP.1943-5487.0000034; http://dx.doi.org/10.1061/(ASCE)CP.1943-5487.0000034

Ghosh, J. K., Bhattacharya, D., Boccardo, P., & Samadhiya, N. K. (2010). A landslide hazard warning system. In *ISPRS Technical Commission Symposium VIII* (pp. 1–10). Kyoto, Japan. Retrieved from http://isprs.org/proceedings/XXXVIII/part8/headline/TS-23/W01OE3_2010030400224 1.pdf

Ghosh, J. K., Bhattacharya, D., Boccardo, P., & Samadhiya, N. K. (2012). A generalized geo-hazard warning system. *Natural Hazards,* Netherlands: Springer. Accepted July 10, 2012. Retrieved July 21, 2012, from http://dx.doi.org.10.1007/s11069-012-0296-0

Ghosh, J. K., Bhattacharya, D., & Samadhiya, N. K. (2009). GEOWARNS: A system to warn geo-deformation failure. In *Proceedings of the FIG Working Week 2009 Surveyors Key Role in Accelerated Development* (Vol. 1, pp. 1–12). Eilat, Israel. Retrieved from http://www.fig.net/pub/Fig2009/papers/ts04b/ts04b_ghosh_bhattacharya_samadhiya_3435.pdf

Gomarasca, M. (Ed.). (2007). *GeoInformation in Europe.* Netherlands: Millpress. ISBN 9789059660618

Hong, Y., Adler, R., & Huffman, G. (2007). An experimental global prediction system for rainfall-triggered landslides using satellite remote sensing and geospatial datasets. *IEEE Transactions on Geoscience and Remote Sensing, 45,* 1671–1680.
http://dx.doi.org/10.1109/TGRS.2006.888436

Kannel. (2010). *Kannel: Open source WAP and SMS gateway.* Author. Retrieved from http://www.kannel.org/

McGinley, M., Bennet, D., & Turk, A. (2006). Design criteria for public emergency warning systems. In *Proceedings of the 3rd International ISCRAM Conference,* Newark, NJ, USA. Retrieved from www.iscram.org/dmdocuments/S2_T1_4_McGinley_etal.pdf

McLoughlin, D. (1985). A framework for integrated emergency management. *Public Administration Review, 45,* 165–172. Retrieved from http://www.jstor.org/pss/3135011
http://dx.doi.org/10.2307/3135011

Montanari, M., Mehrotra, S., & Venkatasubramanian, N. (2007). Architecture for an automatic customized warning system. In *IEEE Conference on intelligence security informatics* (Vol. 1, pp. 32–39). New Brunswick, NJ.

Montoya, L. (2003). Geo-data acquisition through mobile GIS and digital video: An urban disaster management perspective. *Environmental Modelling & Software, 18,* 869–876.

Nayak, S., & Zlatanova, S. (2007). Remote sensing and GIS in disaster monitoring. *Second Gi4DM Symposium GIM International, 21,* 38–39.

Pries, R., Hobfeld, T., & Gia, P. T. (2006). On the suitability of the short message service for emergency warning systems. In *Proceedings of the IEEE 63rd Vehicular Technology Conference* (Vol. 2, pp. 991–995). Melbourne, Australia.

Roy, D. K. (2009). SATCOM—Early warning system for landslides. *Electronics For You,* 128–138. Retrieved from http://efylinux/efyhome/cover/January2009/Landslide-Monitoring.pdf

Samarajiva, R. (2005). Mobilizing information and communications technologies for effective disaster warning: Lessons from the 2004 Tsunami. *New Media and Society, 7,* 731–747. http://dx.doi.org/10.1177/1461444805058159

Siddiqui, A. (2005, March 9–11). *Web-GIS applications in disaster management—Application to the Tsunami.* In National Seminar on GIS application in rural development with focus on disaster management. Hyderabad. Retrieved from http://faculty.kfupm.edu.sa/crp/bramadan/crp514/Lectures/9%20-%20Web-GIS.pdf

Sorensen, J. H. (2000). Hazard warning systems: Review of 20 years of progress. *Natural Hazards Review, 1,* 119–125. http://dx.doi.org/10.1061/(ASCE)1527-6988(2000)1:2(119)

Tobita, J., Fukuwa, N., & Mori, M. (2009). Integrated disaster simulator using WebGIS and its application to community disaster mitigation activities. *Journal of Natural Disaster Science, 30,* 71–82. Retrieved from http://www.jsnds.org/contents/jnds/30_2_3.pdf

Waidyanatha, N., Gow, G., & Anderson, P. (2007). Hazard warnings in Sri Lanka: Challenges of internetworking with common alerting protocol. In *4th International ISCRAM Conference* (Vol. 1, pp. 281–293). Delft, The Netherlands.

Wang, Y. (2002). Mapping extent of floods: What we have learned and how we can do better. *Natural Hazards Review, 3,* 68–73. http://dx.doi.org/10.1061/(ASCE)1527-6988(2002)3:2(68)

Wattegama, C. (2007). *ICT for disaster management.* Retrieved April 15, 2008, from www.apdip.net/publications/iespprimers/eprimer-dm.pdf

Xu, W., & Zlatanova, S. (2007). Ontologies for disaster management. In *Geomatics solutions for disaster management, lecture notes in geoinformation and cartography* (pp. 185–200). Heidelberg: Springer-Verlag Berlin. Retrieved from http://www.gdmc.nl/zlatanova/thesis/html/refer/ps/WX_SZ_2007.pdf

Yang, M. D., Lin, C. C., Chen, S. C., & Su, T. C. (2007, November 27–29). A Web-GIS disaster management system applied in central Taiwan. In *2nd International Conference on urban disaster reduction.* Retrieved from http://ncdr.nat.gov.tw/2icudr/2icudr_cd/PDF/5_1_5.pdf

Zhou, C. H., Lee, C. F., Li, J., & Xu, Z. W. (2002). On the spatial relationship between landslides and causative factors. *Geomorphology, 43,* 197–207. http://dx.doi.org/10.1016/S0169-555X(01)00130-1

Zlatanova, S. (2009). Geoinformation for disaster management. *GIM International* (Vol. 23/7, ISPRS column). Retrieved from http://www.gim-international.com/issues/articles/id1391-Geoinformation_for_Disaster_Manage-ment.html

Zlatanova, S., & Dilo, A. (2010, February). A data model for operational and situational information in emergency response: The Dutch case. In *Proceedings of the Gi4DM Conference—Geomatics for Disaster Management* (p. 4). Torino. Retrieved from www.gdmc.nl/publications/2010/Operational_situational_information_emergency.pdf

11

Behavior of solid matters and heavy metals during conductive drying process of sewage sludge

Jianping Luo[1], Anfeng Li[1]*, Dan Huang[1], Jianju Ma[1] and Jiqiang Tang[2]

*Corresponding author: Anfeng Li, National Engineering Research Center for Urban Environmental Pollution Control, Beijing Municipal Research Institute of Environmental Protection, No. 59, Beiyingfang Street, Xicheng, Beijing 100037, China

E-mail: cee1219@hotmail.com

Reviewing editor: Sanjay Kumar Shukla, Edith Cowan University, Australia

Abstract: Behavior of solid matters and heavy metals during conductive drying process of sewage sludge was evaluated in a sewage sludge disposal center in Beijing, China. The results showed most of solid matters could be retained in the dried sludge after drying. Just about 3.1% of solid matters were evaporated with steam mainly by the form of volatile fatty acids. Zn was the dominant heavy metal in the sludge, followed by Cu, Cr, Pb, Ni, Hg, and Cd. The heavy metals in the condensate were all below the detection limit except Hg. Hg in the condensate accounted for less than 0.1% of the total Hg. It can be concluded that most of the heavy metals are also retained in the dried sludge during the drying process, but their bioavailability could be changed significantly. The results are useful for sewage sludge utilization and its condensate treatment.

Subjects: Environment & the City; Environmental Sciences; Technology

Keywords: sludge drying; heavy metals; solid matters; condensate; sewage sludge; disposal; utilization; mass balance; total organic carbon; total phosphorus

1. Introduction

With development of cities and swell in population, sewage sludge as a by-product in wastewater treatment is increasing sharply. The amount of sewage sludge was 30.4 million tons (moisture content of 80%) in 2010 in China, and it increased to about 40.8 million tons (moisture content of 80%) in 2013. This was expected to 47 million tons in 2015 (Chyxx net, 2014). Disposal of sewage sludge has become an environmental challenge.

ABOUT THE AUTHOR

Anfeng Li, male, professor, 1976, deputy director for environmental engineering design division, Beijing Municipal Research Institute of Environmental Protection. His main research direction is the research and development and application of the technology of advanced treatment of wastewater, wastewater reuse, and waterbody remediation. With the rapidly increasing amount of sewage in China, he began to focus on the reuse of the sludge.

PUBLIC INTEREST STATEMENT

With the sharply increasing amount of sewage sludge, its disposal has become a worldwide environmental challenge. Thermal drying is one of the promising approaches in sludge disposal. Due to the abundant organic matter, N, P, and other plant nutrients, the application of dried sludge in agriculture is an attractive disposal option. However, it is considered to be a secondary pollution source of the soil and groundwater. In this study, behavior of solid matters and heavy metals during indirect conductive drying process of sewage sludge was evaluated to provide useful information on safe application of the sludge in agriculture. It is found that solid matter, total phosphorus, heavy metals, and total organic carbon could be almost retained in the dried sludge after drying. Furthermore, the drying process played an important role in heavy metal bioavailability due to pH increase, dehydration, pyrolyzation, and volatization of organic matter.

The application of sewage sludge in agriculture is an attractive option of disposal due to the abundant organic matter, N, P, and other plant nutrients. It has been an important disposal pathway in Australia, Europe, and China (Kakati, Ponmurugan, Rajasekaran, & Gnanamangai, 2013; Kelessidis & Stasinakis, 2012; Pritchard, Penney, McLaughlin, Rigby, & Schwarz, 2010). However, sewage sludge also contains many pathogens, heavy metals, persistent organic contaminants, and these sludge-derived contaminants in soil have the potential to be accumulated by plants and animals (Harrison, McBride, & Bouldin, 1999; Singh & Agrawal, 2010; Vrkoslavová et al., 2010). Sludge drying has become a necessity for agricultural application to reduce the volume of sludge, kill pathogens, and control toxic contaminants' bioavailability (Tunçal, Jangam, & Güneş, 2011). Now, sludge drying can be operated mainly by three modes: convective drying, conductive drying, and solar drying (Bennamoun, Arlabosse, & Léonard, 2013). In these processes, a large amount of steam would be evaporated. This could result in significant change of sludge properties, and consequently influence the speciation, transformation, and migration of contaminants in the sludge (McLaren & Clucas, 2001; Weng et al., 2014). Meantime, volatilization of contaminants especially for volatile ones can be expected to occur due to higher temperature maintained in convective drying or conductive drying (Chou, Lin, Hsien, Wey, & Chang, 2012; Cousins, Hartlieb, Teichmann, & Jones, 1997; Tunçal et al., 2011; Weng, Ji, Chu, Cheng, & Zhang, 2012). However, behavior of contaminants during drying process is still not very clear, and there has been little published research on behavior of metals during sludge drying. Moreover, attention is also needed to treat the condensate as a by-product of convective drying or conductive drying due to its little-known water quality characteristics now.

In this study, behavior and distribution of heavy metals and other solid matters during conductive drying of sewage sludge were investigated to provide useful information on sludge utilization and treatment of the condensate.

2. Materials and methods

2.1. Sample collection and pre-treatment
Dewatered sludge, dried sludge, and condensate samples were obtained from a sewage sludge disposal center in Beijing, China. The center receives dewatered sludge from different municipal wastewater treatment plants in Beijing, such as Qinghe, Jiuxianqiao, Beixiaohe. Dewatered sludge with the moisture of about 77.7% was treated by indirect conductive drying at 250°C with heated oil, and the dried sludge's moisture was about 29.5%. The dewatered sludge was viscous and odorless with a few visible impurities, while the dried sludge was granular with bad smell.

Sludge and condensate samples were continuously collected one time per day with seven replicates one time for one week for analysis of heavy metals, and general parameters such as pH, total organic carbon (TOC), and total phosphorus (TP) were also detected. The collected sludge samples including dewatered sludge and dried sludge were immediately air-dried, ground, and passed through a 100-mesh nylon sieve, then stored in dry glass bottles at room temperature. The condensate samples were stored in the sealed polyethylene bottle at 4°C in the dark and analyzed within one week.

2.2. Analytical methods
Sludge pH was measured in 1:10 (w/v) sludge and water suspension using a pH meter (Orion 3-Star, Thermo, USA). TP of the sludge was determined by molybdenum antimony colorimetric method, while TP of the condensate was detected by ammonium molybdate spectrophotometric method (721, Sunny Hengping scientific instrument Co., China). The moisture of sludge was determined by drying at 105°C to constant weight. TOC of the sludge was determined by oxidation with potassium dichromate (TOC-Torch, Teledyne-Tekmar, USA), while TOC of the condensate was analyzed by combustion oxidation non-dispersive infrared absorption method. Concentration of volatile fatty acids (VFAs) in the condensate was measured by titrimetric analysis with sodium hydroxide preceded by alkali distillation for ammonia removal and high purity nitrogen for hydrogen sulfide removal. Atomic absorption spectrometry (AAS) (GBC932 & GBC908AA, GBC, Australia) was used to determine the

total metal concentrations of Cu, Zn, Pb, Cr, Ni, and Cd in the sludge (GB/T17138-1997, 1998). 0.5 g of the air-dried sludge was heated with 10-mL HCl (1.19 g/mL) at room temperature (25 ± 1°C). And when about 3-mL liquid was left, then 5-mL HNO_3 (1.42 g/mL), 5-mL HF (1.49 g/mL), and 3-mL $HClO_4$ (1.68 g/mL) were added and the mixture was heated at a middle temperature for 1 h (170 ± 2°C). Hg in the sludge was determined by atomic fluorescence spectrometry (AFS-2202E, Kechuang hai-guang instrument Co., China) (GB/T22105-2008, 2008). 0.5 g of the air-dried sludge was digested with 10-mL (1 + 1) aqua regia in the boiling water bath for 2 h. Cu, Zn, Pb, Cr, Ni, and Cd in the condensate were determined by AAS with 100-mL water samples being digested with 5-mL HNO_3 (1.42 g/mL), then 5-mL HNO_3 (1.49 g/mL), and 2-mL $HClO_4$ (1.68 g/mL) were used for Cd, Cu, Zn, and Pb (GB7475-87, 1987), 5-mL HNO_3 (1.49 g/mL) and 2-mL H_2O_2 (30%) for Cr (HJ 757-2015, 2015), and 5-mL HNO_3 (1.49 g/mL) for Ni (GB11912-89, 1989). Hg in the condensate was detected by AFS with 50-mL water samples being digested with 5-mL (1 + 1) HNO_3 (1.42 g/mL)-$HClO_4$ (1.68 g/mL), and 10-mL $KMnO_4$ (4%) (Ministry of Environmental Protection of the People's Republic of China, 2002).

The mass balance chart was calculated and developed with the content and weight of elements before and after drying (Steger & Meisner, 1996).

3. Results and discussion

3.1. Mass balance of the drying process

The mass balance during the conductive drying process of sewage sludge in this study is shown in Figure 1. 500.0 tons per day (t/d) of dewatered sludge with a solid content of 22.3% was fed into the dryer. 153.2 t/d of dried sludge with a solid content of 70.5% and 346.8 t/d of evaporated fraction were produced. Therefore, about 97% of solid matters were retained in the dried sludge and only 3.5 t were pyrolyzed and evaporated with steam during the drying process. About 98.1% of the evaporated fraction (340.3 t/d) could be condensed and the residual 1.9% (6.5 t/d) was considered as non-condensate. More than 95% of the non-condensate was water vapor. Thus, most of the evaporated solid matters were transferred into the condensate.

The concentration and gross amount of VFAs in the condensate were 6,840.3 mg/L and 2.3 t/d (Table 1 and Figure 1), respectively, which accounted for about 67.0% of the evaporated solid matters. It was indicated that VFAs should be one main component of the evaporated fraction and condensate besides water, which were mainly formed through pyrolyzation of the lipids and proteins in the sludge (Bougrier, Delgenès, & Carrère, 2008; Deng et al., 2009; Shanableh & Jones, 2001). Due to the large amount of VFAs, the condensate was acidic, and the pH was 5.2. Accordingly, the sludge pH increased from 6.7 to 7.0 after drying, and about 6.9% of TOC was evaporated and concentrated in the condensate. About 50% of sulfide in the dewatered could be volatilized into the condensate, and the residual was retained in the dried sludge (Table 1 and Figure 1). There was no significant content difference of TP concentration between dewatered and dried sludge (Table 1). The amount of TP in dried sludge accounted for about 99.1% of that in the dewatered sludge, and the amount of TP in condensate was negligible.

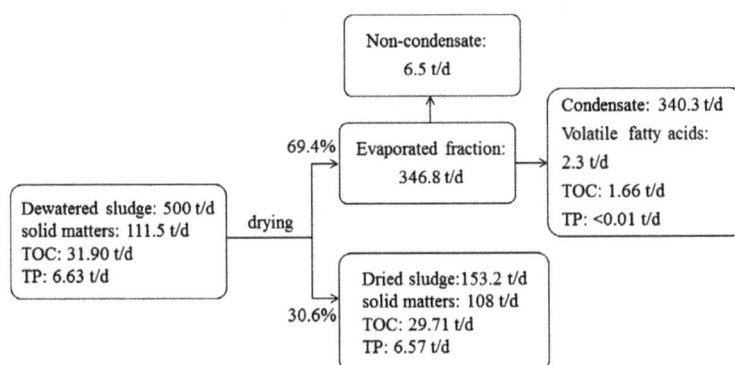

Figure 1. Mass balance during conductive drying process of sewage sludge.

Table 1. General parameters of the dewatered sludge, dried sludge, and condensate

Samples	pH	TOC (g/kg DS)	TP (g/kg DS)	Solid content (%)	VFAs (mg/L)	Sulfide (S²⁻) (mg/kg DS)
Dewatered sludge	6.7 ± 0.2	286.1 ± 10.72	59.5 ± 3.93	22.3 ± 0.59	–	0.031 ± 0.009
Dried sludge	7.0 ± 0.1	275.1 ± 7.70	60.8 ± 0.96	70.5 ± 1.28	–	0.019 ± 0.002
Condensate[a]	5.2 ± 0.1	4,860.6 ± 79.72[a]	8.9 ± 1.18[a]	–	6,840.3 ± 104.49	0.008 ± 0.001

[a]Units of TOC, TP, and Sulfide in condensate are mg/L.

3.2. Concentrations of heavy metals in the dewatered sludge

The highest concentrations of 469.5 and 82.7 mg/kg DS were observed for Zn and Cu in the dewatered sludge, respectively, and followed by Cr, Pb, Ni, Hg, and Cd (Table 2). Metal concentrations in the dewatered sludge were all below Chinese permitted limits of pollutants for sludge to be used for agriculture (GB18918-2002, 2003) in neutral or alkaline soil (pH ≥ 6.5). Only Hg concentration was higher than the limit for agriculture in acid soil (pH < 6.5). The concentration order of heavy metals corresponded well with the results discovered by Dai, Xu, Chen, Yang, and Ke (2007) in sludge from six wastewater treatment plants in Beijing. However, concentrations of the heavy metals became significantly lower. Two reasons may explain the lower level of the metals in the sludge. One is that galvanized pipe has been partially replaced by polyvinyl chloride polymer, polypropylene, and stainless steel tubes in wastewater pipeline in Beijing. This could reduce the release of Zn into wastewater before treatment. The other reason is that use of clean production technologies and separation of industrial wastewater from domestic sewerage networks are very universal, and sewage sludge is typically domestic in origin. Meantime, stricter discharged standard of heavy metals in China has been enforced recently and leads to lower concentration in discharge.

3.3. Behavior of heavy metals during drying process

Like TP, no significant difference of metal concentrations was found in dewatered and dried sludge (Table 2). This was different from composting, digestion, and liquefaction processes. Because during these processes, a mass of organic matters could be decomposed and mineralized, concentrations of heavy metals would increase accordingly (Dong, Liu, Dai, & Dai, 2013; Wagner, Bacon, Knocke, & Switzenbaum, 1990; Yuan et al., 2011). By rough calculation of mass flow, amounts of heavy metals in the dried sludge accounted for about 88–111% of those in the dewatered sludge (Table 3). Meanwhile, the heavy metals were hardly detected in condensate except Hg. Moreover, Hg concentration was very low (Table 2). Only less than 0.1% of Hg in the dewatered sludge could be migrated

Table 2. Total concentrations of heavy metals in dewatered sludge, dried sludge (mg/kg DS), and condensate (mg/L)

Heavy metal	Dewatered sludge	Dried sludge	Condensate	pH < 6.5[b]	pH ≥ 6.5[b]
Zn	469.5 ± 24.7	503.7 ± 31.6	<0.005[a]	2,000	3,000
Cu	82.7 ± 5.1	78.1 ± 4.3	<0.001[a]	800	1,500
Cr	35.1 ± 2.0	32.9 ± 3.6	<0.03[a]	600	1,000
Pb	27.5 ± 2.3	30.6 ± 2.6	<0.01[a]	300	1,000
Ni	12.6 ± 1.4	11.5 ± 0.8	<0.05[a]	100	200
Hg	6.5 ± 0.3	6.6 ± 0.5	0.0014	5	15
Cd	0.14 ± 0.01	0.16 ± 0.02	<0.001[a]	5	20

[a]Lower than detection limit.

[b]Chinese permitted limits for agricultural application of sludge in acid soil (pH < 6.5) and neutral or alkaline soil (pH ≥ 6.5).

Table 3. Mass balance of heavy metals during drying (g/d)							
Samples	Zn	Cu	Cr	Pb	Ni	Hg	Cd
Dewatered sludge	52,349.3	9,221.1	3,913.7	3,066.3	1,404.9	724.8	15.6
Dried sludge[a]	54,399.6 (103.9%)	8,434.8 (91.5%)	3,553.2 (90.8%)	3,304.8 (107.8%)	1,242.0 (88.4%)	712.8 (98.3%)	17.3 (110.9%)
Condensate	<1.70[b]	<0.34[b]	<10.21[b]	<3.40[b]	<17.02[b]	0.48	<0.34[b]

[a]Mass percentages of the heavy metals in dried sludge to ones in dewatered sludge are shown in brackets.
[b]Lower than detection limit.

to the condensate. It was indicated heavy metals during the drying process were difficult to be volatilized from the sludge, and would be almost retained in the dried sludge.

The condensate was also found to be severely lack of trace elements such as Mn and Co, which could significantly affect the activity and settleability of activated sludge in the biological treatment system (data not shown). Therefore, it is necessary to take appropriate measures in condensate biological treatment to avoid trace element deficiency and secondary pollution by volatile metals such as Hg (Zorpas, Vlyssides, Zorpas, Karlis, & Arapoglou, 2001).

Although metals were retained in dried sludge during drying process, their bioavailability could be changed significantly. As well known, the speciation of heavy metals in sludge includes exchangeable/acid soluble fraction, reducible fraction, oxidizable fraction, and residual fraction (Tessier, Campbell, & Bisson, 1979). First, the pH value of the dried sludge increased in this study, probably due to release of a large amount of VFAs. With pH increasing, the mobility and bioavailability of heavy metals might decrease (Wang, Li, Ma, Qian, & Zhai, 2006). Second, dehydration during drying process was an important factor causing the conversion of acid soluble and reducible fractions into the residual form of the heavy metals due to precipitation of heavy metals with minerals and change of iron/manganese amorphous oxyhydroxides to stable crystalline forms (Weng et al., 2014). Third, organic matter and sulfide were found to exhibit a significant positive correlation not only with exchangeable and reducible fractions of heavy metals but also with oxidizable fraction (Wang et al., 2006). Thus, the volatilization of sulfide and pyrolyzation of organic matter, and evaporation of resulting VFAs were one important factor causing conversion of exchangeable, reducible, and oxidizable fractions into residual one. Overall, these changes of sludge and heavy metal speciation indicated that the heavy metals in the dried sludge might be more stable than in the dewatered sludge (Obrador, Rico, Alvarez, & Novillo, 2001; Zorpas et al., 2001).

4. Conclusion

Solid matter, TP, heavy metals, and TOC were almost retained in the dried sludge after indirect conductive drying process of sewage sludge. However, due to the increase in pH, the dehydration and volatilization of sulfide, pyrolyzation of organic matter, and evaporation, the indirect conductive drying process might play an important role in reducing heavy metal bioavailability. Furthermore, as numerous VFAs and little metals during the thermal drying could be transferred into the condensate, appropriate pre-treatment might to be used for its biological treatment.

Funding
This work was supported by the Science & Technology Foundation of Beijing Municipal Research Institute of Environmental Protection [grant number 2013-A-07].

Author details
Jianping Luo[1]
E-mail: luo0409@163.com
Anfeng Li[1]
E-mail: cee1219@hotmail.com
Dan Huang[1]
E-mail: wshd1988@126.com
Jianju Ma[1]
E-mail: majianju@hotmail.com
Jiqiang Tang[2]
E-mail: bshbdl2008@126.com
[1] National Engineering Research Center for Urban Environmental Pollution Control, Beijing Municipal Research Institute of Environmental Protection, Beijing 100037, China.
[2] Beijing Cement Plant Co. Ltd, Beijing 102202, China.

References

Bennamoun, L., Arlabosse, P., & Léonard, A. (2013). Review on fundamental aspect of application of drying process to wastewater sludge. *Renewable and Sustainable Energy Reviews, 28*, 29–43. http://dx.doi.org/10.1016/j.rser.2013.07.043

Bougrier, C., Delgenès, J. P., & Carrère, H. (2008). Effects of thermal treatments on five different waste activated sludge samples solubilisation, physical properties and anaerobic digestion. *Chemical Engineering Journal, 139*, 236–244. http://dx.doi.org/10.1016/j.cej.2007.07.099

Chou, J. D., Lin, C. L., Hsien, Y. L., Wey, M. Y., & Chang, S. H. (2012). Copper emission during thermal treatment of simulated copper sludge. *Environmental Technology, 33*, 17–25. http://dx.doi.org/10.1080/09593330.2010.490853

Chyxx net. (2014). Retrieved from http://www.chyxx.com/industry/201406/252298.html (in Chinese).

Cousins, I. T., Hartlieb, N., Teichmann, C., & Jones, K. C. (1997). Measured and predicted volatilisation fluxes of PCBs from contaminated sludge-amended soils. *Environmental Pollution, 97*, 229–238. http://dx.doi.org/10.1016/S0269-7491(97)00096-1

Dai, J., Xu, M., Chen, J., Yang, X., & Ke, Z. (2007). PCDD/F, PAH and heavy metals in the sewage sludge from six wastewater treatment plants in Beijing, China. *Chemosphere, 66*, 353–361. http://dx.doi.org/10.1016/j.chemosphere.2006.04.072

Deng, W. Y., Yan, J. H., Li, X. D., Wang, F., Zhu, X. W., Lu, S. Y., & Cen, K. F. (2009). Emission characteristics of volatile compounds during sludges drying process. *Journal of Hazardous Materials, 162*, 186–192. http://dx.doi.org/10.1016/j.jhazmat.2008.05.022

Dong, B., Liu, X., Dai, L., & Dai, X. (2013). Changes of heavy metal speciation during high-solid anaerobic digestion of sewage sludge. *Bioresource Technology, 131*, 152–158. http://dx.doi.org/10.1016/j.biortech.2012.12.112

GB/T17138-1997. (1998). *Soil quality-determination of copper, zinc-flame atomic absorption spectrophotometry*. Beijing: Ministry of Environmental Protection of the PRC.

GB/T22105-2008. (2008). *Soil quality-analysis of total mercury, arsenic and lead contents-atomic fluorescence spectrometry*. Beijing: General Administration of Quality Supervision, Inspection and Quarantine of the PRC.

GB11912-89. (1989). *Water quality-Determination of nickel-Flame atomic absorption spectrophotometric method*. Beijing: Standardization Administration of the People's Republicof China.

GB18918-2002. (2003). *Discharge standard of pollutants for municipal wastewater treatment plant*. Beijing: Standard Press of China.

GB7475-87. (1987). *Water quality-Determination of copper, zinc, lead and cadmium-Atomic absorption spectrometry*. Beijing: Standardization Administration of the People's Republicof China.

Harrison, E. Z., McBride, M. B., & Bouldin, D. R. (1999). Land application of sewage sludges: An appraisal of the US regulations. *International Journal of Environment and Pollution, 11*(1), 1–36. http://dx.doi.org/10.1504/IJEP.1999.002247

HJ 757-2015. (2015). *Water quality—Determination of chromium-Flame atomic absorption spectrometry*. Beijing: Ministry of Environmental Protection of the People's Republic of China.

Kakati, J. P., Ponmurugan, P., Rajasekaran, N., & Gnanamangai, B. M. (2013). Effect of textile effluent treatment plant sludge on the growth metabolism of Green gram (*Vigna radiata* L.). *International Journal of Environment and Pollution, 51*, 79–90. http://dx.doi.org/10.1504/IJEP.2013.053181

Kelessidis, A., & Stasinakis, A. S. (2012). Comparative study of the methods used for treatment and final disposal of sewage sludge in European countries. *Waste Management, 32*, 1186–1195. http://dx.doi.org/10.1016/j.wasman.2012.01.012

McLaren, R. G., & Clucas, L. M. (2001). Fractionation of copper, nickel, and zinc in metal-spiked sewage sludge. *Journal of Environment Quality, 30*, 1968–1975. http://dx.doi.org/10.2134/jeq2001.1968

Ministry of Environmental Protection of the People's Republic of China. (2002). *Methods for water and wastewater analysis* (4th ed.). Beijing: China Environmental Science Press.

Obrador, A., Rico, M. I., Alvarez, J. M., & Novillo, J. (2001). Influence of thermal treatment on sequential extraction and leaching behaviour of trace metals in a contaminated sewage sludge. *Bioresource Technology, 76*, 259–264. http://dx.doi.org/10.1016/S0960-8524(00)00101-2

Pritchard, D. L., Penney, N., McLaughlin, M. J., Rigby, H., & Schwarz, K. (2010). Land application of sewage sludge (biosolids) in Australia: Risks to the environment and food crops. *Water Science and Technology, 62*, 48–57. http://dx.doi.org/10.2166/wst.2010.274

Shanableh, A., & Jones, S. (2001). Production and transformation of volatile fatty acids from sludge subjected to hydrothermal treatment. *Water Science and Technology, 44*, 129–135.

Singh, R. P., & Agrawal, M. (2010). Variations in heavy metal accumulation, growth and yield of rice plants grown at different sewage sludge amendment rates. *Ecotoxicology and Environmental Safety, 73*, 632–641. http://dx.doi.org/10.1016/j.ecoenv.2010.01.020

Steger, M. T., & Meisner, W. (1996). Drying and low temperature conversion—A process combination to treat sewage sludge obtained from oil refineries. *Water Science and Technology, 34*, 133–139. http://dx.doi.org/10.1016/S0273-1223(96)00707-X

Tessier, A., Campbell, P. G., & Bisson, M. (1979). Sequential extraction procedure for the speciation of particulate trace metals. *Analytical Chemistry, 51*, 844–851. http://dx.doi.org/10.1021/ac50043a017

Tunçal, T., Jangam, S. V., & Güneş, E. (2011). Abatement of organic pollutant concentrations in residual treatment sludges: A review of selected treatment technologies including drying. *Drying Technology, 29*, 1601–1610. http://dx.doi.org/10.1080/07373937.2011.602307

Vrkoslavová, J., Demnerová, K., Macková, M., Zemanová, T., Macek, T., Hajšlová, J., & Pulkrabová, J. (2010). Absorption and translocation of polybrominated diphenyl ethers (PBDEs) by plants from contaminated sewage sludge. *Chemosphere, 81*, 381–386. http://dx.doi.org/10.1016/j.chemosphere.2010.07.010

Wagner, D. J., Bacon, G. D., Knocke, W. R., & Switzenbaum, M. S. (1990). Changes and variability in concentration of heavy metals in sewage sludge during composting. *Environmental Technology, 11*, 949–960. http://dx.doi.org/10.1080/09593339009384947

Wang, C., Li, X. C., Ma, H. T., Qian, J., & Zhai, J. B. (2006). Distribution of extractable fractions of heavy metals in sludge during the wastewater treatment process. *Journal of Hazardous Materials, 137*, 1277–1283. http://dx.doi.org/10.1016/j.jhazmat.2006.04.026

Weng, H. X., Ji, Z. Q., Chu, Y., Cheng, C. Q., & Zhang, J. J. (2012). Benzene series in sewage sludge from China and its release characteristics during drying process. *Environmental Earth Sciences, 65*, 561–569. http://dx.doi.org/10.1007/s12665-011-1100-2

Weng, H. X., Ma, X. W., Fu, F. X., Zhang, J. J., Liu, Z., Tian, L. X., & Liu, C. (2014). Transformation of heavy metal speciation during sludge drying: Mechanistic insights. *Journal of Hazardous Materials, 265*, 96–103. http://dx.doi.org/10.1016/j.jhazmat.2013.11.051

Yuan, X., Huang, H., Zeng, G., Li, H., Wang, J., Zhou, C., & Zhu, H. (2011). Total concentrations and chemical speciation of heavy metals in liquefaction residues of sewage sludge. *Bioresource Technology, 102*, 4104–4110. http://dx.doi.org/10.1016/j.biortech.2010.12.055

Zorpas, A. A., Vlyssides, A. G., Zorpas, G. A., Karlis, P. K., & Arapoglou, D. (2001). Impact of thermal treatment on metal in sewage sludge from the Psittalias wastewater treatment plant, Athens, Greece. *Journal of Hazardous Materials, 82*, 291–298. http://dx.doi.org/10.1016/S0304-3894(01)00172-8

12

Use of nano-silica in cement based materials

Paratibha Aggarwal[1]*, Rahul Pratap Singh[1] and Yogesh Aggarwal[1]

*Corresponding author: Paratibha Aggarwal, Civil Engineering Department, National Institute of Technology, Kurukshetra, Kurukshetra, India

E-mail: paratibha@rediffmail.com

Reviewing editor: Raja Rizwan Hussain, King Saud University, Saudi Arabia

Abstract: The research nowadays is mainly focusing on the basic science of cementitious material at nano/atomic level. Further, researchers are continuing to improve the durability and sustainability of concrete and have realized significant increment in mechanical properties of cementitious materials by incorporating nano-silica. The review paper summarizes the effect of nano-silica addition on mechanical, durability and microstructure characteristics of paste, mortar and concrete. It provides the current development of application of nano-silica in paste, mortar and concrete. Finally, the future trend/potential and implication of nano-silica in cement-based materials is discussed.

Subjects: Concrete & Cement; Engineering & Technology; Nanoscience & Nanotechnology; Technology

Keywords: nanoparticles; concrete; durability; mechanical properties; microstructure; SEM; XRD

ABOUT THE AUTHORS

Paratibha Aggarwal is working as a professor in Civil Engineering Department of National Institute of Technology, Kurukshetra, India.

Rahul Pratap Singh is pursuing his PhD in Civil Engineering Department of National Institute of Technology, Kurukshetra, India.

Yogesh Aggarwal is working as an assistant professor in Civil Engineering Department of National Institute of Technology, Kurukshetra, India.

Their areas of interest include use of nano-silica in concrete, its study of mechanical and durability properties. High-performance concrete, self-compacting concrete, their performance and durability aspects incorporating industrial by-products. Cement is one of the most energy-consuming materials widely used globally. Efforts are being made mainly to use supplementary cementitious materials in mortars and concrete directed towards the reduction of carbon footprint. The present research work is one such effort towards attaining the above goal.

PUBLIC INTEREST STATEMENT

Concrete is one of most commonly used material on earth after water. Use of nano-materials, particularly nano-silica as supplementary cementitious material, in manufacturing of paste, mortar, and concrete offer the potential of producing materials with new and interesting properties, such as enhanced strength and durability properties. In this article, a review of use of nano-silica in paste, mortar and concrete is presented which provides an insight in the use of nano-silica in recent past. It also provides the current development of application of nano-silica with future trend/potential and implication of nano-silica in cement-based materials.

1. Introduction

The importance of the cementitious materials in the construction industry is very significant; however, their variety of applications must not hide their complexity. They are indeed composite materials with multi-scale internal structures which have evolved over centuries. More specifically, the cement paste matrix is basically a porous material composed of calcium hydroxide (portlandite), aluminates and unhydrated cement (clinker) embedded into an amorphous nanostructured hydration product, the so-called C–S–H (calcium silicate hydrate) gel (Gaitero, Campillo, & Guerrero, 2008). This gel is the dominant hydration product of the cement paste, not only because it is the most abundant component (50–70% by volume), but also because of its exceptionally good mechanical properties.

Nanoparticles have a high surface-area-to volume ratio. In this way, nanoparticles with 4-nm diameter have more than 50% of its atoms at the surface and are thus very reactive (Wiesner & Bottero, 2007). The behaviour of such materials is mainly influenced by chemical reactions at the interface, and by the fact that they easily form agglomerates. When higher surface area is to be wetted, it decreases free dispersant water in aqueous systems available in the mixture. Therefore, the use of nanoparticles in mortars and concretes significantly modify their behaviour not only in the fresh, but also in the hardened conditions, as well as the physical/mechanical and microstructure development (Senff, Labrincha, Ferreira, Hotza, & Repette, 2009).

In recent years, the use of nanoparticles has received particular attention in many fields of applications to fabricate materials with new functionalities. When ultrafine particles are incorporated into Portland cement paste, mortar or concrete, materials with different characteristics from conventional materials were obtained (Lea, 1998; Li, Xiao, Yuan, & Ou, 2004; Neville, 1996). The performance of these cementitious-based materials is strongly dependent on nano-sized solid particles, such as particles of calcium–silicate–hydrates (C–S–H), or nano-sized porosity at the interfacial transition zone between cement and aggregate particles. Typical properties affected by nano-sized particles or voids are strength, durability, shrinkage and steel-bond (Collepardi et al., 2005). Nanoparticles of SiO_2 (nano-silica) can fill the spaces between particles of gel of C–S–H, acting as a nano-filler. Furthermore, by the pozzolanic reaction with calcium hydroxide, the amount of C–S–H increases, resulting in higher densification of the matrix, thus improving the strength and durability of the material (Choolaei, Rashidi, Ardjmand, Yadegari, & Soltanian, 2012; Hou et al., 2013; Zapata, Portela, Suárez, & Carrasquillo, 2013). Previous researches (Björnström, Martinelli, Matic, Börjesson, & Panas, 2004; Lea, 1998; Qing, Zenan, Deyu, & Rongshen, 2007) indicate that the inclusion of nanoparticles modifies fresh and hardened state properties, even when compared with conventional mineral additions. Colloidal particles of amorphous silica appear to considerably impact the process of C_3S hydration (Björnström et al., 2004). Nano-silica decreased the setting time of mortar when compared with silica fume (SF) (Qing et al., 2007) and reduced bleeding water and segregation, while improved the cohesiveness of the mixtures in the fresh state (Collepardi, Olagot, Skarp, & Troli, 2002). Nano-silica-added cement paste showed reduction in setting time (Lin, Lin, Chang, Luo, & Cai, 2008; Singh, Agarwal, Bhattacharyya, Sharma, & Ahalawat, 2011; Singh, Bhattacharyya, & Singh, 2012), shortened duration of dormant and induction period of hydration, shortening of time to reach peak heat of hydration and increased production of calcium hydroxide at early ages (Ltifi, Guefrech, Mounanga, & Khelidj, 2011; Senff et al., 2009). When combined with ultrafine fly ash, better performance is assured than that achieved by use of SF (Lea, 1998; Li, Xiao, & Ou, 2004; Li et al., 2004). Besides, the compressive strength of mortar or concrete with SF was improved when compared with formulations without addition (Jo, Kim, Tae, & Park, 2007; Li et al., 2004; Li, Zhang, & Ou, 2006). Nano-silica addition increased the quantity of C–S–H and C–A–H in the paste (Tobón, Payá, Borrachero, & Restrepo, 2012). Addition of nano-silica into cement paste and mortar demand more water to retain its workability (Quercia, Hüsken, & Brouwers, 2012). To avoid adverse effects on workability, Berra, Carassiti, Mangialardi, Paolini, and Sebastiani (2012) suggested delayed addition of water, stating that instead of adding all the mixing water at a time, certain amount of water should be added later on. Nano-silica samples showed lesser strength loss after being exposed to elevated temperatures (Lim, Mondal, & Cohn, 2012). Mortar containing high volume of fly ash showed higher residual strength after being exposed to 700°C, and dehydration of C–S–H produced calcium silicate, which acts as new binding material to retain residual strength (Rahel, Hamid, & Taha, 2012). The addition of nano-silica modified the porosity of the cement paste and increased the average chain length of silicate chains (Gaitero et al., 2008; Porro et al., 2005).

In particular, the developments in nano-science have had a great impact on concrete industry. Nano-materials have been used in concrete industry over the past decade. Few studies till date are reported with nanoparticles such as nano-silica (nano-SiO_2) and nano-titanium oxide (nano-TiO_2), nano-iron (nano-Fe_2O_3), nano-alumina (nano-Al_2O_3) (Li, 2004; Shekari & Razzaghi, 2011). Additionally, a limited number of investigations are dealing with the manufacture of nano-sized cement particles and the development of nano binders. Thus, limiting the review of literature work to use of nano-silica in cementitious compositions, research has shown that incorporation of nanoparticles in cement matrix could improve durability and mechanical properties of cement-based materials. Nano-silica (nS) in particular has found wide usage in this field because of its high reactivity and very large specific surface area, which results in a high degree of pozzolanic activity. Nano-silica further accelerates the dissolution of C_3S and formation of C–S–H with its activity being inversely proportional to the size, and also provides nucleation sites for C–S–H (Jo, Kim, & Lim, 2007). Even small additions (0.6 wt. % binder) of nS are very efficient and compare to much larger amounts of SF in terms of improvement in mechanical properties of cement-based materials. This is especially pronounced at early ages and for concretes with regular strength grade (Pourjavadi, Fakoorpoor, Khaloo, & Hosseini, 2012). From the nano-indentation studies, it was observed that the nano-silica addition significantly alters the proportions of low stiffness and high stiffness C–S–H (Hou et al., 2013; Mondal, Shah, Marks, & Gaitero, 2010).

2. Nano-material and cement composites

In the construction industry, extensive research is going on to improve the performance of various building materials and development of durable and sustainable concrete is one among them as is clear from Table 1. Among all the nano-materials, nano-silica is the most widely used material in the cement and concrete to improve the performance because of its pozzolanic reactivity besides the pore-filling effect.

2.1. Paste and mortar with nano-silica

Various researchers have investigated the effect of nano-silica on pastes and mortars. Nano-silica-incorporated cement pastes are studied to understand the hydration process and microstructure evolution. Basically, this approach is used for the study of fundamental science behind cement hydration. Mortar studies are used to explore the rheology and mechanical properties.

2.1.1. Fresh properties

Reduction in initial setting time (IST) and final setting time (FST) of pastes was observed on addition of nano-silica (Senff et al., 2009). Also, difference between the IST and FST decreased with increase in nano-silica content (Ltifi et al., 2011; Qing et al., 2007). With regard to the effect of nano-silica on the rheology behaviour of the cementitious mixes' studies on cement paste and mortars, most of the researchers agree in indicating that the addition of nano-silica greatly increases the water demand of cementitious mixes as compared to the control ones. The plasticity loss of mortar with increased torque and yield stress in mixtures with nano-silica was more evident due to higher surface area (Senff, Hotza, Repette, Ferreira, & Labrincha, 2010; Senff et al., 2009). Berra et al. (2012) and Kawashima, Hou, Corr, and Shah (2012) worked on the rheology of cement pastes with various w/b ratios and suggested delayed addition of water by keeping certain amount of water to be added at a later stage. Also, addition of nano-silica into cement paste and mortar resulted in higher demand of water to retain its workability. Direct influence on water amount required in the mixture was observed when nano-silica was incorporated into the mortars in fresh state. This behaviour confirms the fact that additions of high surface area mineral particles to cement mixtures cause the need for higher amounts of water or chemical admixtures in order to keep the workability of the mixture. If the water content is kept constant, as in the actual conditions, an increase of nano-SiO_2 content will promote the packing of particles, decreasing the volume between them and decreasing the free water. Therefore, there is a higher internal friction between solid particles (Ltifi et al., 2011).

Table 1. Summary of use of nano-silica in cementitious compositions

S. No.	Author	Concrete/ mortar	Materials used for comparison	Nano material used	Size of particle (nm)	Fineness (m^2/g)	Water cement ratio	Additive or replacement	Properties
1	Senff et al. (2009)	Paste	–	nS 1.5%, 2%, 2.5%	9	300	0.35	Replacement	Flow table, rehological behaviour, XRD
2	Qing et al. (2007)	Paste 1:0.22:0.025	SF 2%, 3%, 5%	nS 1%, 2%, 3%, 5%	15	160	0.22	Additive	Compressive strength, bond strength, Setting time
3	Ltifi et al. (2011)	Paste, mortar	–	nS 3%, 10%	9	300	0.5	Replacement	Compressive strength, flexure strength
4	Senff et al. (2010)	Mortar	SF 0–20%	nS 0–7%	9	300	0.35–0.59	Additive	Rheological, compressive strength, water absorption
5	Berra et al. (2012)	Paste	–	nS 0.8%, 3.8%	10	345	0.5	Additive	Physical mechanical stability test
6	Kawashima et al. (2012)	Mortar	Fly ash	nS 2%, 5%	20–10		0.43–0.45	Additive	Compressive strength, rate of heat of hydration
7	Hou et al. (2013)	Mortar	–	nS 5%	10 nm		0.4	Additive	Hydration heat, morphology, CH content
8	Stefanidou (2012)	Paste	–	nS 0.5%, 1%, 2%, 5%	14	200	0.34–0.51	Additive	Flexural and compressive strength, microstructure
9	Li et al. (2004)	Mortar	Nano-Fe_2O_3 (3–10%) Silica fume (3–15%)	nS 3%, 5%, 10%	15 ± 5	15 ± 5	0.5	Replacement	Compressive and flexure strength, microstructure
10	Jo et al. (2007)	Mortar 1:2:45	SF 5%, 10%, 15%	nS 3%, 6%, 10%, 12%	40	60	0.5	Additive	Compressive strength, SEM, rate of heat of evolution
11	Zapata et al. (2013)	Mortar	SF	nS 1.5–6%	25	109	0.35, 0.4	Replacement	Compressive strength, SEM, XRD
12	Oltulu and Şahin (2013)	Mortar	Fly ash	nS 0.5%, 1.25%, 2.5%	12	200	0.4	Replacement	Compressive strength
13	Aly et al. (2012)	Mortar	Waste glass 20%, 40%	nS 3%	5	500	–	Replacement	Compressive, fracture, flexural and impact strength, DTA/TGA, SEM
14	Gaitero et al. (2008)	Paste	–	nS 20%, 40%, 45%, 90%	20, 30, 120, 1,400		0.4	Additive	Compressive strength, flexure strength
15	Zhang et al. (2012)	Concrete	SF	nS 0.5%, 1%, 2%	12 and 7	200.1 and 321.6	0.45	Replacement	Compressive strength, rate of heat of hydration, porosity
16	Zhang and Islam (2012)	Concrete	Fly ash, slag and SF	nS 1%, 2%	12	200.1	0.45	Replacement	Compressive strength, rate of heat of hydration, Setting time
17	Pourjavadi et al. (2012)	Concrete	SAP 0.1%, 0.3%	nS 0.5%, 1%	19	172	0.45	Replacement	Compressive strength, flexure strength
18	Li (2004)	Concrete	Fly Ash 50%	nS 4%	10 ± 5	640 ± 50	0.28	Replacement	Hydration heat, pore size
19	Naji Givi et al. (2010)	Concrete	–	nS 0.5%, 1%, 1.5%, 2%	15 and 80	160 ± 12 560 ± 32	0.4	Replacement	Compressive, flexure, and split tensile strength
20	Heidari and Tavakoli (2013)	Concrete	Ground ceramic powder 10–40%	nS 0.5%, 1%		200 ± 30	0.5	Replacement	XRD, compressive strength, water absorption

(Continued)

Table 1. (Continued)

S. No.	Author	Concrete/ mortar	Materials used for comparison	Nano material used	Size of particle (nm)	Fineness (m²/g)	Water cement ratio	Additive or replacement	Properties
21	Zhang and Li (2011)	Concrete	Popypropylene fibre	nS 1%, 3%	10 ± 5	640 ± 50	0.42	Replacement	Compressive and flexure strength, pore structure, chloride permeability
22	Jalal et al. (2012)	Concrete	SF 2%, 10%	nS 2%, 10%	15 ± 3	165 ± 17	0.38	Replacement	Chloride penetration, water absorption, electrical resistivity
23	Ji (2005)	Concrete	Fly ash	nS	15 ± 5	160 ± 20	0.49–0.5	Replacement	Water permeability, SEM
24	Kong et al. (2012)	Mortar	–	nS 0.75%, 1%	100–200 and 200–400	142.9 and 157.8	0.3	Additive	Microstructure, compressive strength

Presence of nano-silica made cement paste thicker and accelerated the hydration process (Qing et al., 2007). Spherical morphology of fly ash helps increase the flowability of cementitious materials, whereas nanoparticles increase stiffness due to their higher specific surface area, thereby reducing fluidity of fly ash cement-blended nano-silica mortar with increase in the amount of nano-silica (Kawashima et al., 2012). The pozzolanic activity of nano-silica and CH adsorption of colloidal nano-silica (CNS) was investigated. It was observed that pozzolanic reaction of nano-silica was complete within 7 days of hydration (Hou et al., 2013).

2.1.2. Mechanical properties

With regard to the influence of nano-silica on the mechanical strength development of cementitious materials, the addition of nano-silica to Portland cement (PC) pastes was found to increase the compressive strength to an extent that was dependent on the nano-silica content, water-to-binder weight ratio (w/b), and curing time. Paste compressive strength was studied (Berra et al., 2012; Qing et al., 2007; Stefanidou, 2012) along with bond strength (Qing et al., 2007) and flexural strength (Stefanidou, 2012). As a general observation, increase in paste strength was observed, with increase in content of nano-silica at early ages along with increase in pozzolanic activity. An increase of approximately 17–41% and 20–25% in compressive strength was observed at the 3rd and 28th days (Qing et al., 2007), 7–11% at the 7th day (Berra et al., 2012) and an average increase of about 25% was observed on addition of 0.5–2% nano-silica (Stefanidou, 2012). The increase in gain of strength and optimum nano-silica content were observed to be 5% (Qing et al., 2007), 0.8% (Berra et al., 2012) and 0.5% (Stefanidou, 2012). The bond strength increase was observed between 16–43% at 7 days and 26–88% at 28 days (Qing et al., 2007). The flexural strength at the ages of 3 days was also observed to be maximum with 1–2% nano-silica content (Stefanidou, 2012).

Li et al. (2004) investigated cement mortars with nano-SiO_2 or nano-Fe_2O_3 to explore their super mechanical and smart potentials. Compressive strength increase in mortar mixes was observed 5.7–20.1% (7 days) and 13.8–26% (28 days). Jo et al. (2007) observed 53.67–63.9% (7 days) and 52.5–62.7% (28 days) increase in mortar mixes compressive strength and suggested the requirement of using higher content of nano-silica must be accompanied by adjustments to water and superplasticizer dosage in the mix in order to ensure that the specimens do not suffer excessive self-desiccation and cracking. Same results were observed by Ltifi et al. (2011), and 6.9–16.9% increase in compressive strength at 90 days was reported (Zapata et al., 2013). On addition of nano-silica to fly-ash concretes (Hou et al., 2013; Kawashima et al., 2012; Oltulu & Şahin, 2013), almost same results were observed with early age strength gain as high as 60%, which became equal at later stage to that of various mixes (Kawashima et al., 2012) and 58–66% i.e. average of 63% increase on strength (Oltulu & Şahin, 2013). Also, flexural strength increase was reported as 28% at 7 days and 19.2–27% at 28 days (Li et al., 2004) and 42–55% at 28 days, along with fracture energy and impact strength increase at 28 days (Aly et al., 2012).

2.1.3. Durability properties

Jo et al. (2007) observed by examining the rate of heat of evolution that nano-scale silica behaves not only as a filler to improve microstructure, but also as an activator to promote pozzolanic reaction. Gaitero et al. (2008) revealed reduced calcium leaching of nano-silica-added cement paste ascribing it to the densification of the paste, transforming of portlandite into C–S–H by means of pozzolanic reaction and modification of internal structure of C–S–H gel, all of which make the cement paste more stable and more strongly bonded. Higher values of water absorption and apparent porosity (Senff et al., 2010) were observed along with unrestrained shrinkage and weight loss of mortars with increase in nano-silica content (highest at 7% nano-silica wt. %). For fly ash replaced cement-based materials, CH generated by cement hydration is critical for later stage pozzolanic reaction. Nano-silica addition has great influence on the CH content of fly ash–cement paste. Also, depletion of $Ca(OH)_2$ was more severe when the nano-silica dosage and fly ash replacement ratio are high (Kawashima et al., 2012). Hou et al. (2013) observed acceleration of cement hydration and maturation of gel structure in CNS added paste, achieved through an acceleration of the dissolution of cement particles and a preferred hydration and hydrates precipitation on CNS particle surface. Although CNS can accelerate cement hydration to a great extent in the early age, the later hydration of cement is hindered.

2.2. Concrete with nano-silica

Nano-silica incorporation into cement concrete is the direct application approach of nanomaterials. Researchers have worked on the mechanical and durability properties and microstructure analysis of concrete with nano-silica as discussed below.

2.2.1. Fresh properties

Reduced setting times were observed by various researchers on incorporation of nano-silica in concrete which is same as observed for pastes and mortar (Zhang & Islam, 2012; Zhang, Islam, & Peethamparan, 2012). Also, decrease in initial and final setting time was observed on incorporation of nS in various quantities, with increase in viscosity and yield stress reported (Pourjavadi et al., 2012).

2.2.2. Mechanical properties

Concrete strength is influenced by lots of factors like concrete ingredients, age, ratio of water to cement materials, etc. Nano-silica incorporation into concrete resulted in higher compressive strength than that of normal concrete to a considerable level. Li et al. (2004) reported 3-day compressive strength increase by 81% and also at later stages, same trend was observed with 4% nano-silica in high volume fly ash concrete. Naji Givi, Abdul Rashid, Aziz, and Salleh (2010) also reported higher compressive strength at all ages, for nano-silica blended concretes up to maximum limit of 2% with average particle size of 15 and 80 nm. Same results were obtained for split tensile and flexural strength. Pourjavadi et al. (2012) reported that negative effect of super absorbent polymer reduced compressive strength by addition of nano silica, but same results were not observed for flexural strength. An increase of about 23–38% and 7–14% at 7 days and 28 days, respectively, in compressive strength of nano-silica concrete was reported, whereas low increase of 9.4% (average) was reported for flexural strength. Zhang and Islam (2012), Zhang et al. (2012) used GGBFS, fly ash and slag and increase in compressive strength was observed as 22% (3 days) and 18% (7 days) and 30% (3 days) and 25% (7 days) of concretes with GGBFS and fly ash and slag, respectively. Heidari and Tavakoli (2013) incorporated nano-silica in ground ceramic concrete and improvement in strength at early stage was observed.

2.2.3. Durability properties

Durability properties of concrete include aspects such as permeability, pore structure and particle size distribution, resistance to chloride penetration, etc. Investigations on nano-silica concrete for its permeability characteristics showed that the addition of nano-silica in concrete resulted in reduction in water absorption, capillary absorption, rate of water absorption, and coefficient of water absorption and water permeability than normal concrete. The pore structure determines the transport properties of cement paste, such as permeability and ion migration. Reduction in water absorption, capillary absorption, rate of water absorption and water permeability has been observed by various researchers (Li, 2004; Zhang & Li, 2011; Zhang et al., 2012). Pore size distribution in concrete was refined and porosity lowered even at short time, curing on addition of 4% nano-silica (Li, 2004; Zhang & Li, 2011). Also, increasing nano-silica dosage decreased capillary porosity (Zhang et al., 2012). Water absorption capacity of nano-silica concretes decreased with incorporation of nano-silica (Jalal, Pouladkhan, Norouzi, & Choubdar, 2012; Zhang et al., 2012). Enhancement of resistance to chloride penetration of concretes with addition of nano-silica was reported (Jalal et al., 2012; Zhang & Li, 2011). Zhang and Islam (2012) studied the behaviour of high-volume fly ash and slag concretes with nano-silica addition and reported that the addition of nano-silica reduced the length of dormant period during hydration and also accelerated the hydration. Chloride ion penetration was also reduced with the addition of nano-silica into fly ash and slag concrete.

3. Microstructure analysis

XRD and SEM observations have been reported by a number of researchers (Aly et al., 2012; Hou et al., 2013; Ji, 2005; Jo et al., 2007; Kong et al., 2012; Li et al., 2004; Pourjavadi et al., 2012; Qing et al., 2007; Senff et al., 2009; Stefanidou, 2012). Li et al. (2004) observed from SEM images that the nanoparticles were not only acting as a filler, but also as an activator to promote hydration process and to improve the microstructure of the cement paste if the nanoparticles were uniformly dispersed. Ji (2005) also revealed through ESEM test that microstructure of concrete with nano-silica was more uniform and compact than that of the normal concrete. Qing et al. (2007) showed from XRD powder patterns of NS and SF that strong broad peaks of NS and SF were centred on 23° and 22° (2θ), respectively, which was in keeping with the strong broad peak that is characteristic of amorphous SiO_2. The results showed that both NS and SF were in an amorphous state. SEM examination was performed to verify the mechanism predicted by compressive strength test (Jo et al., 2007) and nano-silica particles were found to influence hydration behaviour and lead to the differences in the microstructure of the hardened paste. The microstructure of the mixture containing nano-SiO_2 revealed a dense, compact formation of hydration products and a reduced number of $Ca(OH)_2$ crystals. Qing et al. (2007) showed that NS can reduce the size of CH crystals at the interface more effectively than SF. Senff et al. (2009) also showed that nano-silica addition contributed to the increased production of CH at early stage compared to samples without nano-silica. Aly et al. (2012) showed through SEM micrographs show that the densest mortar structure was observed for the specimen with a hybrid combination of waste glass powder (WG) and colloidal nano-silica (CS). Stefanidou's (2012) observation recorded a denser structure in nano-modified samples. Also, Kong et al. (2012), through SEM observation, recorded an obvious microstructure improvement of the hardened cement paste (HCP) and the ITZ in mortar by adding nano-silica, regardless of its agglomerate size. It was found that C–S–H gels from pozzolanic reaction of the agglomerates cannot function as binder. The gels from cement hydration did not penetrate into the pozzolanic gels. Pourjavadi et al. (2012) reported that the addition of 1% nano-silica reduced the porosity of hardened cement paste because of super pozzolanic performance and production of higher amounts of C–S–H gel. In addition, the microstructure was considerably improved due to the micro and nano-filling effects. Crystals of portlandite were reduced in size and quantity as a consequence of the pozzolanic reaction and crystal growth control by nano-silica.

4. Conclusions and summary

The present paper reviews the current state of the field of nanotechnology in concrete and recent key advances. Current status of nano-silica opens up widely for research in cementitious compositions. Applications of nanotechnology have the potential to make breakthrough in materials technology. Nano-silica application in paste, mortar and concrete is a good way of enhancing their properties. It has been observed that optimum quantity of nano-silica to be used is still contradictory and it is for the researcher to decide the optimum quantity for his/her own material. Using nanotechnology in future will make it possible to design materials for their specific purpose of application. New developments have taken place in the nano-engineering and nano modification of concrete; however, current challenges need to be solved before the full potential of nanotechnology can be realized in concrete applications, including proper dispersion, compatibility of the nanomaterials in cement.

Some of the outcomes from the literature reviewed can be summarized as:

Direct influence on water amount required in the mixture was observed when nano-silica was incorporated into the mortars in fresh state. This behaviour confirms the fact that additions of high-surface area mineral particles to cement mixtures cause the need for higher amounts of water or chemical admixtures in order to keep the workability of the mixture.

Compressive strengths increase with increase in nano-silica content, which acts as activator to promote hydration and also to improve the microstructure of cement paste if nanoparticles were uniformly dispersed. The compressive strength is enhanced with nano-SiO_2 addition, especially at early stages, and the pozzolanic activity of nano-silica is much greater than that of SF. It was observed that nano-silica-blended concretes have higher strength as compared to non-blended concretes. Compressive strength is higher at all stages for nano-silica-blended concretes.

Nano-Silica was observed to have no positive effect on the strength gain of fly ash-replaced cement-based material at later stages. Flexural and split tensile strengths also improved by increasing the silica nano-particle content. Fly ash-based cements have low initial pozzolanic activity, but the addition of a little nano-SiO_2 significantly increases pozzolanic activity. Thus, nano-SiO_2 activates fly ash.

Nano-SiO_2 adsorbs the $Ca(OH)_2$ crystals reducing the size and amount of the $Ca(OH)_2$ making the interfacial transition zone of aggregates and binding paste matrix denser. The nano-SiO_2 particles fill the voids of the C–S–H gel structure and act as nucleus to tightly bond with C–S–H gel particles, making binding paste matrix denser, resulting in an increase in long-term strength and durability of concrete. Nano-scale SiO_2 behaves not only as filler to improve mortar cement microstructure, but also as a promoter of pozzolanic reaction.

Future research should address the following issues.

(1) Physical state and dispersion of nano-silica into the concrete is a major issue requiring thorough study.

(2) The optimum quantity of nano-silica for concrete or cement paste needs to be determined for certain percentage, which depends on the type of nano-silica, i.e. colloidal, dry powder, etc., and also the average particle size of nano-silica. A relationship needs to be established between optimum quantity and characteristics of nano-silica.

(3) Most of the research works are limited to cement pastes and mortars, with only a few researchers having worked extensively on mechanical properties and permeability of the concrete incorporating nano-silica as is clear from the review paper. Durability properties still need to be investigated further on carbonation, corrosion resistance, acid resistance, sulphate resistance.

(4) Optimization, fresh, mechanical, microstructural and durability properties of concrete along with mathematical modelling of concrete behaviour requires extensive research.

Additionally, introduction of these novel materials into the public sphere through civil infrastructure will necessitate an evaluation and understanding of the impact they may have on the environment and human health.

Funding
The authors received no direct funding for this research.

Author details
Paratibha Aggarwal[1]
E-mail: paratibha@rediffmail.com
Rahul Pratap Singh[1]
E-mail: rahul9588@gmail.com
Yogesh Aggarwal[1]
E-mail: yogesh.24@rediffmail.com
[1] Civil Engineering Department, National Institute of Technology, Kurukshetra, Kurukshetra, India.

References
Aly, M., Hashmi, M. S. J., Olabi, A. G., Messeiry, M., Abadir, E. F., & Hussain, A. I. (2012). Effect of colloidal nano-silica on the mechanical and physical behaviour of waste-glass

cement mortar. *Materials and Design, 33*, 127–135. http://dx.doi.org/10.1016/j.matdes.2011.07.008

Berra, M., Carassiti, F., Mangialardi, T., Paolini, A. E., & Sebastiani, M. (2012). Effects of nanosilica addition on workability and compressive strength of Portland cement pastes. *Construction and Building Materials, 35*, 666–675. http://dx.doi.org/10.1016/j.conbuildmat.2012.04.132

Björnström, J., Martinelli, A., Matic, A., Börjesson, L., & Panas, I. (2004). Accelerating effects of colloidal nano-silica for beneficial calcium–silicate–hydrate formation in cement. *Chemical Physics Letters, 392*, 242–248. http://dx.doi.org/10.1016/j.cplett.2004.05.071

Choolaei, M., Rashidi, A. M., Ardjmand, M., Yadegari, A., & Soltanian, H. (2012). The effect of nanosilica on the physical properties of oil well cement. *Materials Science and Engineering: A, 538*, 288–294. http://dx.doi.org/10.1016/j.msea.2012.01.045

Collepardi, M., Olagot, J. J. O., Skarp, U., & Troli, R. (2002, September 9–11). Influence of amorphous colloidal silica on the properties of self-compacting concretes. In *Challenges in concrete constructions–innovations and developments in concrete materials and constructions* (pp. 473–483). Dundee.

Collepardi, S., Borsoi, A., Olagot, J. J. O., Troli, R., Collepardi, M., & Cursio, A. Q. (2005, July 5–7). Influence of nano-sized mineral additions on performance of SCC. In *Proceedings of the 6th International Congress, Global Construction, Ultimate Concrete Opportunities*. Dundee.

Gaitero, J. J., Campillo, I., & Guerrero, A. (2008). Reduction of the calcium leaching rate of cement paste by addition of silica nanoparticles. *Cement and Concrete Research, 38*, 1112–1118. http://dx.doi.org/10.1016/j.cemconres.2008.03.021

Heidari, A., & Tavakoli, D. (2013). A study of the mechanical properties of ground ceramic powder concrete incorporating nano-SiO2 particles. *Construction and Building Materials, 38*, 255–264. http://dx.doi.org/10.1016/j.conbuildmat.2012.07.110

Hou, P., Kawashima, S., Kong, D., Corr, D. J., Qian, J., & Shah, S. P. (2013). Modification effects of colloidal nanoSiO$_2$ on cement hydration and its gel property. *Composites Part B: Engineering, 45*, 440–448. http://dx.doi.org/10.1016/j.compositesb.2012.05.056

Jalal, M., Pouladkhan, R. A., Norouzi, H., & Choubdar, G. (2012). Chloride penetration, water absorption and electrical resistivity of high performance concrete containing nano silica and silica fume. *Journal of American Science, 8*, 278–284.

Ji, T. (2005). Preliminary study on the water permeability and microstructure of concrete incorporating nano-SiO2. *Cement and Concrete Research, 35*, 1943–1947. http://dx.doi.org/10.1016/j.cemconres.2005.07.004

Jo, B. W., Kim, C. H., & Lim, J. H. (2007). Investigations on the development of powder concrete with nano-SiO$_2$ particles. *KSCE Journal of Civil Engineering, 11*, 37–42. http://dx.doi.org/10.1007/BF02823370

Jo, B. W., Kim, C. H., Tae, G. H., & Park, J. B. (2007). Characteristics of cement mortar with nano-SiO$_2$ particles. *Construction and Building Materials, 21*, 1351–1355. http://dx.doi.org/10.1016/j.conbuildmat.2005.12.020

Kawashima, S., Hou, P., Corr, D. J., & Shah, S. P. (2012). Modification of cement-based materials with nanoparticles. *Cement and Concrete Composites, 36*, 8–15.

Kong, D., Du, X., Wei, S., Zhang H., Yang, Y., Shah, S. P. (2012). Influence of nano-silica agglomeration on microstructure and properties of the hardened cement-based materials. *Construction and Building Materials, 37*, 707–715. http://dx.doi.org/10.1016/j.conbuildmat.2012.08.006

Lea, O. I. (1998). *Chemistry of cement and concrete* (4th ed.). London: Arnold.

Li, G. (2004). Properties of high-volume fly ash concrete incorporating nano-SiO$_2$. *Cement and Concrete Research, 34*, 1043–1049. http://dx.doi.org/10.1016/j.cemconres.2003.11.013

Li, H., Xiao, H. G., & Ou, J. (2004). A study on mechanical and pressure-sensitive properties of cement mortar with nanophase materials. *Cement and Concrete Research, 34*, 435–438. http://dx.doi.org/10.1016/j.cemconres.2003.08.025

Li, H., Xiao, H. G., Yuan, J., & Ou, J. (2004). Microstructure of cement mortar with nano particles. *Composites Part B: Engineering, 35*, 185–189. http://dx.doi.org/10.1016/S1359-8368(03)00052-0

Li, H., Zhang, M., & Ou, J. (2006). Abrasion resistance of concrete containing nano-particles for pavement. *Wear, 260*, 1262–1266. http://dx.doi.org/10.1016/j.wear.2005.08.006

Lim, S., Mondal, P., & Cohn, I. (2012). Effects of nanosilica on thermal degradation of cement paste. In *NICOM 4 – 4th International Symposium on Nanotechnology in Construction*. Greece.

Lin, D. F., Lin, K. L., Chang, W. C., Luo, H. L., & Cai, M. Q. (2008). Improvements of nano-SiO$_2$ on sludge/fly ash mortar. *Waste Management, 28*, 1081–1087. http://dx.doi.org/10.1016/j.wasman.2007.03.023

Ltifi, M., Guefrech, A., Mounanga, P., & Khelidj, A. (2011). Experimental study of the effect of addition of nano-silica on the behaviour of cement mortars Mounir. *Procedia Engineering, 10*, 900–905. http://dx.doi.org/10.1016/j.proeng.2011.04.148

Mondal, P., Shah, S. P., Marks, L. D., & Gaitero, J. J. (2010). Comparative study of the effects of microsilica and nanosilica in concrete. *Transport Research Recording*. doi:10.3141/2141-02.

Naji Givi, A. N., Abdul Rashid, S. A., Aziz, F. N. A., & Salleh, M. A. M. (2010). Experimental investigation of the size effects of SiO$_2$ nano-particleson the mechanical properties of binary blended concrete. *Composites Part B: Engineering, 41*, 673–677. http://dx.doi.org/10.1016/j.compositesb.2010.08.003

Neville, A. M. (1996). *Properties of concrete* (4th ed.). Harlow: ELBS with Addison Wesley Longman.

Oltulu, M., & Şahin, R. (2013). Effect of nano-SiO2, nano-Al2O3 and nano-Fe2O3 powders on compressive strengths and capillary water absorption of cement mortar containing fly ash: A comparative study. *Energy and Buildings, 58*, 292–301. http://dx.doi.org/10.1016/j.enbuild.2012.12.014

Porro, A., Dolado, J. S., Campillo, I., Erkizia, E., de Miguel, Y. R., Sáez de Ibarra, Y., & Ayuela, A. (2005). Effects of nanosilica additions on cement pastes. In R. K. Dhir, M. D. Newlands, & L. J. Csetenyi (Eds.), *Applications of nanotechnology in concrete design* (pp. 87–98). London: Thomas Telford.

Pourjavadi, A., Fakoorpoor, S. M., Khaloo, A., & Hosseini, P. (2012). Improving the performance of cement-based composites containing superabsorbent polymers by utilization of nano-SiO$_2$ particles. *Materials and Design, 42*, 94–101. http://dx.doi.org/10.1016/j.matdes.2012.05.030

Qing, Y., Zenan, Z., Deyu, K., & Rongshen, C. (2007). Influence of nano-SiO$_2$ addition on properties of hardened cement paste as compared with silica fume. *Construction and Building Materials, 21*, 539–545. http://dx.doi.org/10.1016/j.conbuildmat.2005.09.001

Quercia, G., Hüsken, G., & Brouwers, H. J. H. (2012). Water demand of amorphous nano silica and its impact on the workability of cement paste. *Cement Concrete Research, 42*, 344–357. http://dx.doi.org/10.1016/j.cemconres.2011.10.008

Rahel, I., Hamid, R., & Taha, M. R. (2012). Fire resistance of high-volume fly ash mortars with nanosilica addition. *Construction and Building Material, 36*, 779–786.

Senff, L., Hotza, D., Repette, W. L., Ferreira, V. M., & Labrincha, J. A. (2010). Mortars with nano-SiO$_2$ and micro-SiO$_2$ investigated by experimental design. *Construction and Building Materials, 24*, 1432–1437. http://dx.doi.org/10.1016/j.conbuildmat.2010.01.012

Senff, L., Labrincha, J. A., Ferreira, V. M., Hotza, D., & Repette, W. L. (2009). Effect of nano-silica on rheology and fresh properties of cement pastes and mortars. *Construction and Building Materials, 23*, 2487–2491. http://dx.doi.org/10.1016/j.conbuildmat.2009.02.005

Shekari, A. H., & Razzaghi, M. S. (2011). Influence of nano particles on durability and mechanical properties of high performance concrete. *Procedia Engineering, 14*, 3036–3041. http://dx.doi.org/10.1016/j.proeng.2011.07.382

Singh, L. P., Agarwal, S. K., Bhattacharyya, S. K., Sharma, U., & Ahalawat, S. (2011). Preparation of silica nanoparticles and its beneficial role in cementitious materials. *Nanomaterial Nanotechnology, 1*, 44–51.

Singh, L. P., Bhattacharyya, S. K., Singh, & P. (2012). Granulometric synthesis and characterisation of dispersed nanosilica powder and its application in cementitious system. *Advances in Applied Ceramics, 111*, 220–227. http://dx.doi.org/10.1179/1743676112Y.0000000002

Stefanidou, M. (2012). Influence of nano-SiO_2 on the Portland cement pastes. *Composites Part B: Engineering, 43*, 2706–2710. http://dx.doi.org/10.1016/j.compositesb.2011.12.015

Tobón, J. I., Payá, J. J., Borrachero, M. V., & Restrepo, O. J. (2012). Mineralogical evolution of Portland cement blended with silica nanoparticles and its effect on mechanical strength. *Construction and Building Materials, 36*, 736–742. http://dx.doi.org/10.1016/j.conbuildmat.2012.06.043

Wiesner, M. R., & Bottero, J. Y. (2007). *Environmental nanotechnology: Applications and impacts of nanomaterials.* New York, NY: McGraw-Hill.

Zapata, L. E., Portela, G., Suárez, O. M., & Carrasquillo, O. (2013). Rheological performance and compressive strength of super plasticized cementitious mixtures with micro/nano-SiO_2 additions. *Construction and Building Materials, 41*, 708–716. http://dx.doi.org/10.1016/j.conbuildmat.2012.12.025

Zhang, M. H., & Islam, J. (2012). Use of nano-silica to reduce setting time and increase early strength of concretes with high volumes of fly ash or slag. *Construction and Building Materials, 29*, 573–580. http://dx.doi.org/10.1016/j.conbuildmat.2011.11.013

Zhang, M. H., Islam, J., & Peethamparan, S. (2012). Use of nano-silica to increase early strength and reduce setting time of concretes with high volumes of slag. *Cement and Concrete Composites, 34*, 650–662. http://dx.doi.org/10.1016/j.cemconcomp.2012.02.005

Zhang, M. H., & Li, H. (2011). Pore structure and chloride permeability of concrete containing nano-particles for pavement. *Construction and Building Materials, 25*, 608–616. http://dx.doi.org/10.1016/j.conbuildmat.2010.07.032

13

Engineer's estimate reliability and statistical characteristics of bids

Fariborz M. Tehrani[1]*

*Corresponding author: Fariborz M. Tehrani, Department of Civil and Geomatics Engineering, California State University, Fresno, 2320 E. San Ramon Avenue, M/S EE94, Fresno, CA 93740, USA

E-mail: ftehrani@csufresno.edu

Reviewing editor: Amir H. Alavi, Michigan State University, USA

Abstract: The objective of this report is to provide a methodology for examining bids and evaluating the performance of engineer's estimates in capturing the true cost of projects. This study reviews the cost development for transportation projects in addition to two sources of uncertainties in a cost estimate, including modeling errors and inherent variability. Sample projects are highway maintenance projects with a similar scope of the work, size, and schedule. Statistical analysis of engineering estimates and bids examines the adaptability of statistical models for sample projects. Further, the variation of engineering cost estimates from inception to implementation has been presented and discussed for selected projects. Moreover, the applicability of extreme values theory is assessed for available data. The results indicate that the performance of engineer's estimate is best evaluated based on trimmed average of bids, excluding discordant bids.

Subjects: Engineering Economics; Engineering Project Management; Statistics for Business, Finance & Economics; Transportation Engineering

Keywords: cost estimate; estimate reliability; bid analysis; contingency; construction cost; competitive bid; normal distribution; theory of extreme values; Weibull distribution; public sector procurement

1. Introduction

Strategic planning and programming of infrastructure projects require reliable cost estimating methods to secure appropriate funding. The expected accuracy of cost estimates varies throughout the life cycle of a project from inception to implementation. Thus, cost estimating is a constant challenge for long-term planning.

ABOUT THE AUTHORS

Fariborz M. Tehrani is an assistant professor, professional civil engineer (PE), sustainability professional (ENV SP), and project management professional (PMP) with academic and professional backgrounds in engineering design, management, and education. His research and practice experiences focus on resilient and sustainable infrastructures. This work presents research on resilient project delivery methods which emphasize the probabilistic approach to resource management.

PUBLIC INTEREST STATEMENT

This paper highlights sources of uncertainties in estimating the true cost of selected highway maintenance projects. Presented methodology traces these sources in the preparation of engineer's estimate at various stages of the project development. Further, statistical analysis has been utilized to show the reliability of engineering estimate in respect to bid values, and particularly the lowest bid amount.

Sources of uncertainty in a project's cost can be categorized to inherent variability and prediction error. The latter, also known as modeling uncertainties, can be reduced through improvements in prediction models, inclusion of more data, and enhancement of statistical analysis. An unclear scope of the work alongside undefined means and methods of construction are known examples of errors which can be reduced through a detailed design of projects (Bajari, McMillan, & Tadelis, 2003). Project studies aim to clarify project deliverable items and recognize all principal factors that may affect the final cost of the project, e.g. site conditions, economic trends, and environmental issues. Obviously, prediction errors beyond certain levels of risk or confidence, e.g. occurrence of major disasters during the life time of a routine maintenance project, cannot be taken into account in cost estimates. Such risks are often transferred by means of insurance, indemnification, force majeure clauses, etc.

Some uncertainties pertain to the nature of the project nature and remain nearly consistent throughout the project development cycle. As a result, the main objective in studying inherent variability is simply acknowledging them and incorporating their effect in project estimates rather than reducing them. These types of uncertainties are caused by random variables, which in practical terms could not be controlled. For instance, minor variations in site conditions could not be avoided nor could they be wholly anticipated. The same is true for fluctuations in construction material costs. Cost estimates may also include strategies to cover minor and limited effects of unknown random phenomena. The most common strategy for covering these unexpected costs is the inclusion of contingencies in the cost estimate. Appropriate contingencies protect owners from minor unpredictable events that may increase the cost and cause over-run. However, contingencies should not be used to cover inaccurate estimates or poorly defined projects. The contingency amount is a function of uncertainty in project. It could be evaluated and represented as a percentage of contract cost, either itemized or total cost, or defined through probabilistic models, e.g. PERT or Monte Carlo (Ergin, 2005).

Although an engineer's estimate at the end of the design phase is the most detailed estimate accompanying plans and specifications, but, it may not represent the final and real cost of the project. Rather, the real cost of the project is developed through the construction phase. These estimates, along with a contractor's estimate, known as bids, at their best are fundamentally expert opinions on the final cost of a project. In this arena, the bid process is also a factor in long-term planning of projects.

Competitive bidding is the most common method of procurement in the US public sector. Transportation departments award most projects to the lowest bidders as unit price contracts to comply with government statutes, e.g. Federal Acquisition Regulation System (Schexnayder, Weber, & Fiori, 2003). In spite of the appropriateness or fairness of this method, a bid may not necessarily represent the bidder's opinion on the contract cost. Strategic manipulation of bids and the behavior of the bidder in response to a competitive environment of bidding are main sources of bid variation (Gaver & Zimmerman, 1977). Therefore, bid analysis is often necessary to assure the reliability of bids, particularly the lowest bid, in comparison to the engineer's estimate. Furthermore, some agencies implement alternative awarding processes to select the lowest reliable bid, as opposed to the lowest bid. Examples of these alternate procedures include the selecting of the average bid (e.g. in Italy and Taiwan), the average bid in middle 80% (e.g. in Peru), as well as the lowest bid after excluding discordant bids (Salem Hiyassat, 2001).

2. Bidding and the extreme value theory

Consider the random variable B as a representation of rational bids from m bidders on a project with a known distribution (Rothkopf, 1969):

$$X = (B_1, B_2, \dots, B_m) \tag{1}$$

Suppose that bidders provide positive cost estimates representing their true estimate of a project cost as an independent random variable. Assume that the distribution X is normal. This assumption is not a general necessity, but rather complies with observations made in this study. Other distribution functions could be analyzed likewise. Furthermore, assume that the probability of negative numbers in this distribution may be disregarded and does not require the use of the log-normal distribution function.

$$f_X(B) = \frac{1}{\sigma\sqrt{2\pi}} \exp\left[\frac{-(B-\mu)^2}{2\sigma^2}\right] \tag{2}$$

$$F_X(B) = \int_{-\infty}^{B} \frac{1}{\sigma\sqrt{2\pi}} \exp\left[\frac{-(B-\mu)^2}{2\sigma^2}\right] dx \tag{3}$$

Assume that there are n realizations of the sample random variable X with minimum values of Y, which represent the number of projects with identical bidder distributions:

$$Y = \min(X_1, X_2, \ldots, X_n) \tag{4}$$

Therefore, Y is also a random variable and its distribution function can be derived from the initial random variable X:

$$F_Y(y) = 1 - \left[1 - F_X(y)\right]^n \tag{5}$$

$$f_Y(y) = n\left[1 - F_X(y)\right]^{n-1} f_X(y) \tag{6}$$

Similar formulation could be provided for the maximum value of X. Figure 1 shows the normal distribution and extreme values distribution for a given number of projects.

As n grows larger, the distribution of the extreme Y value converges to a particular functional form, Weibull distribution, which does not depend on the initial distribution. Rather, the tail behavior of the initial distribution, an exponential function, defines the extreme value distribution. Distribution of smallest values is provided as (Rothkopf, 1969):

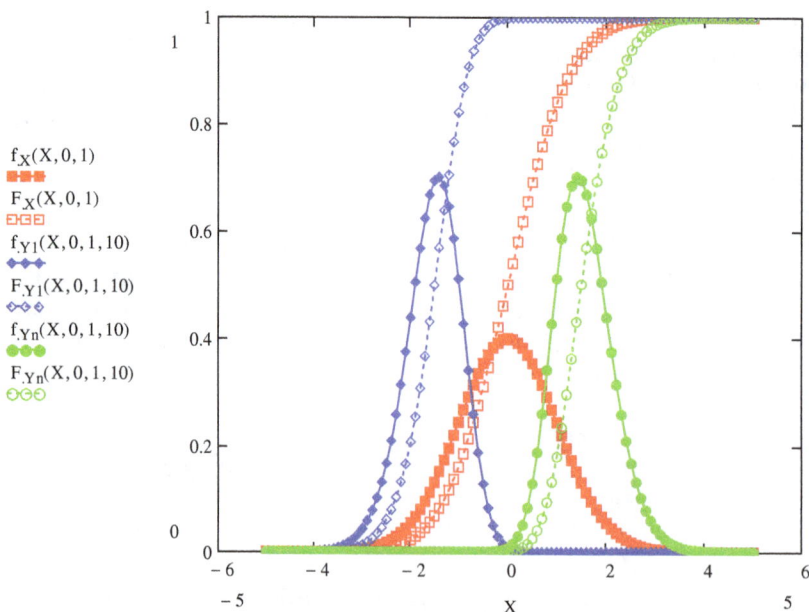

Figure 1. Normal and extreme value distributions.

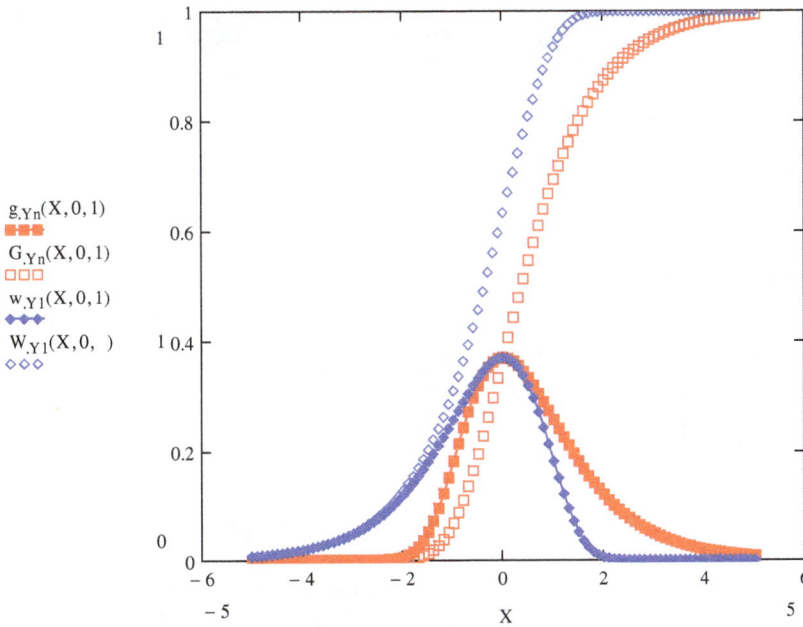

$g_{.Yn}(X,0,1)$
$G_{.Yn}(X,0,1)$
$w_{.Y1}(X,0,1)$
$W_{.Y1}(X,0,)$

Figure 2. Extreme value distributions.

$$F_Y(y) = 1 - \exp\left\{-\exp\left[\alpha(y-u)\right]\right\} \tag{7}$$

$$f_Y(y) = \alpha \cdot \exp\left\{\alpha(y-u) - \exp\left[\alpha(y-u)\right]\right\} \tag{8}$$

where u and a are characteristics of the smallest value of X (mode of Y) and inverse measure of dispersion of Y, respectively. Other moments of smallest values are defined below:

Mean: $$\bar{Y} = u - \frac{\gamma}{\alpha} \tag{9}$$

Standard Deviation: $$\sigma_Y = \frac{\pi}{\alpha\sqrt{6}} \tag{10}$$

Median: $$\hat{Y} = u - \frac{0.3665}{\alpha} \tag{11}$$

Skew: $$g_Y = -1.1396 \tag{12}$$

In the above equations, g is the Euler's constant equal to 0.577216.

Figure 2 shows the distribution of extreme values as defined in the above equations.

3. Selected engineer's estimates and bids

Projects are unique endeavors, characterized by specific scope, schedule, and cost. Furthermore, the engineer's estimate and the bid values for a project are subject to additional economy-driven parameters of the market at the time of delivery. Therefore, statistical data-sets should be carefully selected based on the common characteristics of projects. Thus, large number of sample projects with different scopes or advertising time may not provide more reliable outcomes, regardless of resulted statistical confidence levels. The main data-set in this research contains 22 projects with identical scope within a single program. This set of data is incorporated to show the development of cost estimate throughout the life cycle of the project design. These projects were developed and delivered over a four-year period. Due to wide range of their advertising date, their bid values may not represent the same market characteristics. To compare bid values, a subset of the above

Table 1. Sample project estimates and bids

Project	Engineer's estimate	Bid 1	Bid 2	Bid 3	Bid 4	Bid 5	Bid 6	Bid 7	Bid 8
1	$676,530	$528,600	$572,414	$616,692	$703,020				
2	$870,970	$493,660	$640,607	$642,174					
3	$998,285	$794,131	$871,768	$924,824	$1,017,242	$1,044,540			
4	$1,271,040	$1,376,810	$1,516,227	$1,659,646					
5	$933,000	$774,925	$891,248	$983,970	$993,100	$1,094,350			
6	$567,372	$437,460	$473,360	$516,498	$523,474	$540,800	$567,373	$581,432	$679,302
7	$824,960	$709,632	$753,716	$812,188	$838,700				
8	$1,158,380	$1,034,765	$1,053,133	$1,106,080	$1,125,000	$1,126,436	$1,147,703		
9	$1,044,294	$845,995	$910,106	$1,059,240	$1,118,800				
10	$324,263	$316,702	$434,031	$444,444					
Total	$8,669,094	$7,312,680							

referenced data-set is considered, as shown in Table 1. Sample projects in this data subset have been advertised through a four-month time frame. All projects received three to eight bids and were awarded to the lowest bidder. The standard deviation of the sample is 33% and the Chi-test result confirms that the engineer's estimate distribution over the sample projects follows a normal distribution at 99.44% confidence level. The skew is as small as −0.05 and median is 1.01.

4. Development of cost estimate

Figure 3 shows the development of the cost estimate during different stages of project development. Each line in this graph reflects real data for a sample project in the main data-set introduced in the previous section. Thus, this graph contains more projects than the subset listed in Table 1. All projects have the same scope of the work, but were advertised in the earlier years. All values are normalized to the final cost of project. Lower and higher bound curves in this graph show the level of accuracy in cost estimation through four specific milestones. Table 2 summarizes these observations with expected values reported by AbouRizk, Babey, and Karumanasseri (2002) and Rothwell (2004). The cost estimate at project initiation varies within ±80% of the final cost. The cost estimate at this stage is often based on the average unit cost of major construction activities obtained from historical records. The main sources of inaccuracy at this stage are uncertainties in definition of project scope, method, and schedule. Specifically, the project initiation takes place nearly three years before construction begins. Although some adjustments could be made to account for inflation, alternative construction methods, and materials, and other minor items, unjustified

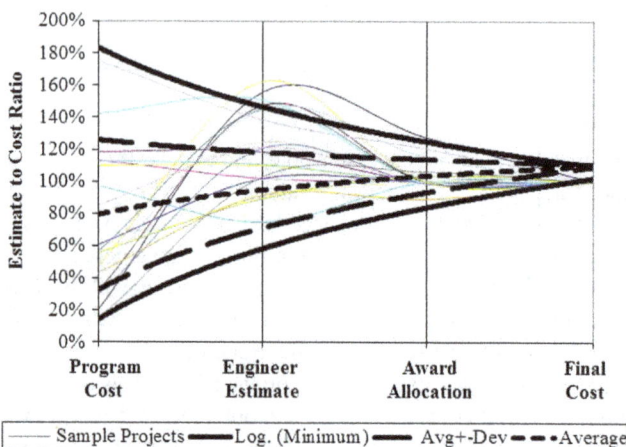

Figure 3. Typical cost estimate development.

Table 2. Summary of observed vs. expected variations in cost estimate

Project stage	Programming	Design	Bid/Award
Estimate type	Conceptual	Engineer's estimate	Bid analysis
Observed extreme values	±80%	±40%	±20%
Observed average ± Std. Dev.	−65% ~ +25%	−30% ~ +20%	−10 ~ +15%
Expected accuracy (Municipal Projects)[a]	±30%	±10%	±10%
Expected accuracy (Roadway Projects)[a]	±90%	−50% ~ +60%	−50% ~ +60%
Expected accuracy (AACEI)[b]	−50% ~ +100%	−15% ~ +20%	−10% ~ +15%
Suggested contingency (AACEI)[b]	50%	15%	5%

[a]After AbouRizk et al. (2002).
[b]After Rothkopf (1969).

contingencies should not be included in the estimate. Such high contingencies do not improve the accuracy and would only shift the estimate to higher amounts, causing unnecessary blockage of funds.

The major qualitative improvement in cost estimate occurs in design stage, where engineer's estimate varies within ±40% of final cost. Preparation of plans and specifications provides the opportunity for the engineer to narrow down the scope and method of the work. The average construction period of sample projects is between 4 and 8 months after the engineer prepares the estimate. Therefore, the engineer's estimate should also include adjustments for inflation and unforeseen conditions within the scope of the project.

Award allocation is the assumed cost of project for contracting purposes. Bidders provide their bids in the advertising period, from two to three months before beginning construction. Further increase or decrease in the project cost is often due to minor changes in the quantities of the work or additional minor tasks which are necessary to perform the project.

5. Actual cost of the project

Although the contract cost is derived from the lowest bid, this number may not represent the true cost of the project in general conditions, which were assumed in the engineer's estimate. Bidders consider availability of their resources at the time and place of construction in preparation of their bids. For instance, preoccupation with other projects, lack of experience in the specific environment or location, overheads, and size of the business tend to raise the bid amount. Also, desire to expand business, access to mobilized resources in the vicinity of project location, or at the certain time frame are usual causes for lower bids. Moreover, minor errors might also change the bid amount without justifying bid withdrawal by bidder or bid rejection by the owner. In this respect, the lowest bid might simply come from the bidder who has specific advantages in accessing required resources, has specific desire to get the project, or simply has provided an erroneous bid. Statistical analysis of bids is the tool to search for the true cost of the project within bid results, by recognizing non-project-related parameters.

Table 3 summarizes the results of a sample study on various hypotheses on the best estimate on project cost. Three error measures are provided for each hypothesis. The error is defined as the difference between a reference value, engineer's estimate in this study, and proposed cost estimate. The weighted average of these errors includes the cost of each project and dollar amount of the difference. Therefore, this measure is the most useful value to evaluate the entire program considering both over- and under-estimated bids. The un-weighted average of errors represents simple average of individual projects. Further, the least square method represents the probable error measures in the program considering the independence of projects and orthogonal nature of error values for different projects. This measure is simply a test to validate consistency of results.

Table 3. True cost hypothesis vs. error measures							
True cost hypothesis vs. error measures	The lowest bid (%)	The second low bid (%)	Average third low bids (%)	Average bids (%)	Average of non-extreme bids (%)`	Trimmed mean of bids (%)	Median bid (%)
Weighted-signed average (Program)	−15.65	−6.37	−6.97	−3.19	−2.14	−2.68	−1.84
Signed average (Projects)	−16.31	−5.30	−6.64	−2.69	−1.05	−1.29	−0.80
Least square root (Probable)	20.81	17.98	16.92	14.83	15.89	15.61	16.00

The low bid concept is evaluated in three different measures. The lowest bid column, representing the common acceptable rule, reveals the large gap between the engineer's estimate and the lowest bid. In other word, taking the lowest bid as the true cost of the project is the poorest hypothesis among the provided methods. The second lowest bid is a more reliable value, especially when the lowest bid is suspected to represent bidders with special characteristics discussed earlier. Substantial lower error measures in Table 3 indicate this point. Average of the three lowest bids is another approach to filter-out special cases in bids. However, this method is not helpful in projects with less than four bids. The results of this method are closer to the second-low-bid approach.

The average bid is a well-known approach to estimate the real cost of the project. This method has provided the lowest least-square-root-error value in the table of results for small number of projects. However, it does not perform well in presence of very low or very high bids. Therefore, another approach is taken by removing the highest and lowest bids before taking the average. In an alternative approach, all low and high bids, beyond 68% middle population (Average ± standard deviation), are removed before evaluating the average bid amount. The results indicate that trimming extreme values reduces the error. For comparison, the error values for the median bid, at 50% percentile, are also provided.

6. Statistical characteristics of bids
Figure 4 shows the distribution of bids for sample projects, representing the ratio of each bid to the average bid of the same project. This distribution is close to the normal distribution with 99% confidence.

Further analysis reveals that bids are distributed with average standard deviation of 11.6%. Considering each bid as an expert opinion on the project cost, and disregarding the engineer's estimate, the variation of cost estimate is practically higher than 8 at 99% confidence level. In this distribution, the lowest bid is within 0.004–4.237 percentile of bids, which is at the average of 1.34 percentile of sample bids for the first data-set. The average ratios of the lowest bid to engineer's estimate is 84%. Figure 5 shows the distribution of these ratios for sample projects. The confidence level for normal distribution of lowest bid ratio to engineer's estimate is 90.2% only, which is much less than bids distribution in Figure 4.

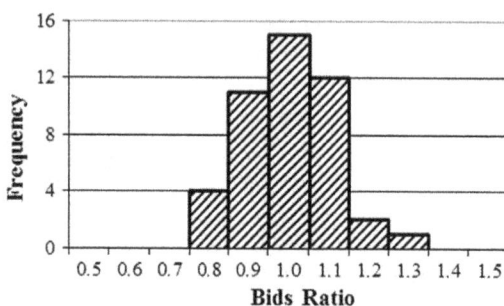

Figure 4. Bids' distribution.

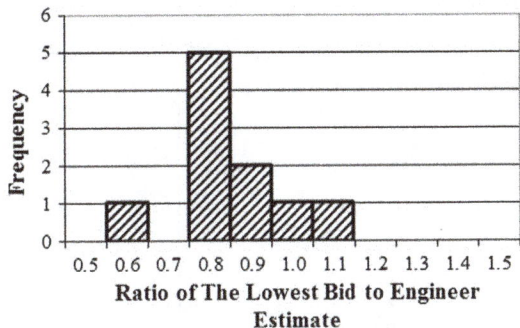

Figure 5. Lowest bids' distribution using ratio to engineer's estimate.

Considering the ratio of lowest bid to average bid, an alternative distribution would be obtained as shown in Figure 6. This distribution has the average of 86%, with 99.9% confidence level for normal distribution. The median is the same as average, at 86%. Standard deviation is as low as 0.048 and skew is 0.33. Using Equations (5) and (6), the distribution of extreme values could be obtained as shown in Figure 7. This graph indicates that the lowest bid would fall within 69–96% of average bid at 95% confidence level. Further, assuming no discrepancies between the average bid and engineer's estimate, the expected value of lowest bid is 18% below the engineer's estimate, or 82% of engineer's estimate. This number is very close to the average ratio of the lowest bid to average bids, 84%, evaluated from distribution shown in Figure 5. For large number of projects, the distribution would converge to the graphs shown in Figure 8. This graph shows how distributions of the smallest and largest numbers become distant for large number of bids.

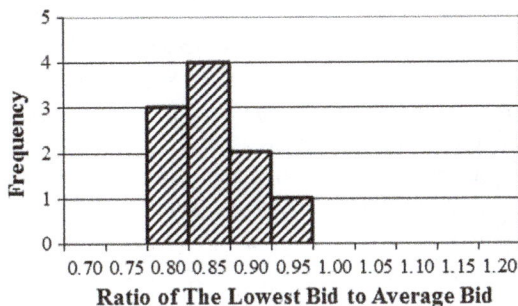

Figure 6. Lowest bids' distribution using ratio to average bid.

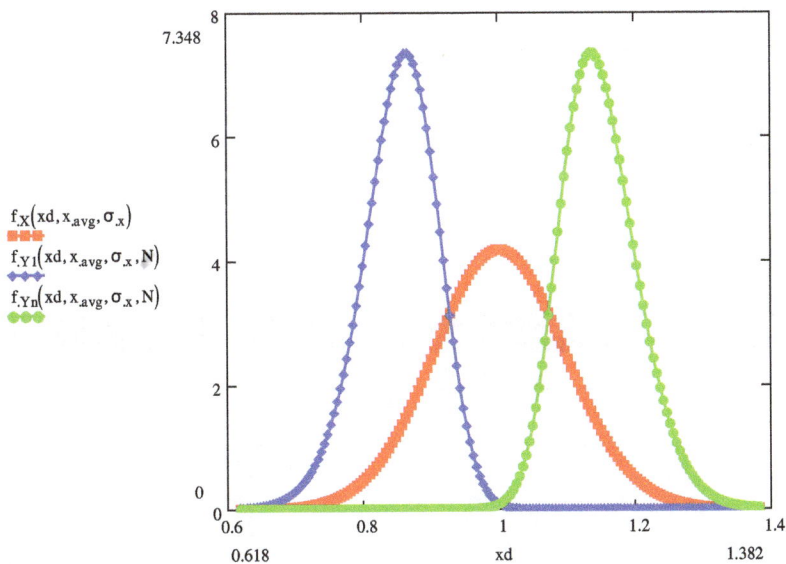

Figure 7. Extreme value distributions for selected projects.

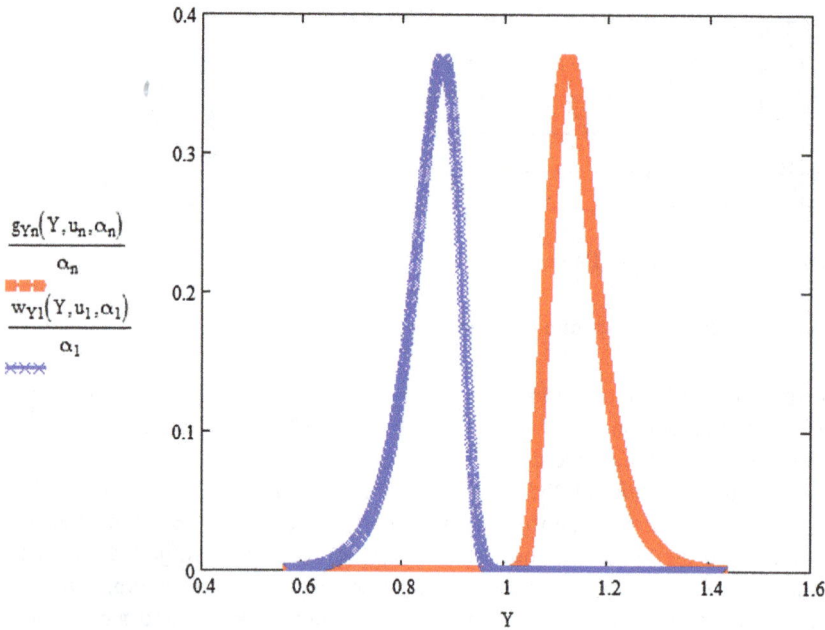

Figure 8. Extreme value distributions for sample bids and large number of projects.

Assuming 95% confidence level, the standard deviation of bids would be 9.6%. The engineer's estimate at this standard deviation would be within ±20% of the average bid with the same confidence level. Suppose that average bid is the same as engineer's estimate, mean ratio of 1.0. Then, rational lowest bid ratio could be derived for different values of n as shown in Table 4. It might be helpful to realize that Figure 9 shows the ratio of bids to engineer's estimate. The shape of this distribution is an important tool to understand behavior of bidders. The confidence of level of normality of this distribution is only 80%, which would reject the normal hypothesis. Further, the skew of distribution is 0.4 which would confirm the visual observation of distribution being skewed to values smaller than engineer's estimate. Nearly 64% of bids are below engineer's estimate in sample projects.

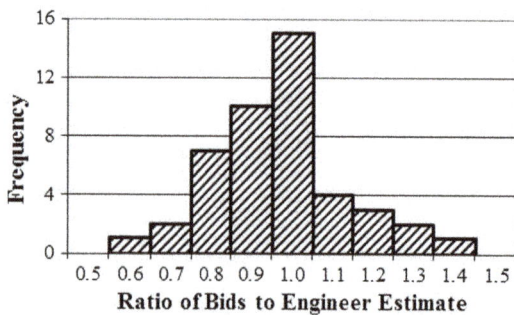

Figure 9. Bid to engineer's estimate ratio distribution.

Table 4. The lowest bid ratio limits at 95% confidence level					
Distribution	Normal				Weibull
n-Value	1	2	5	10	Very large
Average (%)	100	94.6	88.9	85.3	88
Lower bound (%)	80.9	78.8	76.1	74.1	75
Upper bound (%)	119.1	110.4	101.7	96.5	101

7. Conclusions

Development of cost estimates in various stages of projects shows that unjustified increase of costs through large contingency does not improve the accuracy of estimates. Distinguishing between errors caused by inaccurate predictions, and variations caused by inherent parameters are important in appropriate use of contingency. Summary of observed variations in cost estimate of sample projects indicates that the accuracy of cost estimates is improved from 80% deviation to 40% deviation as the project makes progress from programming to design. The lowest bid (award amount) may also deviate from the final cost by 20%. These deviations are not symmetric, that is cost under-estimation is more frequent and more considerable than the cost over-estimation throughout the project development. This trend slightly changes after the award, as the over-estimation was observed to be higher than under-estimation for sample projects.

Statistical analysis of bid results for sample projects indicates that normal distribution is appropriate to model the bid ratio to average bids. This confirms the perception of bid values as expert opinions about the cost of a project. Further, performance of engineer's estimate should be evaluated based on trimmed average of bids, excluding discordant and extremely low or high bids, rather than the lowest bid. Thus, the accuracy of the engineer's estimate may not directly correlate with the reliability of the budget, as the budget is ultimately compared with the award amount or the lowest bid.

Implementing theory of extreme values reveals that distribution of lowest bids is independent of initial distribution of bids. Expected range of lowest bids could be derived for desired level of confidence using proposed analytical methods. This conclusion directly relates to the budget estimation for a multi-project program, where the total budget is the sum of the lowest bids for each project.

Validity of provided results is limited to assumptions made for rational competitive bidding. Errors caused by deficient design, bids manipulation, change orders, and similar sources were neither investigated nor considered in analysis. Further, numerical analyses are performed on limited number of projects with specific scope and size. Moreover, market parameters were intentionally filtered out by careful selection of projects advertised in a short time period. Regardless, the presented methodology can be implemented over larger pool of samples to provide appropriate guidelines on accuracy and performance of engineer's estimate and reliability of bids in various situations.

Acknowledgments
Raw data have been extracted from public Bid Summary Results provided by the Office of Contract Awards and Services in California Department of Transportation. The author is grateful to Ms. Kinana Negoro and Ms. Roshanak Farshidpour for their help with data mining and edits on an earlier version of the manuscript.

Funding
The author received no direct funding for this research.

Author details
Fariborz M. Tehrani[1]
E-mail: ftehrani@csufresno.edu
ORCID ID: http://orcid.org/0000-0002-7618-8009
[1] Department of Civil and Geomatics Engineering, California State University, Fresno, 2320 E. San Ramon Avenue, M/S EE94, Fresno, CA 93740, USA.

References
AbouRizk, S. M., Babey, G. M., & Karumanasseri, G. (2002). Estimating the cost of capital projects: An empirical study of accuracy levels for municipal government projects. *Canadian Journal of Civil Engineering, 29,* 653–661. doi:10.1139/l02-046

Bajari, P. L., McMillan, R. S., & Tadelis, S. (2003). Auctions versus negotiations in procurement: An empirical analysis. *Journal of Law, Economics and Organization, 25,* 372–399. doi:10.3386/w9757

Ergin, A. A. (2005). *Determination of contingency for international construction projects during bidding stage* (Master's thesis). Middle East Technical University, Ankara. Retrieved from http://lib.metu.edu.tr/

Gaver, K. M., & Zimmerman, J. L. (1977). An analysis of competitive bidding on BART contracts. *The Journal of Business, 50,* 279–295. Retrieved from http://www.jstor.org/journal/jbusiness http://dx.doi.org/10.1086/jb.1977.50.issue-3

Rothkopf, M. H. (1969). A model of rational competitive bidding. *Management Science, 15,* 362–373. doi:10.1287/mnsc.15.7.362

Rothwell, G. (2004). *Cost contingency as the standard deviation of the cost estimate for cost engineering* (SIEPR Discussion Paper No. 04-05). Stanford, CA: Stanford Institute for Economic Policy Research. Retrieved from https://siepr.stanford.edu/

Salem Hiyassat, M. A. (2001). Construction bid price evaluation. *Canadian Journal of Civil Engineering, 28*, 264–270. doi:10.1139/l00-111

Schexnayder, C. J., Weber, S. L., & Fiori, C. (2003). *Project cost estimating: A synthesis of highway practice* (Part of NCHRP Project 20-7, Task 152). Washington, DC: National Cooperative Highway Research Program; Transportation Research Board. Retrieved from http://www.trb.org

14

Characterization of alum floc in water treatment by image analysis and modeling

Petri Juntunen[1]*, Mika Liukkonen[1], Markku Lehtola[1] and Yrjö Hiltunen[1]

*Corresponding author: Petri Juntunen, Department of Environmental Science, University of Eastern Finland, Kuopio FI-70210, Finland

E-mail: petri.juntunen@kuopionvesi.fi

Reviewing editor: Roberto Revelli, Politechnico Di Torino, Italy

Abstract: In many water treatment plants, flocculation is the key unit concerning the performance of water treatment. For this reason, monitoring the flocculation (i.e. floc size) is a crucial issue to achieve the acceptable performance for the process. Generally, flocculation is monitored by naked eye or using complex, sample-based methods. This is laborious and expensive, however, and should be either automated or alternative methods for estimating the floc quality should be developed if possible. In this paper, we present an online characterization system for estimating the most essential quality parameters of floc using digital images taken in the flocculation unit. In addition, we compare the surface area of the floc particles defined using the images with other measurement data collected from the process, and create a multivariable regression model for it. We also illustrate the dependencies between the floc properties and other process variables using a self-organizing map.

Subjects: Engineering & Technology, Intelligent Systems, Systems & Control Engineering, Water Engineering, Water Science

Keywords: water treatment, water quality, image analysis, flocculation, regression, self-organizing map

ABOUT THE AUTHOR

The University of Eastern Finland (UEF) enjoys a leading national position in the field of forestry research. The research projects of the Process Informatics research group in the Department of Environmental Science have focused on advanced measurement and modeling methods and advanced systems utilized by the process industry. Research has been done within the energy, pulp, chemical, electronics, and water industries. So far we have developed and applied intelligent methods, which can be exploited in offline or online software tools. These tools can be used in process monitoring and optimization, for example. Currently, we are developing advanced online monitoring systems in which we combine novel measurement technology and advanced data processing to extract new kind of information from processes. In addition, several commercial software tools have been developed which have been adopted by service business companies operating in the process industry.

PUBLIC INTEREST STATEMENT

In this article, we introduce a low-cost camera application, which is suitable for characterization of alum floc in water treatment processes. The application is capable to produce different floc describing parameters to produce parameter such as floc size or floc number by image analysis. In addition, the calculated parameters are combined with other process parameters and analyzed by data-driven modeling methods such as multiple linear regression or self-organizing maps. The application enables a fruitful way to achieve more online information about the most essential unit of the water treatment process.

1. Introduction

Flocculation is a critical unit process of drinking water treatment. It involves the formation of floc followed by the aggregation of floc that is amenable to solid/liquid separation with subsequent processes such as sedimentation, flotation, and/or filtration. The most common coagulants used in water treatment are alum and other aluminum-based chemicals. However, the flocculation process is extremely complex because many chemical and physical features of raw water affect the aluminum coagulation and flocculation process. Speaking of chemical features, many organic and inorganic compounds in suspended, colloid, or solved form influence the process. Organic compounds such as fulvic or humic acids play an essential role in the coagulation and flocculation. Furthermore, many inorganic compounds such as the SiO_2^-, OH^-, F^-, PO_4^{3-} or SO_4^{2-} affect the process. As an example of physical parameters, temperature of the water has a remarkable influence on the flocculation process (Kvech & Edwards, 2002; van Benschoten & Edzwald, 1990). Moreover, it is evident that the process conditions itself have a great effect on Al coagulation and flocculation. The dose of the Al chemical is, of course, one of the key parameters, as is the adjusted pH value. Furthermore, hydraulic variables in flocculation units such as velocity or velocity gradient (G-value) affect flocculation. Running variables of the separation processes such as surface load or washing periods influence the performance (Letterman, 1999). Moreover, in water treatment, there are observable cycles and/or episodic events such as rapid changes, which cause the process to behave dynamically (Bratby, 2006; Juntunen, Liukkonen, Pelo, Lehtola, & Hiltunen, 2013). In summary, the coagulation and flocculation processes are both physically and chemically heterogeneous, for which the development of online monitoring applications is extremely challenging, and time-consuming and expensive *in situ* testing such as z-potential metering may be needed.

For these reasons, there is a need for developing new approaches to monitor and manage water treatment and water quality online and to ensure safe drinking water with reasonable costs. Real-time monitoring, which is characterized by a rapid response time, full compatibility with automation, sufficient sensitivity, high rate of sampling, minimal requirements for skill and training, etc. (Mays, 2004) can provide data that can be used in multiple purposes, such as fault monitoring or process assessment. In water treatment, as in many domains, process monitoring and control relies heavily on accurate and reliable sensor information. While many process variables can be measured continuously using relatively simple and cheap physical sensors, the determination of certain quantities of interest requires costly laboratory analyses that cannot be performed online (Valentin & Denoeux, 2001).

One interesting approach to characterize the floc formation is the image analysis of the forming floc. The main advantage of this approach is that we can measure the most essential features of the floc such as the size and the form of the floc. Previously, there have been some studies of the image analysis in ex situ or *in situ* considering for water treatment (Chakraborti, Atkinson, & Van Benschoten, 2000; Wang, Lu, Du, Shi, & Wang, 2009), or waste water treatment cases (Perez, Leite, & Coelho, 2006). Nonetheless, these applications are either laboratory or pilot-scale studies not designed for online use, which is useful if the purpose is to study the chemistry of floc formation or validate the results of image analysis. Considering online use in full-scale water treatment plants, such applications would be complex and expensive. On the other hand, the information contained by digital images can be exploited without knowing the exact value of the measured parameters. In this case, we make interpretations of the image analysis by comparing the calculated image characterization variables to measured process variables.

Advanced statistical methods, including artificial neural networks, have been applied successfully to various problems in different industrial fields, such as the pulp and paper industry (Alonso, Negro, Blanco, & San Pío, 2009; Avelin, Jansson, Dotzauer, & Dahlquist, 2009). For example, self-organizing maps (SOM) have proved to be useful and efficient in modeling water quality in general level (Kalteh, Hjorth, & Berndtsson, 2008; Maier, Morgan, & Chow, 2004). A SOM is an unsupervised learning method to analyze various data-sets, including those with missing values.

The application of techniques of artificial intelligence, which can be used for analyzing and visualizing complex multivariable data, can potentially be useful in analyzing data, such as features of water quality parameters (Juntunen, Liukkonen, Pelo, Lehtola, & Hiltunen, 2011) or outliers (Muruzábal & Muñoz, 1997). Typical for a SOM is that the desired solutions or targets are not given and the network intelligently learns to cluster the data by recognizing different patterns (Alhoniemi, Hollmen, Simula, & Vesanto, 1999; Kohonen, 2001). A SOM possesses advantages over other multivariate approaches, because it can handle the nonlinearities in a system; it can be produced using the data without mechanistic knowledge of the system; it can deal with noisy, irregular, or missing data; it can be easily and quickly updated; and it can interpret information from multiple variables or parameters with its excellent visualization capabilities (Hong, Bhamidimarri, & Charleson, 1998; Liukkonen, Havia, Leinonen, & Hiltunen, 2011; Liukkonen, Hiltunen, Hälikkä, & Hiltunen, 2011). In addition, SOM can combine the different data-sets to a single form which is visually and easily understandable (Kohonen, 2001).

As far as we know, characterization of floc using a method capable of operating online in a real water treatment process has not been possible this far. In this paper, we present a low-cost online characterization system for estimating the size and other features of the floc in the flocculation unit. The system consists of an ordinary systems camera, which is automated to capture images over the flocculation pool. The images are then analyzed to calculate characteristics that indicate the size of the floc and other features such as eccentricity and the number of floc particles. After characterizing the floc, multiple linear regression (MLR) and SOM are used to analyze the reasons for changes in floc properties.

2. Materials and methods

2.1. Case process and data
In our case process, the water is first bank filtrated. After filtration, the water is purified with chemical coagulation in a chemical purification process. The aluminum flocculant is dosed to the process for flocculation. In addition, lime is dosed to the process for adjusting the pH in two separate locations (A and B later in the text). After coagulation, the coagulated floc is separated by sedimentation or flotation followed by sand filtration. In the final stage, water is disinfected by chlorination.

The measurement period was organized during 26 May 2011 to 12 June 2011. The period was divided into two parts. In the first part, which consisted of the first 11 days, the raw water was pumped mainly from the wells. In the latter period, the raw water was pumped from the lake. The process was shut down in two week-end periods. The resolution of the process data is 1 h. Thirteen of the process variables were selected manually for further investigation (Table 1).

2.2. Measurement system
A low-cost industrial camera system was installed to learn whether digital images could be used in monitoring the floc properties in the flocculation unit and in characterizing the features of the floc.

Table 1. The process variables used in the analysis

Level of the wells A*	Derivative of lime feed A**
Level of the wells B*	Lime feed B**
Surface/ground water feed ratio	Derivative of lime feed B**
Flow from wells A*	pH in flocculation
Flow from wells B*	pH of treated water
Total flow to process	Pressure cut in filtration
Lime feed A**	

*A and B in the context of wells refer to two separate well groups in different locations.

**A and B in the context of lime feed refer to two separate dosing points of lime.

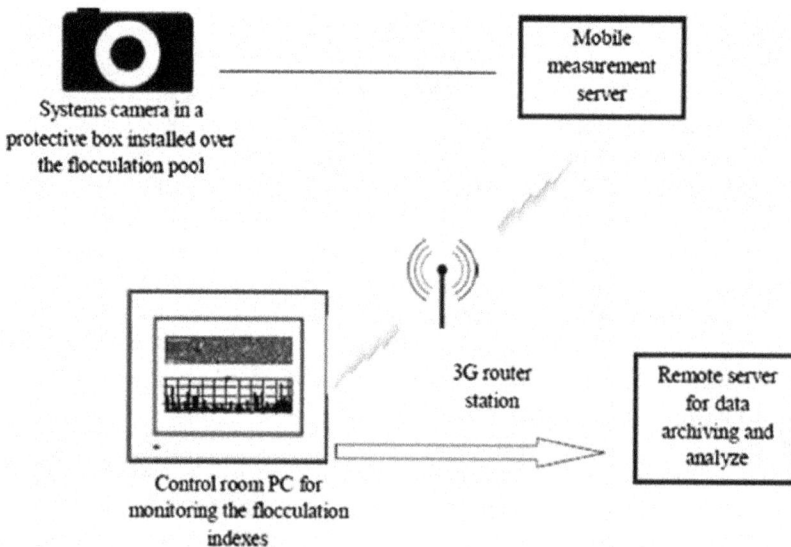

Figure 1. An example implementation of the industrial camera system for monitoring the flocculation in the flocculation unit.

The system includes a commercial systems camera (Nikon D3100+18–55 mm standard objective) and a mobile measuring and analyzing unit for capturing online images from the process. The systems camera is located over the pool of the flocculation unit. The automatically taken images (every 30 min in this case) were recorded on a measurement server and transferred to a remote PC using a wireless connection. In a more sophisticated application, the images could be forwarded to a control room PC, for example, to be viewed and analyzed. The proposed implementation of the system in a water treatment plant is shown in Figure 1.

2.3. Image data and analysis

Images were taken from the process automatically by the system during one month in the summer of 2011. The size of digital images is 4608×3072 pixels. After the measurement period, the system had collected 950 digital images. The resolution of process data is one hour, however, and therefore the floc variables calculated from the digital images were averaged to have the same resolution, and thus the final data-set had 464 rows of data.

There are a large number of image-processing tools, which can be potentially used for preparing digital images for the measurement of the features and structures they reveal (Russ, 2011). Analysis of floc images taken from the process is a relatively difficult problem due to many real-world challenges: the illumination can be poor; floc particles are small and overlap with each other; material is in layers; and so on. Fortunately, computational processing of images makes it possible to solve many of these problems.

Various variables were determined to indicate the size, form, and color of the floc particles. A computer program for analyzing floc was coded on the Matlab (Version 7.10) platform and the image-processing toolbox (Version 7.0). The stages for determining the floc characteristics in a single image are as follows:

(1) Upload digital color image to analysis.

(2) Convert image to grayscale.

(3) Convert grayscale image to binary form (zeros and ones) to detect floc particles using a fixed threshold value (.65).

(4) Find components (i.e. floc objects) connected by pixels using the neighborhood of four pixels.

(5) Calculate desired object properties based on the pixel data.

The following floc properties were calculated for each image:

(1) Average *surface area* of floc particles

(2) The *number* of floc particles

(3) Average *equivalent diameter* of floc particles: the diameter of a circle with the same area as the region

(4) *Eccentricity* of particles: the eccentricity (ratio of the distance between the foci of an ellipse and its major axis length) of the ellipse that has the same second-moments as the region. The value of the eccentricity is between 0 and 1, where 0 is a circle and 1 is a line segment.

(5) Average *perimeter* of floc particles

(6) The average red (*R*), green (*G*), and blue (*B*) levels of the color model in digital images.

(7) *Color index* from the indexed color images

2.4. Modeling methods

2.4.1. Multivariate linear regression
In *MLR* (Cohen, 1968), the purpose is to model the relationship between two and more explanatory variables and a response variable by fitting a linear equation to observed data samples. In principle, the MLR model with observations and variables is given in Equation 1:

$$Y = e + a_0 + a_1X_1 + a_2X_2 + ... + a_nX_n \tag{1}$$

where Y is the value of the response variable, X is the value of the predictor (explanatory) variable, $a_0, ... , n$ equals the unknown coefficient to be estimated, and e signifies the uncontrolled factors and experimental errors of the model. In this case, Y is the average surface area of floc particles, and X_n are the independent explanatory variables used in the model (see Table 1). The fitting works by minimizing the sum of the squares of the vertical deviations from each data point to the line that fits best for the observed data, which is also called least-squares fitting.

In *variable selection* based on MLR, sequential forward selection method was used. In this approach, the variables are included in progressively larger subsets, so that the goodness of the model is maximized. To select p variables from the set P:

(1) Search for the variable that gives the best value for the selected criterion.

(2) Search for the variable that gives the best value with the variable(s) selected in Stage 1.

(3) Repeat Stage 2 until p variables have been selected.

2.5. Self-organizing map
Multivariate measurement data may be difficult to interpret using standard data processing methodology. Advanced descriptive methods such as SOM (Kohonen, 2001) can be useful in detecting nonlinear relationships between data variables. SOM is an unsupervised neural network methodology, which can be used to transform an *n*-dimensional input vector into a lattice (or discrete feature map), which is usually two-dimensional. In SOM, the input vectors sharing common features are projected to the same areal units (i.e. neurons) of the map. Each neuron of the feature map has an *n*-dimensional weight vector, which works as a link between the input and output spaces in such a way that input vectors with common characteristics are assigned to the same or neighboring neurons. This will preserve the topological order of the original input data. The lattice of neurons reflects variations in the statistics of the data-set and highlights common characteristics that approximate to the distribution of the data points. Each neuron includes an *n*-dimensional reference vector (prototype vector), which describes its common properties. The array of neurons (the map) can be illustrated as a rectangular, hexagonal, or even irregular organization, the size of which can be altered depending on the application; the more neurons, the more details are represented. In view of its

ability to compress information, the SOM is an ideal means of analyzing large data-sets typical of industrial processes.

In summary, training of a SOM advances as follows:

(1) Initialize the map.

(2) Find a best-matching unit (BMU) for the input vector using Euclidean distance.

(3) Move the reference vector of the BMU toward the input vector.

(4) Move the reference vectors of the neighboring neurons toward the input vector.

(5) Repeat Steps 2–4 for all input vectors in sequence.

(6) Repeat Steps 2–5 using a smaller learning rate factor (fine-tuning phase).

(7) Find the final BMUs for the input vectors.

The power of a SOM lies in presenting the most characteristic features of a multidimensional data-set in a low-dimensional display. This enables easier detection of multivariate interactions in large data-sets. During the learning phase, the input data vectors are mapped one by one into particular neurons based on the minimal n-dimensional (Euclidean) distances between the input vectors and the reference vectors of the map neurons. Each time a new input vector is located, the reference vectors of the activated neurons are updated, eventually leading to a self-organizing network. In this unsupervised methodology, the SOM can be constructed without previous a priori knowledge (Kohonen, 2001).

In the case study, variance scaling was used for pre-processing the data. A 15×15 hexagonal grid was used in the map, and linear initialization, batch training algorithm, and a Gaussian neighborhood function were used in training.

3. Results and discussion

Monitoring the floc properties is an important but difficult issue in water treatment plants, because of process dynamics, seasonal and episodic variations, and complex chemical reactions, which are partly unknown. A low-cost online system for estimating the size and other features of the floc in the flocculation unit of a water treatment plant has been presented. Here, we present the results from analyzing the collected digital images and the corresponding measurement data from the process. The analysis stage consisted of the following phases. First, the digital images were analyzed to get average values for floc properties. Next, the obtained data were coded to a SOM together with the measurement data obtained from the process. In addition, a multivariate regression model was created for the average surface area of floc determined from the images.

The average floc surface areas (in pixels) calculated from digital images and two sample images are presented in Figure 2. As can be seen, the index indicates both short- and long-term changes during the measurement period. In addition, it can be seen that the floc size is obviously smaller when ground water is treated (period 1) than when surface water is used (period 2).

Furthermore, the calculated floc variables are shown in Figures 3 and 4. Again, we can observe a change in the behavior of several variables when the ground water period is changed into the surface water period. Especially, the average red value of the RGB space drops remarkably after this transition.

In Figure 5, the SOM component planes of the process and image data are presented. As can be seen, there seems to be plenty of similarity between the patterns of eccentricity and the derivatives of lime feed A and B. In other words, the form of the floc particles seems to correlate with the lime feed. Furthermore, the lime feed seems to correlate with the calculated surface area of floc. It can

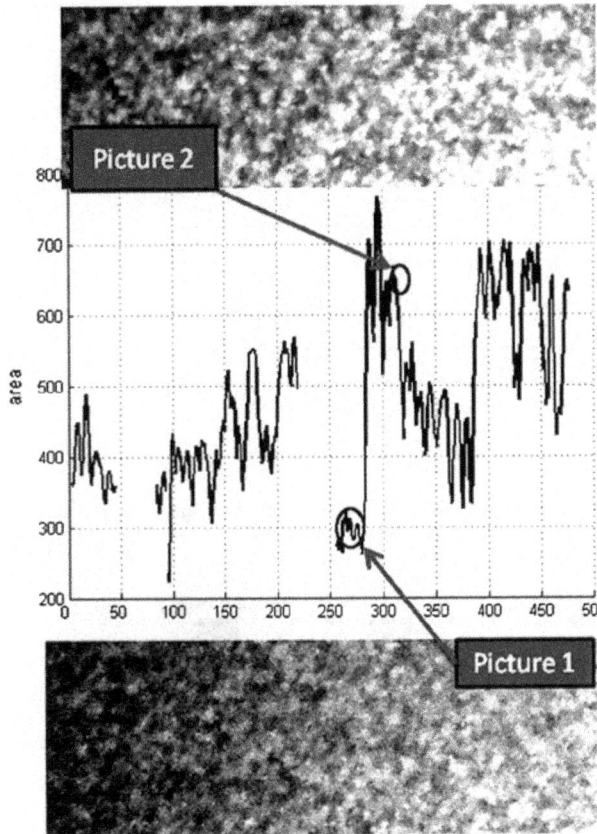

Figure 2. The calculated average values of the floc area (in pixels) and two samples from the test period. number (in pixels).

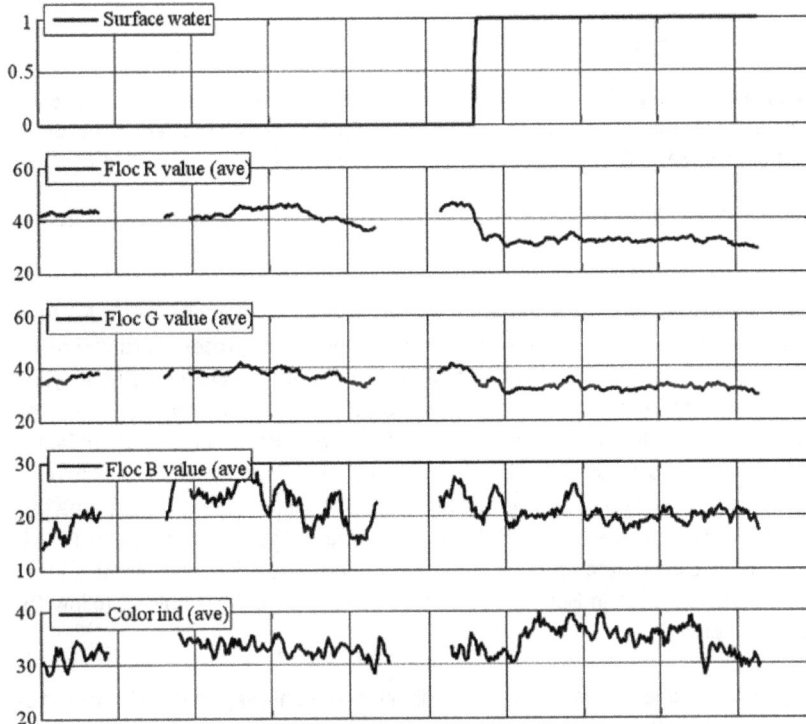

Figure 3. The ratio of surface water during the test period and the calculated average values of the floc diameter, eccentricity, perimeter, and number (in pixels).

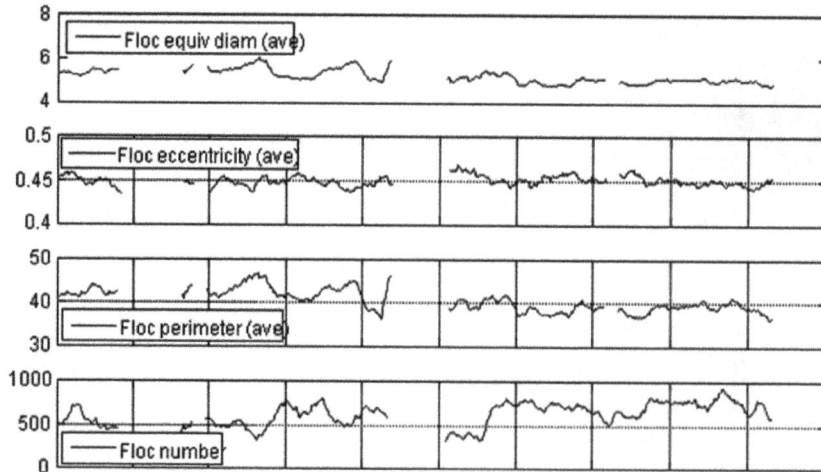

Figure 4. The ratio of surface water during the test period and the calculated average values of the floc R, G, and B levels and the color index value.

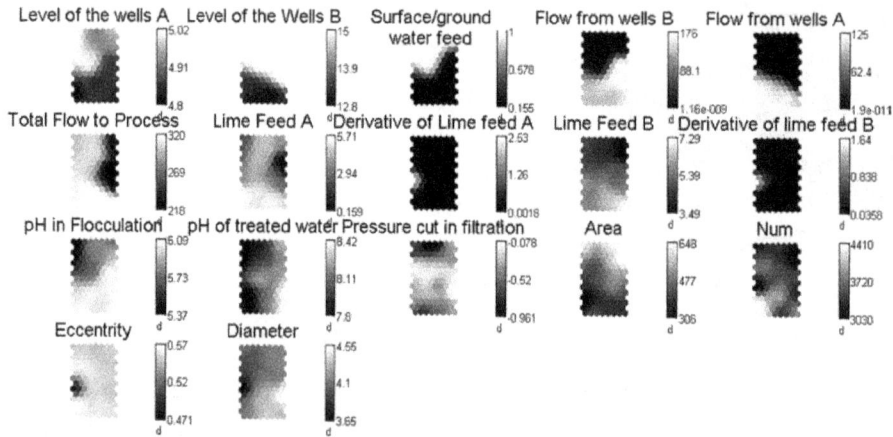

Figure 5. The SOM component planes of process and image data. Darker color tones correspond to low values of variables. Note the similarity between the patterns of eccentricity and the derivatives of lime feed A and B: in case of low eccentricity of flocs, the rate of lime feed has been generally high.

also be seen that the area of floc particles becomes larger, in general, when surface water is treated (see Figure 2).

Next, we present the regression model created for the average surface area of flocs. In Figure 6, the results of variable selection for the average surface area of flocs using process data and multi-variate linear regression are shown. The so-called "cumulative" correlation is presented here, which is produced by the variable selection procedure. In other words, the first correlation value is the correlation of a regression model including only the topmost (i.e. the most important) variable, the second value is the correlation of a model including the two most important variables, and so on. It is notable that the first two of the variables selected are lime feed variables, which confirms that there is a connection between the lime feed and the surface area of the floc particles forming in flocculation. The five variables presented produce a relatively good model for the average surface area ($R \approx .82$). This model can be seen in Figure 7. As can be seen, the model is able to follow the baseline of the calculated value.

Validation of the results of image analysis is of course difficult. The only way in practice would be to use naked eye, and calculate the number of particles manually, for example. The other option would be to use a microscope in parallel with the systems camera, calculate the floc properties manually from the microscope images, and see if the results would match with those given by the

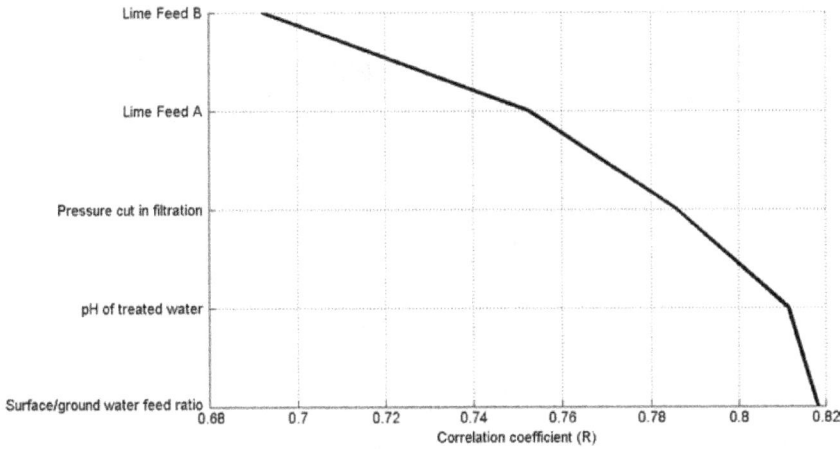

Figure 6. Results of variable selection for the average surface area of floc using process data and mulvariate linear regression.

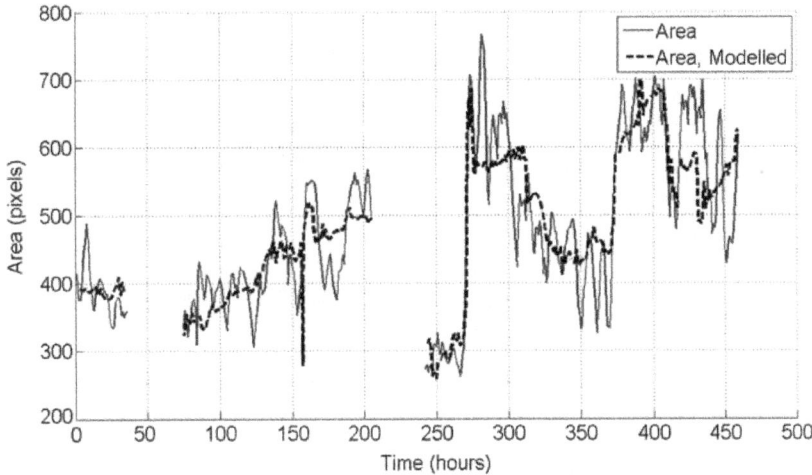

Figure 7. The regression model for the surface area of floc ($R = .82$). The output of the model is the calculated mean area of the floc. The inputs of the model are presented in Figure 6.

approach presented here. This would be extremely difficult in a large scale, however. Therefore, in practice, the only reasonable way to get comparison data for images is to use naked eye (e.g. see Figure 2).

Furthermore, the same method is used for the number and R component (see Figures 8 and 9).

Furthermore, in this case, following conclusions can be made from Figures 6 to 9:

- *Size* and *form* of floc particles: pH important

- *Number* of particles: pH and water source important

- *Color* of floc particles: water source important. For example, the ground water wells were identified as the potential sources of iron, because there is an observable drop in the R value when surface water is taken into use

On the other hand, when monitoring the process we should be more interested in changes occurring in the floc quality, after all, and not so much on the absolute values of its different properties. The quality parameters calculated from digital images offer one possible tool for monitoring these changes, and it can be used (1) for diagnosing the reasons for changes when combined with corresponding process data and (2) as the basis of a warning system, which would alarm when either

Model for the *number* of floc particles

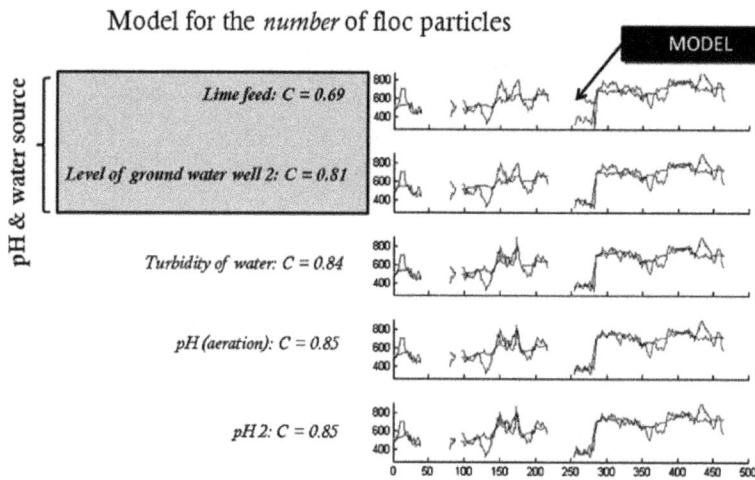

Figure 8. The regression model for the number of floc (R = .85) with the five most affecting variables.

Model for R component (*redness*) of images

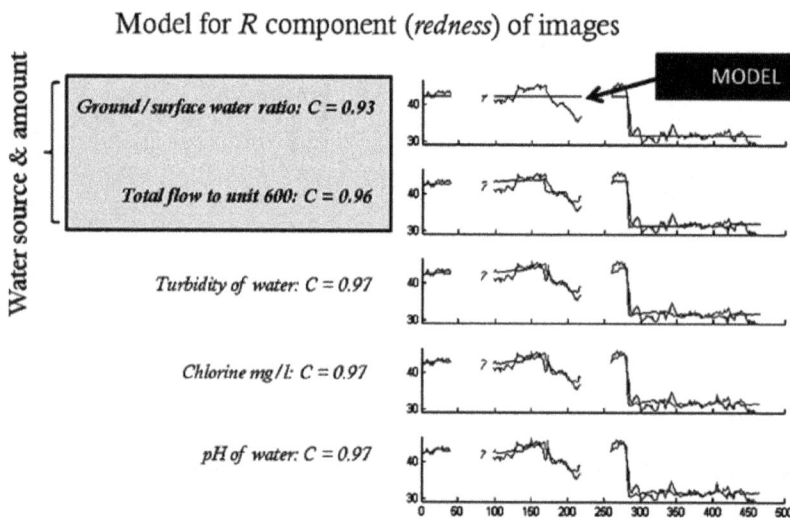

Figure 9. The regression model for the R component (R = .85) with the five most affecting variables.

sudden or long-term changes occur in the process. The system also has potential in real-time use, because the processing of images and calculation of floc property parameters are not computationally heavy. It would be possible to install the monitoring system into the control room of a water treatment plant, as presented in Figure 1, where an online image would be shown together with a long-term floc property indexes.

Models for floc variables vary from moderate to good; there are clear connections between floc properties and the process. The results show that especially the average surface area and eccentricity of the floc seem to be the most interesting quality parameters in this case. The source of the raw water (i.e. ground or lake water) seems to have an effect on the size of the floc particles. The changes in the surface area can be explained using five variables, two of which are variables describing lime feed. In addition, the results suggest that the eccentricity of the floc particles seems to provide an indicator of the changes in the process. Especially, the lime feed seems to have an influence on the formation or breaking of the floc particles. This kind of information is extremely important for the planning of the pilot tests. When we can select the right parameters and range, resources, such as time and money, can be saved.

Furthermore, the most important issue for the water treatment plant is: what kind of floc is best for the plant? This should be evaluated case-specifically by comparing with quality variables of purified water, but the usual problem is the lack of proper measurement data. However, the camera analysis is a potential way of studying this.

4. Conclusions

A low-cost inspection system for estimating floc properties in the flocculation unit of a water treatment plant has been developed. The benefits of the system include:

- Only a single calibration is required (focusing of the camera).
- Camera and other instruments are all commercially available, which ensures low material costs.
- The system is compact and easy to install.
- Since the enclosure is pressurized and sealed, the system is applicable to dusty, dirty, and humid industrial environments, as long as the window through which the images are captured is kept clean.

The main conclusions from testing the industrial camera system in the flocculation unit and analyzing the data are as follows:

- Image analysis enables the monitoring of different floc properties and therefore indirect estimation of floc quality.
- The system enables both online and long-term monitoring, because it provides online information on the process, and trend lines can be used to monitor changes occurring during a longer period of time.
- The system can be programmed to alarm in case there are unwanted trends in any of the quality parameters. A warning signal could be delivered to process operators, so that they could check the condition of the floc by naked eye.
- Based on the preliminary results of data analysis, it seems that there are dependencies between the surface area of the floc and certain process measurements, which suggests that it is possible to create data-based models for floc quality.
- Quality models can reveal interesting factors affecting flocculation and help in studying physical phenomena behind the complex process.

Funding

The writing of this paper was supported by Maa- ja Vesitekniikan tuki Ry. The material on which it is based was produced in the POLARIS project financed by the Finnish Funding Agency for Technology and Innovation (Tekes), which the authors thank for its financial support. In addition, Mika Liukkonen is grateful to the Finnish Cultural Foundation for financial support.

Author details

Petri Juntunen[1]
E-mail: petri.juntunen@kuopionvesi.fi
Mika Liukkonen[1]
E-mail: mika.liukkonen@uef.fi
Markku Lehtola[1]
E-mail: markku.lehtola@kuopionvesi.fi
Yrjö Hiltunen[1]
E-mail: yrjö.hiltunen@uef.fi
[1] Department of Environmental Science, University of Eastern Finland, Kuopio FI-70210, Finland.

References

Alhoniemi, E., Hollmen, J., Simula, O., & Vesanto, J. (1999). Process monitoring and modeling using the self-organizing map. *Integrated Computer-Aided Engineering, 6*, 3–14.

Alonso, A., Negro, C., Blanco, A., & San Pío, I. (2009). Application of advanced data treatment to predict paper properties. *Mathematical and Computer Modelling of Dynamical Systems, 15*, 453–462. http://dx.doi.org/10.1080/13873950903375445

Avelin, A., Jansson, J., Dotzauer, E., & Dahlquist, E. (2009). Use of combined physical and statistical models for online applications in the pulp and paper industry. *Mathematical and Computer Modelling of Dynamical Systems, 15*, 425–434. http://dx.doi.org/10.1080/13873950903375403

Bratby, J. (2006). *Coagulation and flocculation in water and wastewater treatment.* Cornwall: IWA.

Chakraborti, R. K., Atkinson, J. F., & Van Benschoten, J. E. (2000). Characterization of alum floc by image analysis. *Environmental Science & Technology, 34*, 3969–3976. http://dx.doi.org/10.1021/es990818o

Cohen, J. (1968). Multiple regression as a general data-analytic system. *Psychological Bulletin, 70*, 426–443. http://dx.doi.org/10.1037/h0026714

Hong, Y. S., Bhamidimarri, S., & Charleson, T. (1998). A genetic adapted neural network analysis of performance of the nutrient removal plant at Rotorua. In *Proceedings of Institute of Professional Engineers New Zealand (IPENZ) Annual Conference. Simulation and control section, 2,* 43.

Juntunen, P., Liukkonen, M., Pelo, M., Lehtola, M., & Hiltunen, Y. (2011). Cluster analysis of a water treatment process by self-organizing maps. In *Watermatex 2011: Conference Proceedings.* London: IWA.

Juntunen, P., Liukkonen, M., Pelo, M., Lehtola, M., & Hiltunen, Y. (2013, April). Dynamic soft sensors for detecting factors affecting turbidity in drinking water. *Journal of Hydroinformatics, 15,* 416.

Kalteh, A., Hjorth, P., & Berndtsson, R. (2008). Review of the self-organizing map (SOM) approach in water resources: Analysis, modelling and application. *Environmental Modelling and Software, 23,* 835–845. http://dx.doi.org/10.1016/j.envsoft.2007.10.001

Kohonen, T. (2001). *Self-organizing maps* (3rd ed.). Berlin Heidelberg, NY: Springer-Verlag. http://dx.doi.org/10.1007/978-3-642-56927-2

Kvech, S., & Edwards, M. (2002). Solubility controls on aluminium in drinking water at relatively low and high pH. *Water Research, 36,* 4356–4368. http://dx.doi.org/10.1016/S0043-1354(02)00137-9

Letterman, R. D. (Ed.). (1999). *Water quality & treatment— Handbook of community water supplies.* Denver, CO: AWWA.

Liukkonen, M., Havia, E., Leinonen, H., & Hiltunen, Y. (2011). Quality-oriented optimization of wave soldering process by using self-organizing maps. *Applied Soft Computing, 11,* 214–220. http://dx.doi.org/10.1016/j.asoc.2009.11.011

Liukkonen, M., Hiltunen, T., Hälikkä, E., & Hiltunen, Y. (2011). Modeling of the fluidized bed combustion process and NOx emissions using self-organizing maps: An application to the diagnosis of process states. *Environmental Modelling & Software, 26,* 605–614.

Maier, H. R., Morgan, N., & Chow, C. W. K. (2004). Use of artificial neural networks for predicting optimal alum doses and optimal water quality parameters. *Environmental Modeling and Software, 15,* 105–124.

Mays, L. (2004). *Water supply systems security.* Tempe, AZ: McGraw-Hill.

Muruzábal, J., & Muñoz, A. (1997). On the visualization of outliers via self-organizing maps. *Journal of Computational and Graphical Statistics, 6,* 355–382.

Perez, Y. G., Leite, S. G. F., & Coelho, M. A. Z. (2006). Activated sludge morphology characterization through an image analysis procedure. *Brazilian Journal of Chemical Engineering, 23,* 319–330.

Russ, J. C. (2011). *The image processing handbook* (6th ed.). Boca Raton, FL: CRC Press.

Valentin, V., & Denoeux, T. (2001). A neural based software sensor for coagulation control in a water treatment plant. *Intelligent Data Analysis, 5,* 23–39.

van Benschoten, J. E., & Edzwald, J. K. (1990). Chemical aspects of coagulation using aluminium salts-I. Hydrolytic reactions of alum and polyaluminium chloride. *Water Research, 24,* 1519–1526. http://dx.doi.org/10.1016/0043-1354(90)90086-L

Wang, Y., Lu, J., Du, B., Shi, B., & Wang, D. (2009). Fractal analysis of polyferric chloride-humic acid (PFC-HA) flocs in different topological spaces. *Journal of Environmental Sciences, 21,* 41–48. http://dx.doi.org/10.1016/S1001-0742(09)60009-7

15

Integration of a prototype wireless communication system with micro-electromechanical temperature and humidity sensor for concrete pavement health monitoring

Shuo Yang[1], Keyan Shen[2], Halil Ceylan[1]*, Sunghwan Kim[1], Daji Qiao[2] and Kasthurirangan Gopalakrishnan[1]

*Corresponding author: Halil Ceylan, Department of Civil, Construction & Environmental Engineering (CCEE), Iowa State University of Science and Technology, Ames, IA 50011-3232, USA

E-mail: hceylan@iastate.edu

Reviewing editor: Amir H. Alavi, Michigan State University, USA

Abstract: In recent years, structural health monitoring and management (SHMM) has become a popular approach and is considered essential for achieving well-performing, long-lasting, sustainable transportation infrastructure systems. Key requirements in ideal SHMM of road infrastructure include long-term, continuous, and real-time monitoring of pavement response and performance under various pavement geometry-materials-loading configurations and environmental conditions. With advancements in wireless technologies, integration of wireless communications into sensing device is considered an alternate and superior solution to existing time- and labor-intensive wired sensing systems in meeting the requirements of an ideal SHMM. This study explored the development and integration of a wireless communications sub-system into a commercial off-the-shelf micro-electromechanical sensor-based

ABOUT THE AUTHOR

Halil Ceylan is an associate professor in the Department of Civil, Construction and Environmental Engineering (CCEE), Director of Program for Sustainable Pavement Engineering and Research (PROSPER), and Pavement Research Engineer at the Institute for Transportation (InTrans) at Iowa State University (ISU). He received his PhD in Civil Engineering in 2002 from University of Illinois at Urbana-Champaign, USA. He has extensive experience in non-destructive testing and health monitoring of transportation infrastructure systems; mechanistic-based pavement analysis and design; engineering applications of intelligent data mining and computational intelligence techniques. He has been involved with over 65 research projects most of which are related to pavement systems and pavement materials. He has authored over 200 publications and delivered over 150 technical presentations which includes over 90 invited lectures and presentations at a number of universities and technical conferences and meetings. He has served as editorial board member/editor of nine international journals and as a member of more than 20 national and international committees and organizations.

PUBLIC INTEREST STATEMENT

Frequent monitoring of health of any system (human or man-made) is important to ensure that it is functioning optimally and to identify any corrective measures. Concrete pavement systems are no different. Traditionally, the structural health of such systems have been monitored using embedded sensors connected to data acquisition systems through long wires. However, the use of wired sensors can be very time-consuming and costly, especially if a large number of sensors have to be installed. Recent advancements in wireless technologies and development of miniature (micro-scale) sensing devices have the potential to enable the promise of smart pavement health monitoring systems. This paper describes the development and field performance of a wireless micro-electromechanical sensor system (MEMS) for monitoring concrete pavement temperature and relative humidity.

concrete pavement monitoring system. A success-rate test was performed after the wireless transmission system was buried in the concrete slab, and the test results indicated that the system was able to provide reliable communications at a distance of more than 46 m (150 feet). This will be a useful feature for highway engineers performing routine pavement scans from the pavement shoulder without the need for traffic control or road closure.

Subjects: Engineering & Technology; Concrete & Cement; Pavement Engineering; Transportation Engineering

Keywords: pavement; structural health monitoring; wireless; sensor; temperature; humidity; concrete; MEMS

1. Introduction

Like many advanced technologies, wireless sensor technologies were initially developed and deployed for military and industrial purposes (SILICON LABS, 2015). In recent years, these kinds of technologies are extensively applied in civil engineering infrastructure to measure the changes in material and geometric properties for serviceability assessment, which is referred to as structural health monitoring (SHM). Over past decades, SHM has been widely used in civil engineering infrastructure to monitor structural integrity failures such as cracks, structural deterioration, and steel corrosion. An early warning could avoid unnecessary costs to the maintenance programs. Moreover, continuously measured data can contribute to improved modeling and analytics resulting in prolonged system service life and reduced life cycle costs (Barroca et al., 2013; Buenfeld, Davis, Karmini, & Gilbertson, 2008; McCarter & Vennesland, 2004). Wired sensor systems are widely used in traditional SHM to detect structural damage (Lynch, 2002). However, the use of wired sensors can be very time-consuming and costly if a large number of sensors have to be installed to improve quality of measured data in SHM (Cepero, 2013; Cho et al., 2008; Lynch, Sundararajan, Law, Kiremidjian, & Carryer, 2003). Furthermore, in cases where wires are buried in concrete, wires may be corroded or damaged (Cepero, 2013; Cho et al., 2008; Lynch et al., 2003). Due to these drawbacks, the use of wireless technologies is considered to be a promising substitute to provide better functionality at a lower price especially when a higher spatial density of sensors is desired (Kim et al., 2007). In addition, micro-electromechanical sensors and systems (MEMS) technology has been investigated for SHM since MEMS make it possible for systems of all kinds to be smaller, faster, more energy-efficient, and less expensive (Ceylan et al., 2011, 2013).

Numerous studies have been conducted to apply wireless sensor technologies in bridge system SHM (Maser, Egri, Lichtenstein, & Chase, 1996; Loh, Zimmerman, & Lynch, 2007; Lynch, 2007; Lynch & Loh, 2006). However, only few recent studies have investigated these technologies in pavement SHM applications. For instance, Lajnef, Chatti, Chakrabartty, Rhimi, and Sarkar (2013) focused on development of a wireless strain sensing system for asphalt pavement SHM to detect fatigue damage. The sensor system developed in this study contained a low-power consumption wireless integrated circuit sensor interfaced with a piezoelectric transducer. This piezoelectric ceramic transducer was designed with an array of ultra-low power floating gate (FG) computational circuits and it could generate power to supply FG analog processor in the sensor under stress. Each sensor node could store the data and then periodically transmit them to radio frequency (RF) reader mounted on a moving vehicle (Lajnef, Rhimi, Chatti, Mhamdi, & Faridazar, 2011).

The objective of this current study is to investigate the feasibility of developing wireless-based MEMS for concrete pavement SHM. A wireless network system was integrated with an off-the-shelf MEMS sensor, originally designed and configured for wired data acquisition. The field performance of commercial wired MEMS sensors was evaluated in a newly constructed concrete pavement under actual traffic load and weather conditions to identify the system requirements for development of the wireless MEMS sensor system. A preliminary design of the prototype wireless system with robust packaging was developed to improve the survivability of MEMS sensors. The wireless system utilized

XBee-PRO modules interfaced with Arduino boards (http://pdf1.alldatasheet.com/datasheet-pdf/view/530828/TI/TPL5000.html) to build the transmission system based on ZigBee protocol. Detailed discussions and findings pertaining to the development of wireless-based MEMS are discussed.

2. Evaluation of commercial off-the-shelf wired MEMS sensors

2.1. Sensor description

Temperature and moisture content are significant factors in the hydration process between cementitious materials and water, which in turn influence early-age concrete properties (Norris, Saafi, & Romine, 2008; Saafi & Romine, 2005). Anomalies in the hydration process may result in insufficient strength and durability since the development of early concrete strength mainly depends on the moisture diffusion and hydration temperature (Choi & Won, 2008; Every, Faridazar, & Deyhim, 2009; Ye, Zollinger, Choi, & Won, 2006). Furthermore, concrete pavement can be subjected to deformation due to different temperature and moisture gradients throughout the concrete slab, commonly referred to as curling (temperature) and warping (moisture/humidity) behaviors. This, when combined with heavy traffic loading, could lead to cracking of slabs. Considering the significant impact of temperature and moisture conditions on the overall concrete pavement response and performance (Ruiz, Rasmussen, Chang, Dick, & Nelson, 2005), these two properties were selected for investigation in this study. Continuous wireless monitoring and communication of temperature and moisture changes within in-service concrete pavements can provide an early warning and alert the highway engineers/agencies regarding their potential for structural integrity failure. This can enable the selection of appropriate pavement distress mitigation and preventive strategies resulting in sustainable and durable pavement systems.

The Sensirion SHT71 digital humidity sensor, classified as a commercial off-the-shelf MEMS device that can simultaneously measure relative humidity (RH) as well as temperature, was evaluated in this study. Note that moisture content measured inside concrete is typically expressed as RH which refers to the ratio of moisture content of air compared to saturated moisture level at the same temperature and pressure (Ye et al., 2006).

The commercial MEMS digital humidity sensor integrates sensor elements coupled with signal processing circuitry on a silicon chip by MEMS technology to provide a fully calibrated digital output. A unique capacitive sensor element consisting of paired conductors is built out of the capacitor of MEMS sensor to capture humidity while another band-gap sensor measures temperature. These conductors are separated by a polymer dielectric that can absorb or release water proportional to the relative environmental humidity, and thus can change the capacitance of the capacitor (Sensirion Inc., 2014, http://www.sensirion.com/en/technology/humidity/). An electronic circuit calculates RH by measuring the capacitance difference. Additionally, the capacitance for the chip of this MEMS sensor is formed by a "micro-machined" finger electrode system with different protective and polymer cover layers, which can simultaneously protect the sensor from interference as well. However, in order to continuously monitor and store measurement data, MEMS sensors have to be connected with a data reader of evaluation kit EK-H4 (see Figure 1) and a computer which require power (battery) supply all the time.

2.2. Field instrumentation and findings

A set of four wired commercial MEMS RH/temperature (RH/T) sensors were instrumented in a newly constructed jointed plain concrete pavement (JPCP) in a US-30 highway section near Ames, Iowa, USA. The instrumented JPCP, constructed at 8:00 am on 24 May 2013, consists of 254 mm (10 inch) thick concrete slab with approximately 6 m (20 feet) transverse joints spacing. The passing lane and travel lane widths for this JPCP are 3.7 m (12 feet) and 4.3 m (14 feet), respectively. A 152 mm (6 inch) thick Hot-Mix Asphalt (HMA) shoulder was constructed approximately 28 days after the paving of concrete. A set of wireless temperature sensors and longitudinal strain gauges were installed in the same section along with the commercial MEMS RH/T sensors for another series of investigations, not part of this study.

(a)

(b)

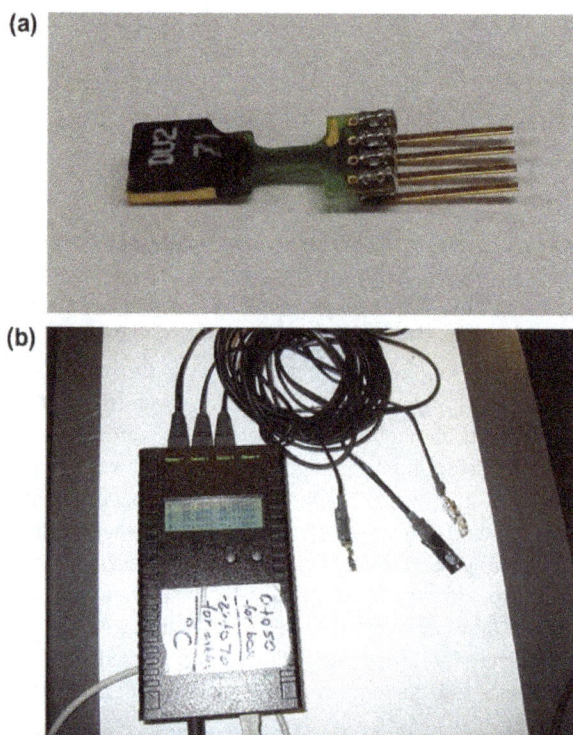

Figure 1. Sensors evaluated in this study: (a) Sensirion SHT71 sensor; and (b) Sensirion evaluation kit EK-H4.

Before the paving of concrete took place, the RH/T sensors were tied on to short wood sticks installed on top of the base course. As seen in Figure 2, all the cables/wires from the sensors were tied together and then placed in a polyvinyl chloride (PVC) pipe buried underground to protect them from damage during concrete paving operations. The cables in the PVC pipe were connected to a data acquisition system (DAS) equipment (laptop, data logger, evaluation kit, and batteries) in a plastic shield box placed near the drainage ditch away from the HMA shoulder (see Figure 3). The installation of these wired sensors required great care, was time consuming and labor-intensive.

Figure 4 illustrates measured temperature and RH profiles captured by wired MEMS RH/T sensors one month after concrete paving. This figure shows measurements from MEMS RH/T sensor No. 3 (installed at 51 mm (2 inch) below pavement surface and 711 mm (28 inch) away from shoulder) and MEMS RH/T sensor No. 4 (installed at 2.5 mm (0.1 inch) below pavement surface and 203 mm (8 inch) away from shoulder). Among the four sensors installed before paving, two sensors (No. 3 and 4) remained functional in measuring temperature inside concrete while one sensor (No. 4) measured only RH of concrete.

Figure 2. Sensor instrumentation on US-30 highway section near Ames, Iowa before paving of concrete.

Figure 3. Data acquisition system.

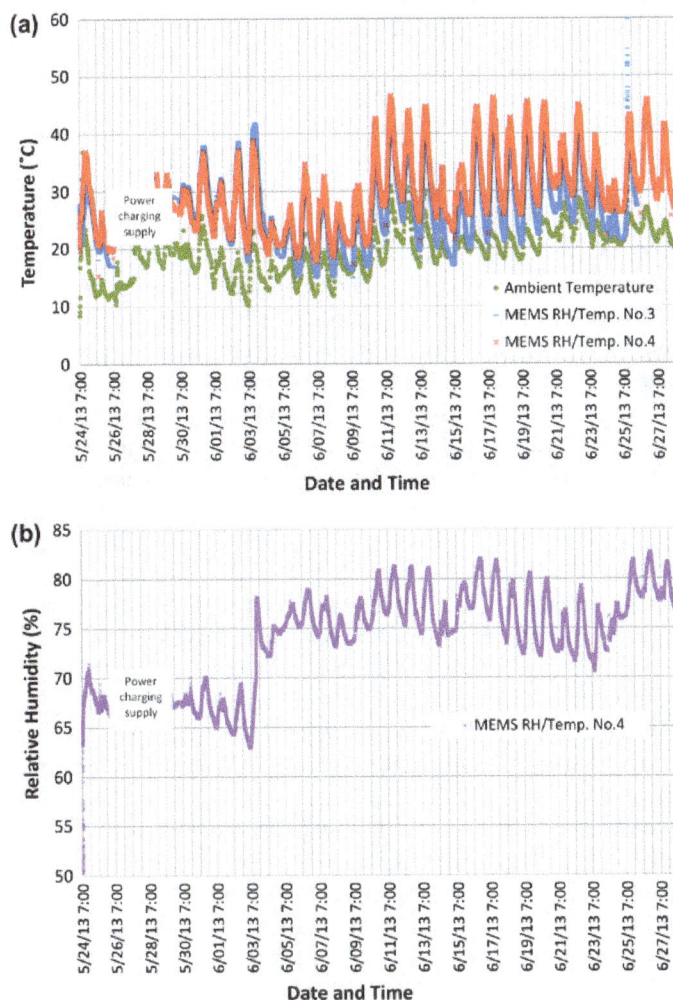

**Figure 4. Commercial off-the-shelf wired MEMS RH/Temperature sensor measurements:
(a) Temperature profile; and (b) Relative humidity profile.**

The other two sensors, No. 1 and 2 (installed at 216 mm (8.5 inch) and 140 mm (5.5 inch) below pavement surface and 711 mm (28 inch) away from shoulder), malfunctioned several hours after concrete paving operations. This could probably be attributed to damages incurred to the wires/cables from concrete paver and vibrator operations. The sensor itself could also have been damaged because of the high alkali environment prevailing during concrete hydration. Data could not be acquired from 26 May to 28 May 2013 since the battery (power supply) for the DAS was not recharged. These practical constraints and limitations of wired sensor systems with respect to continuous monitoring and storage of measured data highlight the need for a self-powered, wireless sensor system.

2.3. Lessons learned from field evaluation

The on-site experiences from US-30 highway sensor installation and monitoring and the identified limitations of wired sensor systems proved to be resources in identifying the system requirements in the development of wireless MEMS sensor systems. These limitations include time-consuming and labor-intensive installation process, poor sensor survivability caused by cable damage, and complicated sensor packaging required to protect sensor from high alkali environment during concrete hydration.

Critical factors to be considered in wireless MEMS sensor systems include hardware architecture, packaging, embedded software, wireless signal strength, and low-power consumption under on-site conditions. Considering these critical factors, a preliminary wireless system with a robust packaging for MEMS sensors was developed.

3. Development of a prototype wireless communication system

3.1. Overview

The wireless communications system presented in this study is a preliminary design mainly focusing on the wireless transmission function. In this study, an Institute of Electrical and Electronics Engineers (IEEE) standard-based wireless system was utilized because of both its low price and its low power consumption. A MEMS digital humidity sensor was used as sensing unit; this pin-based sensor had no packaging for its sensing element so an additional robust packaging system was also required. The wireless system could be divided into two parts: wireless transmission end and wireless receiving end. The wireless transmission end is used to transfer data from MEMS sensors into wireless transmission devices. The wireless receiving end connected with computer is used to download data without the need for a wire. Microcontrollers and XBee-PRO modules are required for the transmission and receiving ends to communicate with each other.

3.2. Wireless network protocol

Wireless network protocols are used to define or standardize the rules and conventions for communication between devices (Lee, Su, & Shen, 2007). The wireless system implemented in this design was ZigBee, used to construct a decentralized self-healing wireless mesh network. In this mesh network, nodes can find a new route when an original route fails (Texas Instruments, 2013). ZigBee uses an IEEE 802.15.4-based protocol; in addition to ZigBee, there are also other possibilities, including Bluetooth, Wi-Fi, cellular, etc. Table 1 gives a comparison between different wireless technologies by evaluating their total scores derived from weighted scores considering various aspects such as data rate, range, and energy consumption. In this table, the weighted score of each aspect is calculated by multiplying its weight by the score of each specific wireless technology; a higher total score represents a better wireless technology for this application. Based on this table, it can be seen that ZigBee is more energy-efficient, cost-effective, and easier to work with than the other technologies.

3.3. Microcontrollers

The Arduino board is a single-board microcontroller employing an Atmel AVR® 8-bit or 32-bit microcontroller that can be wirelessly programmed by a device utilizing the ZigBee protocol (Atmel, 2014, http://www.atmel.com/products/microcontrollers/avr/). In the present study, Arduino Uno and Arduino Mega 2560, respectively, shown in Figure 5, were used for wireless transmitter and receiver.

Table 1. Comparison of wireless technologies (Al-Khatib et al., 2009)					
Aspects		Score (0–10, with 0 being the worst and 10 being the best)			
Factors	Weight	Bluetooth®	ZigBee®	Wifi®	Cellular
Multi-node network support	100	5	10	10	10
Throughput	60	7	6	8	3
Data rate	60	7	6	10	10
Range	50	6	5	7	10
Ease of implementation	50	6	8	6	4
Power consumption	−80	6	2	8	6
Cost	−100	5	3	7	8
Total score		460	910	390	200

(a)

(b)

Figure 5. Microcontrollers used in this study: (a) Arduino Uno for wireless transmission end; and (b) Arduino Mega 2560 for wireless receiving end.

The Arduino Uno is a microcontroller using an ATmega328 processor with 32 KB of flash memory, 2 KB of static random-access memory (SRAM), and 1 KB of electrically erasable programmable read-only memory (EEPROM). The board has 14 digital input/output pins, 6 analog inputs, a 5-volt linear regulator, a 16 MHz ceramic resonator, a USB connection, a power jack, an In-Circuit Serial Programming header, and a reset button. The Arduino Mega 2560 is similar to the Arduino Uno but it has an ATmega2560 processor with 54 digital input/output pins, 16 analog inputs, 4 hardware serial ports (UARTs), and a 16 MHz crystal oscillator. The Arduino Mega 2560 is compatible with most shields designed for the Arduino Duemilanove or Diecimila and it has 256 KB of flash memory, 8 KB of SRAM, and 4 KB of EEPROM for storing code and data. These two microcontrollers were selected because of their high reliability and low cost. Furthermore, Arduino 1.0.4 (open-source software) can be used for program coding such as setting up a time interval and changing the format of exported data, to control both the Arduino Uno and the Arduino Mega 2560.

3.4. XBee-PRO modules

XBee-PRO RF module (series 1) as shown in Figure 6 is a wireless device, which can offer low-cost wireless connectivity in ZigBee (DIgI, https://www.digi.com/technology/rf-articles/wireless-zigbee) mesh networks. It is reliable in point-to-point, multipoint wireless transmission and it is designed to meet the IEEE 802.15.4 standard. Furthermore, XBee-PRO module also has an easy setup process and the software used to program is called X-CTU, which adjusts its frequency, signal strength, energy consumption, and so on. Additionally, an XBee Explorer Regulated board can be used to help it regulate the voltage input.

3.5. Wireless transmission

The wireless transmission end, as shown in Figure 7, consists of a MEMS sensor (SHT71 digital humidity and temperature sensor), an XBee-PRO module, an XBee Explorer Regulated, an Arduino Uno microcontroller, and 12 × 1.5 V AA batteries. Among these devices, XBee Explorer Regulated is a board that can be pinned on XBee-PRO to help it regulate the voltage input. In the wireless transmission end, both the SHT71 sensor and XBee-PRO with XBee Explorer Regulated were pinned on the digital port and power port on Arduino Uno board. Meanwhile, 12 1.5 V AA batteries were placed in

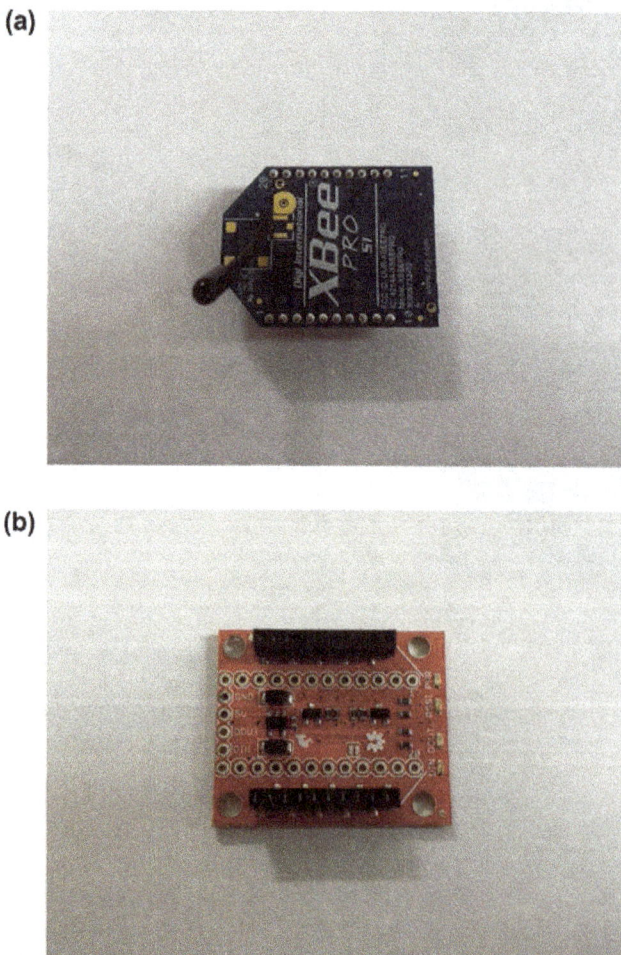

(a)

(b)

Figure 6. Wireless transmission device: (a) XBee-PRO; and (b) XBee Explorer Regulated.

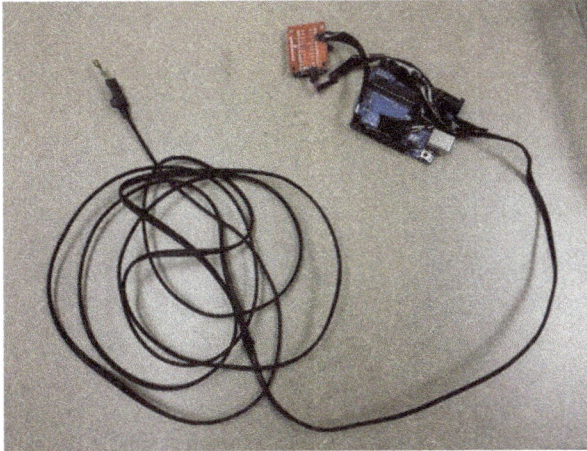

Figure 7. Wireless transmission end without batteries.

a plastic holder connected with the microcontroller to power the entire wireless transmission end through the voltage output pin on the board. Furthermore, because the entire wireless transmission end will be buried in concrete, a robust packaging framework is needed for the wireless transmission system which will be discussed later.

3.6. Wireless reception

The wireless receiving end, as shown in Figure 8, consists of an XBee-PRO module, an XBee Explorer Regulated, and an Arduino Mega 2560 microcontroller. The XBee Explorer Regulated here plays the same role as it was used in the wireless transmission end. However, there were no batteries on the Arduino Mega 2560 because it was powered by computer through a USB cable. The XBee-PRO on the Arduino Mega 2560 was paired with the other XBee-PRO on the Arduino Uno in the wireless transmission end to receive the transmitted data. After that, the data will be stored in a data-storage module with 4096 bytes non-volatile memory on the Arduino Mega 2560.

3.7. Packaging

Robust packaging is required to protect both the sensor and wireless transmission devices like the XBee-PRO module and the microcontroller to ensure that they work properly inside the concrete. The packaging functions include protecting the wireless transmitter during sensor installation and pavement construction processes, protecting the sensor from alkali-cement hydration reaction, and protecting the wireless transmitter under harsh climate and traffic conditions.

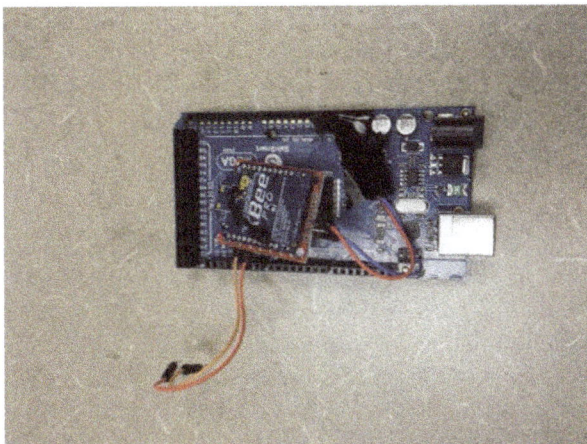

Figure 8. Wireless receiving end.

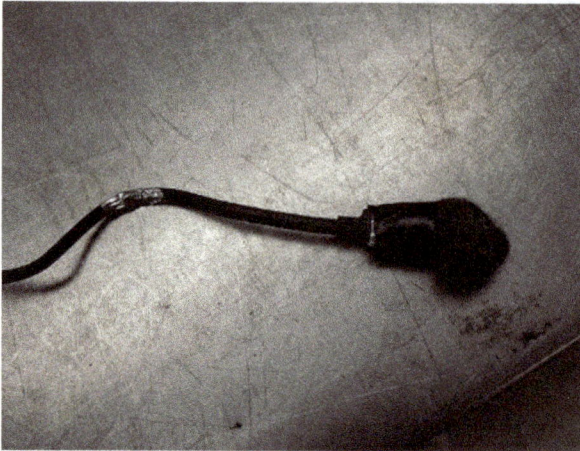

Figure 9. Packaged MEMS sensor.

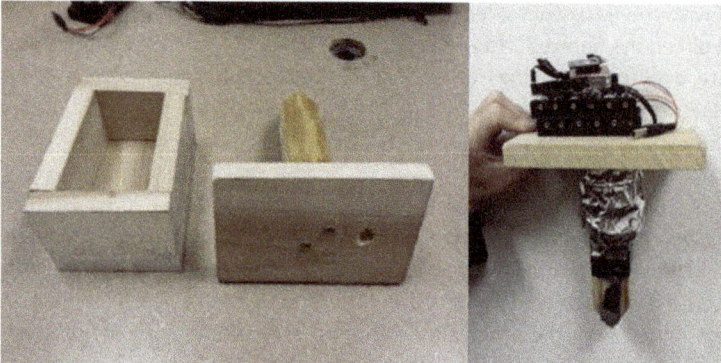

Figure 10. Protective packaging for wireless transmission end.

Two kinds of in-house packaging were designed to protect the sensor, microcontroller and XBee-PRO module, respectively. For the MEMS sensor, a piece of adhesive tape, a protection filter cap, and steel wool were used to make the protective package to prevent direct contact between the raw sensor and fresh concrete. In this packaging, a filter cap was placed on the top of the MEMS sensor using adhesive tape. Steel wool was used to attach the sensor (Figure 9). As for the microcontroller and XBee-PRO module, a small box with the bottom open, consisting of 12 mm thick wood board and a wood board nailed with a 180 mm long sharp-edged wood stick, was prepared. A hole was drilled on the board nailed with the stick to allow the cable from the sensor to go through to connect the Arduino Uno microcontroller. The size of the box was 160 mm × 105 mm × 88 mm (6.3″ × 4.1″ × 3.5″) which was sufficient to place the entire wireless transmission system, as shown in Figure 10. Silicon glue and adhesive tape were used as well to seal the small gap in the box.

4. Evaluation of developed prototype wireless communication system

4.1. Working principle

The data-exchange principle of this wireless system is based on the ZigBee protocol. This system requires no external cables. When it is turned on, the MEMS sensor will sense temperature and RH and transfer that data to the XBee-PRO through the Arduino Uno microcontroller. Then the XBee-PRO, using the wireless transmitter, will transmit data to the paired XBee-PRO at the wireless receiver through an antenna; this data will be stored in the Arduino Mega 2560, so the wireless receiver and a computer must be placed nearby because only the Arduino Mega 2560 microcontroller is used to store data in this wireless system. The data can finally be downloaded to the computer through software called "CoolTerm," (http://download.cnet.com/CoolTerm/3000-2383_4-10915882.html#ixzz2ueTEGqCd) a simple freeware serial port terminal application without terminal emulation that supports data exchange with hardware connected through serial ports (Sparkfun Electronics, 2014, https://learn.sparkfun.com/tutorials/terminal-basics/coolterm-windows-mac-linux). Temperature, relative humidity, and dew point are the data elements exported from the system.

4.2. Comparison between wired and wireless MEMS system

Figure 11 provides an overall system-level comparison between wired MEMS system and wireless MEMS system developed. In the wired MEMS system, the sensor must be connected to the data reader and the computer through cables to continuously monitor concrete properties and the data, so both the data reader and the computer require an electrical power supply. However, the implemented wireless system requires no external cables and can thereby save installation time and reduce the risk of sensor malfunction.

4.3. Evaluation of wireless communication capacity

To test the reliability and survivability of the wireless communication system inside the concrete, both wireless transmitter and receiver were embedded in concrete as shown in Figure 12 to conduct a success-rate test. Success rate refers to the success rate of data transmitted from the transmitter in successfully reaching the receiver. The higher this rate, the more reliable the system will be.

The success-rate test was conducted for wireless MEMS system inside concrete buried underground by increasing horizontal and vertical distances between wireless transmission and receiving ends (see Figure 13). Figure 14 illustrates success rates (in transmitting concrete pavement temperature

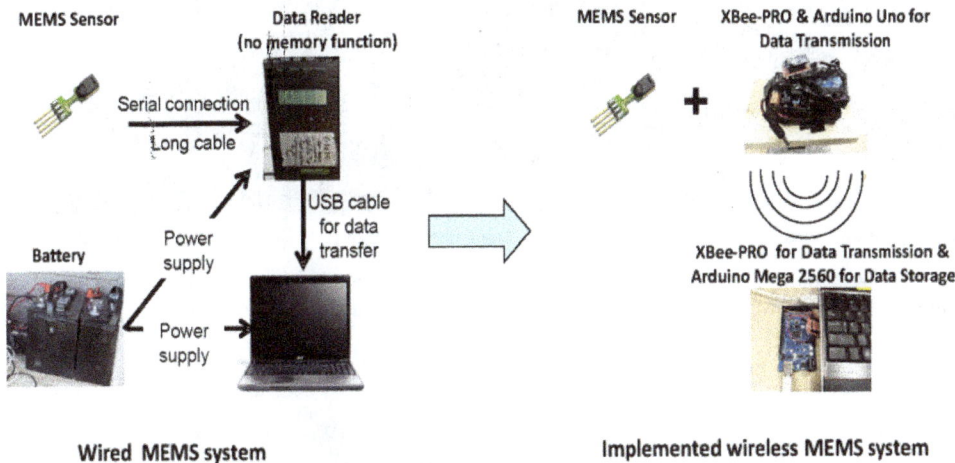

Wired MEMS system   Implemented wireless MEMS system

Figure 11. Comparison between wired MEMS system and implemented wireless-based MEMS system.

Figure 12. Wireless MEMS system inside concrete.

(a)

(b)

(c)

Ground position (0m)   Crouch position (0.75m)   Stand position (1.2m)   Above head position (1.7m)

Figure 13. Success-rate test: (a) wireless MEMS system inside concrete buried underground; (b) measuring horizontal distance; and (c) four positions to measure vertical distance.

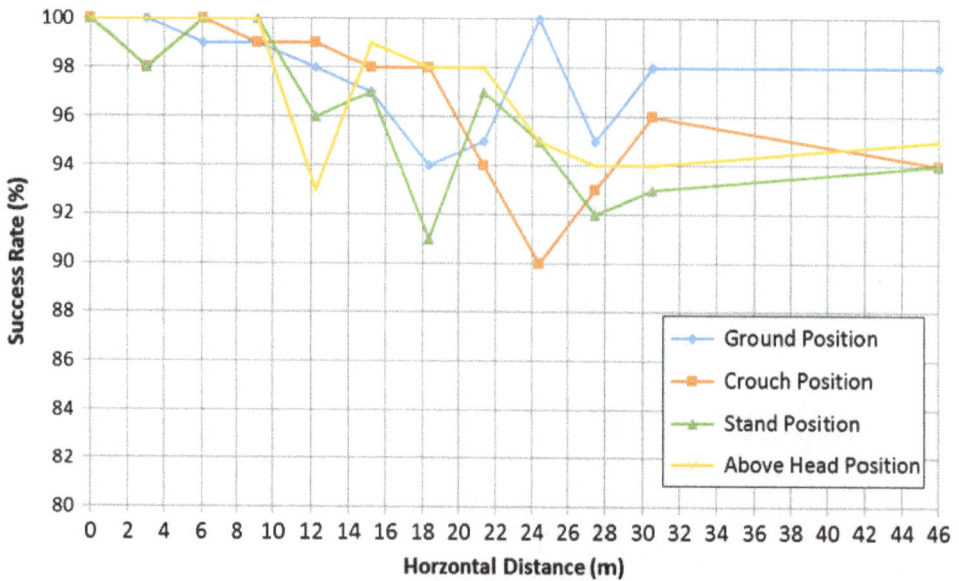

Figure 14. Success-rate test results.

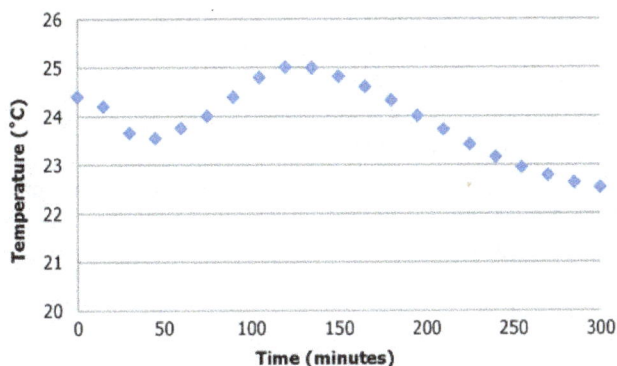

Figure 15. Temperature measurement using the implemented wireless MEMS system.

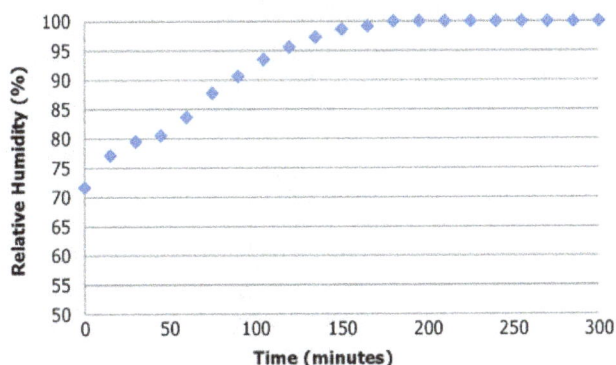

Figure 16. Relative humidity measurement using the implemented wireless MEMS system.

and RH measurements) at different working distances which tend to indicate that the implemented wireless MEMS system can maintain high success rate (greater than 92%) at a horizontal distance of over 46 m (150 feet). Also, the success rate tends to increase as the vertical distance decreases, especially at higher horizontal distances.

The temperature and RH measurements acquired by the wireless sensor system during the success-rate test are presented in Figures 15 and 16. As mentioned before, the implemented wireless communication system was able to transmit temperature and RH measurement when the receiver was positioned approximately 46 m away from transmission end with an almost 100% success rate.

5. Summary, findings, and recommendations

The objective of this study was to investigate the feasibility of implementing wireless-based MEMS for concrete-pavement structural-health monitoring. The system requirements for the wireless MEMS system were derived from field experience using a US-Highway-30 wired MEMS system. In this design, a wireless communication system was integrated with off-the-shelf MEMS sensors originally designed to be wired. The wireless MEMS system developed was capable of providing reliable temperature and RH measurement data over a distance of more than 150 feet from the receiver when embedded in concrete. However, the entire system was still consuming energy from a currently limited energy source. It could work for just a few days at a reasonable data-sampling rate using 12 1.5 AA batteries. The lifetime of these batteries could easily be diminished by harsh environmental factors like high temperatures occurring during concrete hydration; extremes of both temperature and humidity can reduce the lifetime and capacity of such batteries. One attractive, but not necessarily easy solution to this whole issue, is to develop a self-powered system that can utilize electromagnetic wave, wind, solar, thermo-electricity, and physical vibration to power sensor operation (Yildiz, 2009). Among these options, physical vibration could be an ideal energy source for pavement health monitoring applications since it can be obtained from moving vehicles on pavements. Furthermore, future research should focus on increasing memory capacity and making the whole system smaller.

Some recommendations to resolve the aforementioned issues are:

- A power management circuit called Texas Instruments (2014, http://pdf1.alldatasheet.com/datasheet-pdf/view/530828/TI/TPL5000.html) Debuts TPL5000 power timer can be used to control power output of battery; this can possibly extend the current working time to as much as several years under ideal conditions.
- A micro-SD card or QuadRam Shield can be added to the microcontroller to tremendously increase its memory capacity.
- A smaller microcontroller called Arduino Fio with XBee plug can be used to replace the original microcontroller to reduce the overall system size.

Acknowledgments
The authors gratefully acknowledge the Iowa Department of Transportation (DOT) and Iowa State University (ISU) for supporting this study. The contents of this paper reflect the views of the authors who are responsible for the facts and accuracy of the data presented within. The contents do not necessarily reflect the official views and policies of the Iowa DOT and Iowa State University. This paper does not constitute a standard, specification, or regulation.

Funding
This work was supported by the Iowa Department of Transportation [project number TR-637].

Author details
Shuo Yang[1]
E-mail: shuoy@iastate.edu
Keyan Shen[2]
E-mail: kshenyy@gmail.com
Halil Ceylan[1]
E-mail: hceylan@iastate.edu
Sunghwan Kim[1]
E-mail: sunghwan@iastate.edu
Daji Qiao[2]
E-mail: daji@iastate.edu
Kasthurirangan Gopalakrishnan[1]
E-mail: rangan@iastate.edu

[1] Department of Civil, Construction & Environmental Engineering (CCEE), Iowa State University of Science and Technology, Ames, IA 50011-3232, USA.
[2] Department of Electrical and Computer Engineering (ECpE), Iowa State University of Science and Technology, Ames, IA 50011-3232, USA.

References
Al-Khatib, Z., Yu, J. M., Al-Khakani, H. G., & Kombarji, S. (2009). *A wireless multivariable control scheme for a quadrotor hovering robotic platform using IEEE® 802.15.4* (IEEE Published Student Application Papers). Concordia University, Canada. Retrieved from https://www.ieee.org/documents/zaid_student%20application%20paper.pdf

Barroca, N., Borges, L. M., Velez, F. J., Monteiro, F., Górski, M., & Castro-Gomes, J. (2013). Wireless sensor networks for temperature and humidity monitoring within concrete structures. *Construction and Building Materials, 40,* 1156–1166. doi:10.1016/j.conbuildmat.2012.11.087

Buenfeld, N., Davis, R., Karmini, A., & Gilbertson, A. (2008). *Intelligent monitoring of concrete structures.* London: CIRIA.

Cepero, E. D. (2013). *Structural health monitoring inside concrete and grout using the wireless identification and sensing platform (WISP)* (Published doctoral dissertation). Florida International University, Miami, FL.

Ceylan, H., Gopalakrishnan, K., Kim, S., Taylor, P. C., Prokudin, M., & Buss, A. F. (2013). Highway infrastructure health monitoring using micro-electromechanical sensors and systems (MEMS). *Journal of Civil Engineering and Management, 19,* S188–S201. http://dx.doi.org/10.3846/13923730.2013.801894

Ceylan, H., Gopalakrishnan, K., Taylor, P., Shrotriya, P., Kim, S., Prokudin, M., ... Zhang J. K. (2011). *A feasibility study on embedded micro-electromechanical sensors and systems (MEMS) for monitoring highway structures* (IHRB Project No. TR-575). Ames, IA: Iowa State University.

Cho, S., Yun, C. B., Lynch, J. P., Zimmerman, A. T., Spencer, Jr., B. F., & Nagayama, T. (2008). Smart wireless sensor technology for structural health monitoring of civil structures. *Steel Structures, 8,* 267–275.

Choi, S., & Won, M. (2008). *Literature review of curing in Portland cement concrete pavement* (FHWA Publication No. FHWA/TX-09/0-5106-2). Austin, TX: University of Texas.

Every, E. V., Faridazar, F., & Deyhim, A. (2009). Embedded sensors for life-time monitoring of concrete. *Proceedings of 4th International Conference on structural health monitoring on intelligent infrastructure (SHMII-4).* Zurich, Switzerland.

Kim, S., Pakzad, S., Culler, D., Demmel, J., Fenves, G., Glaser, S., & Turon, M. (2007). Health monitoring of civil infrastructures using wireless sensor networks. *Proceedings of 6th International Symposium, Information Processing in sensor networks* (pp. 254–263). Cambridge, MA: IEEE. doi:10.1109/IPSN.2007.4379685

Lajnef, N., Chatti, K., Chakrabartty, S., Rhimi, M., & Sarkar, P. (2013). *Smart pavement monitoring system* (FHWA Publication No. FHWA-HRT-12-072). East Lansing, MI: Michigan State University.

Lajnef, N., Rhimi, M., Chatti, K., Mhamdi, L., & Faridazar, F. (2011). Toward an integrated smart sensing system and data interpretation techniques for pavement fatigue monitoring. *Computer-Aided Civil and Infrastructure Engineering, 26,* 513–523. doi:10.1111/j.1467-8667.2010.00712.x

Lee, J. S., Su, Y. W., & Shen, C. C. (2007). A comparative study of wireless protocols: Bluetooth, UWB, ZigBee, and Wi-Fi. In *Proceedings of 33rd Annual Conference of* the IEEE, IECON (pp. 46–51). Taipei: IEEE. doi:10.1109/IECON.2007.4460126

Loh, K. J., Zimmerman, A. T., & Lynch, J. P. (2007). Wireless monitoring techniques for structural health monitoring. *Proceedings of the International Symposium of applied electromagnetics & mechanics*. Lansing, MI.

Lynch, J. P. (2002). *Decentralization of wireless monitoring and control technologies for smart civil structures* (Technical Report No.140). Stanford, CA: Stanford University.

Lynch, J. P. (2007). An overview of wireless structural health monitoring for civil structures. *Philosophical Transactions of the Royal Society A: Mathematical, Physical and Engineering Sciences, 365*, 345–372. doi:10.1098/rsta.2006.1932

Lynch, J. P., & Loh, K. J. (2006). A summary review of wireless sensors and sensor networks for structural health monitoring. *The Shock and Vibration Digest, 38*, 91–128. doi:10.1177/0583102406061499

Lynch, J. P., Sundararajan, A., Law, K. H., Kiremidjian, A. S., & Carryer, E. (2003). Power-efficient wireless structural monitoring with local data processing. *Proceedings of the International Conference on structural health monitoring and intelligent infrastructure* (Vol.1, pp. 331–338). Tokyo, Japan.

Maser, K., Egri, R., Lichtenstein, A., & Chase, S. (1996). Field evaluation of a wireless global bridge evaluation and monitoring system. *Proceedings of the 11th Conference on engineering mechanics* (Vol. 2, pp. 955–958). Fort Lauderdale, FL.

McCarter, W., & Vennesland, Ø. (2004). Sensor systems for use in reinforced concrete structures. *Construction and Building Materials, 18*, 351–358. doi:10.1016/j.conbuildmat.2004.03.008

Norris, A., Saafi, M., & Romine, P. (2008). Temperature and moisture monitoring in concrete structures using embedded nanotechnology/microelectromechanical systems (MEMS) sensors. *Construction and Building Materials, 22*, 111–120. doi:10.1016/j.conbuildmat.2006.05.047

Ruiz, J. M., Rasmussen, R. O., Chang, G. K., Dick, J. C., & Nelson, P. K. (2005). *Computer-based guidelines for concrete pavements, volume II—Design and construction guidelines and HIPERPAV II user's manual* (Technical Report No. FHWA-HRT-04-122). Austin, TX: Transtec Group.

Saafi, M., & Romine, P. (2005). Preliminary evaluation of MEMS devices for early age concrete property monitoring. *Cement and Concrete Research, 35*, 2158–2164. doi: http://dx.doi.org/10.1016/j.cemconres.2005.03.012

SILICON LABS. (2015). Retrieved from http://www.silabs.com/Support%20Documents/TechnicalDocs/evolution-ofwireless-sensor-networks.pdf

Texas Instruments. (2013). *ZigBee wireless networking overview*. Retrieved from http://www.ti.com/lit/sg/slyb134d/slyb134d.pdf

Ye, D., Zollinger, D., Choi, S., & Won, M. (2006). *Literature review of curing in Portland cement concrete pavement* (FHWA Publication No. FHWA/TX06/0-5106-1). Austin, TX: University of Texas.

Yildiz, F. (2009). Potential ambient energy-harvesting sources and techniques. *The Journal of Technology Studies, 35*(1). Retrieved from http://scholar.lib.vt.edu/ejournals/JOTS/v35/v35n1/yildiz.html

16

Thermal effect of mass concrete structures in the tropics: Experimental, modelling and parametric studies

Herbert Abeka[1], Stephen Agyeman[1]* and Mark Adom-Asamoah[2]

Abstract: This paper is an experimental investigation and analytical simulation of thermal effects on mass concrete structures in the tropics. A study of the temperature rise of a 1.1 m × 1.1 m × 1.1 m experimental mass concrete block, well instrumented with thermocouples to monitor the temperatures distribution was performed. A validated finite element model was used to predict the temperature development of the hydrating experimental mass concrete block. Thermal stress analysis was performed to give an estimate of stresses induced by the thermal gradient of the concrete block section and the crack index was used to quantify the probability of thermal cracking. A parametric study on the effect of the surface area to volume ratio (SVR) of mass concrete was performed to quantify the maximum allowable thermal gradient as well as the induced thermal stresses that may cause thermal cracks. For SVR less than 0.36, thermal cracks may occur at early ages of concrete strength development in the tropics.

ABOUT THE AUTHORS

Ing. Herbert Abeka obtained his Master's degree in Development Planning and Policy, and MPhil Structural Engineering from the Kwame Nkrumah University of Science and Technology (KNUST). He is a Member of the Ghana Institution of Engineers (GhIE) and currently lecturing at civil engineering department of Sunyani Technical University. His research interests includes urban planning and settlements, structural simulations and animations.

Engr. Stephen Agyeman obtained his Master's degree in Road and Transportation Engineering from the KNUST. He is a Member of the Institution of Engineering and Technology (IET), Ghana and currently lecturing at civil engineering department of Sunyani Technical University. His research interest includes building permits, public transportation, traffic simulations and animations, and road traffic accidents analysis.

Prof. Mark Adom-Asamoah obtained his PhD in Civil Engineering from Bristol University. He is a lecturer and Provost for the College of Engineering, KNUST. His research interest includes structural design, structural analysis and earthquake engineering.

PUBLIC INTEREST STATEMENT

Albeit the effects of thermal gradients on mass concrete is well known in developed countries, the maximum allowable temperature differential published value of 20°C (35°F) between the centre of a mass concrete element and its surface is being hotly debated among researchers. However, the application of the fixed temperature differential value by many agencies has been based on time and location where such massive concrete projects have taken place. The novelty of this work is the developed finite element analysis model that can be used by agencies to predict the temperature distribution and the associated stress development in massive concrete structures in the tropics. We conclude that in the tropics, mass concrete structures with SVR less than 0.36 are expected to cause thermal cracking at early ages of cement hydration and a minimum of 5 days are required for de-shuttering of the formwork to prevent the occurrence of thermal shock.

Subjects: Materials Science; Composites; Materials Processing; Civil, Environmental and Geotechnical Engineering; Concrete & Cement; Rock Mechanics; Soil Mechanics; Waste & Recycling; Foundations and Piling; Pavement Engineering

Keywords: mass concrete; hydration; temperature; thermal cracks; finite element model

1. Introduction

Whenever large volume of fresh concrete is poured during the construction of large homogeneous structures such as dams, bridges, water retaining structures and foundations, consideration is always given to the amount of heat that will be generated (Gajda, 2007; Klemczak, 2014). The concrete hydration is an exothermic reaction that can produce high amounts of heat during curing, especially in the first few days or weeks after casting (ACI Committee 318, 2005; Lawrence, 2009). This heat production can produce high temperatures at the centre of the mass concrete due to the insulating effect of the concrete. Since the concrete surface temperatures are lower due to the heat dissipated into the ambient environment, temperature gradients are formed (Khan, Cook, & Mitchell, 1998; Lachemi & Aïtcin, 1997; Lawrence, 2009; Pofale, Tayade, & Deshpande, 2013). These changes in temperature create volumetric changes, i.e. expansion from heating and contraction from cooling in the concrete (Lin & Chen, 2015; Tia, Lawrence, Ferraro, Do, & Chen, 2013). When these volumetric changes are restrained by the supports and the more mature interior concrete, tensile stresses are formed on the concrete's surface (Riding, Poole, Folliard, Juenger, & Schinder, 2012). If the surface tensile stresses become higher than overall tensile strength of the concrete, cracking normally occurs (de Borst & van den Boogaard, 1994; Kim, 2010; Lawrence, 2009). The cracking is even magnified in early age concrete that is still developing its full strength (Cervera, Faria, Oliver, & Prato, 2002; Lee & Kim, 2009).

Past research works on the creation of numerical models for the prediction of temperature distribution in mass concrete mainly focused on using basic heat generation functions for the calculation of adiabatic temperature rise (Ballim, 2004; Chini & Parham, 2005; De Schutter, 2002; De Schutter, Yuan, Liu, & Jiang, 2014; Ilc, Turk, Kavčič, & Trtnik, 2009; Tanabe, Kawasumi, & Yamashita, 1986; van Breugel, 1991). The used of real measured heat of hydration results from calorimetry testing of the cement paste is mostly uncommon in Africa, especially Ghana due to the initial cost in acquiring the instruments (Kim, 2010; Milestone & Rogers, 1981). However, available literature reviews that numerous labs in North America and Europe have calorimeter(s) for measuring the real heat of hydration (Cao, Zavaterri, Youngblood, Moon, & Weiss, 2014). Instead, attempts at modelling hydrating mass concrete have treated the heat generated by the reacting cement as being uniform throughout the concrete mass. Whereas, in reality, the heat generation is a function of the temperature and time history of the concrete at individual locations in the concrete mass (Lawrence, 2009; Radovanovic, 1998).

Although the effects of thermal gradients on mass concrete is well known in developed countries, there is no agreed maximum allowable temperature differential value between the centre of a mass concrete element and its surface. Bobko, Edwards, Seracino, and Zia (2015), have modelled the thermal behaviour of hydrating mass concrete with some degree of success and have fixed the temperature differential at 20°C (35°F). However, in the country where this temperature differential value was developed, a several agencies have established their own guidelines to regulate and control the adverse effect of thermal cracking in mass concrete depending on the time and location where such massive concrete projects are taking place (Edwards, 2013; Lawrence, Tia, Ferraro, & Bergin, 2012; Lawrence, 2009). This confirms the fact that heat generation in mass concrete structures varies for the tropics and the temperate zones for the same type of cement (Do, Lawrence, Tia, & Bergin, 2015). But in the tropics, specifically Ghana, these values do not even exist. This paper therefore formulated a finite element model, taking into consideration the non-homogeneity of heat generation within the mass concrete, the resulting thermal gradients, the associated thermal stresses and strains, and how to accurately predict the distribution of temperature in a hydrating concrete mass in the tropics. Novelty of this work is the developed finite element analysis model that can be used by agencies to predict the temperature distribution and the associated stress development in massive concrete structures in the tropics.

2. Thermal stresses

The thermal stresses that occur during the hardening of mass concrete are extremely complex and difficult to measure. This is due to several factors, chief among which is the complex distribution of temperature changes throughout the volume of the mass concrete. The central region of the mass concrete at early age experiences high but uniform temperatures while the temperature in the outer region decreases as we move closer to the surface (Folliard et al., 2008). Since the maturity of concrete and strength are functions of temperature, the central region of the mass concrete structure will be matured and stronger than the outer region. As the concrete hydrates faster in the middle, large thermal gradients are produced, and strength and maturity are decreased moving outwards towards the surface. Restraint against this contraction will cause tensile stresses and strains to develop, creating the possibility for cracks to occur at or close to the surface of the concrete (Atrushi, 2003; Yuan & Wan, 2002). These cracks are initiated when the tensile stresses exceed the low tensile strength at the surface. The magnitude of the tensile stresses are dependent on the difference in the mass concrete, creep or relaxation of the concrete, the coefficient of thermal expansion, the degree of restraint in the concrete and elastic modulus. The development of cracks will affect the ability of the concrete structure to withstand its design load, and further allow the infiltration of lethal materials which will undermine the integrity and durability of the mass concrete structure (De Schutter et al., 2014; Lawrence, 2009; Lawrence et al., 2012).

The causes of early age thermal cracking may include either internal or external restraint (ACI Committee 207, 2005a; Kim, 2010). Internal restraint is brought about by strain gradients within the material while exterior restraint is brought about by externally applied loads. This degree of restraint varies between 0 and 100% depending on the physical boundary conditions and on the geometry of the structure (Muhammad, 2009). To accurately predict these thermal cracks, thermal properties that need to be modelled include the specific heat, the coefficient of thermal expansion, thermal diffusivity and heat production. Mechanical properties that need to be quantified, in order to simulate a finite element model of the experimental block include the tensile strength, tensile strain and elastic modulus (Atrushi, 2003; Gawin, Pesavento, & Schrefler, 2006a, 2006b; Ulm & Coussy, 1995).

According to de Borst and van den Boogaard (1994), Ishikawa (1991), Jaafar (2007), Lawrence et al. (2012), Noorzaei, Bayagoob, Thanoon, and Jaafar (2006), and Tang, Millard, and Beattie (2015), Finite Element Method (FEM) which is a numerical modelling method is seen as the best predictor of thermal cracks in concrete. It offers a step-by-step approach in solving the problem though it has its own limitations of been costly and impossibly used at site to quickly determine the maximum heat of hydration of concrete (De Freitas, Cuong, Faria, & Azenha, 2013; Tatro & Schrader, 1992; Zhai, Wang, & Wang, 2015).

3. Experimentals—materials and methods

3.1. Concrete mix design (sample preparation)

The two concrete mixes used in this study had a water to cement ratio of 0.42 to allow for complete hydration. Both concrete mixes had 100% Type I Portland cement concrete. The concrete mix designs evaluated in this paper were prepared manually by mixing Type I Portland cement concrete, water, sand (quarry dust) and 20 mm machine crushed aggregates to obtain mixes 1 and 2 using design mix ratios in Table 1. Equipment used were strain gauges, loading fame, signal conditioning

Table 1. Mix designs of concrete used in the large-scale blocks		
Materials	**Mix 1**	**Mix 2**
Cement	485 Kg	500 kg
Sand: quarry dust (1:1)	762 Kg	740 Kg
Aggregates (20 mm) (machine crushed)	1,009 Kg	1,031 Kg
Design mix ratio	1:2:4	1:2:4
Water (w/c = 0.42)	204 Kg	210 Kg

Figure 1. Experimental block geometry.

Note: All dimensions in mm.

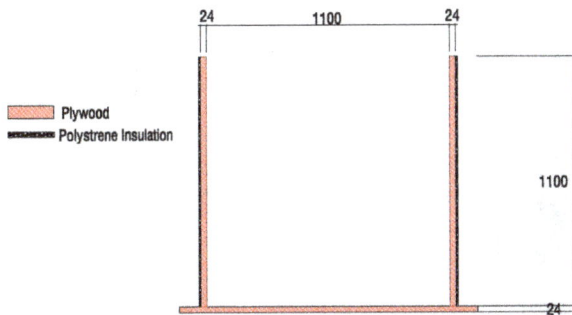

unit, data logger thermocouples and two computers (one for strain and the other for load cell acquisition). Concrete mix 2 was assumed to be fully adiabatic while mix 1 was assumed to be semi adiabatic. The semi adiabatic mix actually simulates the existing field practises in the tropics as achieving fully adiabatic is not possible under the prevailing working conditions. The raw materials information is also found in Table 1.

Two mass concrete blocks of dimensions $1.1\,\mathrm{m} \times 1.1\,\mathrm{m} \times 1.1\,\mathrm{m}$ were built (Figure 1) using $150\,\mathrm{mm} \times 150\,\mathrm{mm} \times 600\,\mathrm{mm}$ beam moulds. Also, a 24 mm thick plywood with a 1 mm layer of polystyrene sheet was used as insulating material for one of the blocks. These blocks were setup to measure the thermal behaviour of concrete under semi adiabatic conditions. However, in an effort to simulate a full adiabatic process a 50 mm thick sand was added to the top surface of the already placed concrete. Figures 2 and 3 are the individual blocks after the concrete had been poured with and without the formwork respectively.

3.2. Instrumentation for data collection

Data logger thermocouples were instrumented at critical positions in the two experimental concrete blocks to monitor the temperature distribution with time. The layout of the thermocouples are presented in Figure 4. The thermocouples data were recorded at various times in order to calibrate the proposed finite element model to be used to estimate the potential for crack-growth in early age mass concrete.

Figure 2. Experimental block with concrete and sand insulation.

Figure 3. Early age concrete block without formwork.

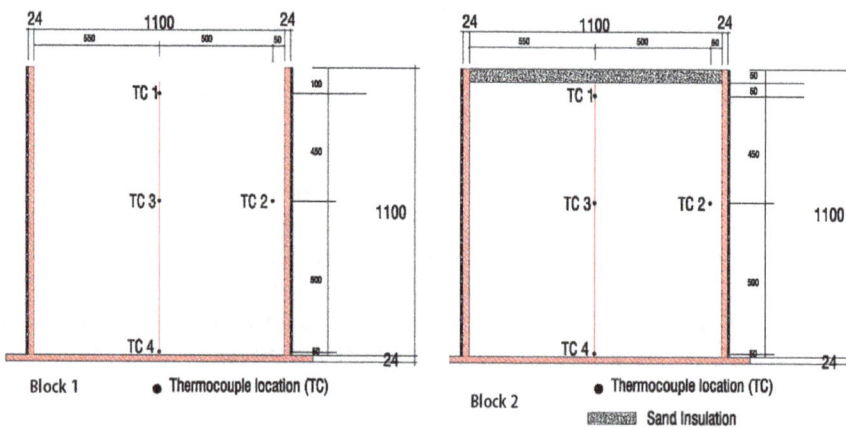

Figure 4. Thermocouple locations in Block 1 and Block 2.

Note: All dimensions in mm.

3.3. Temperature profiles

The location of the temperature sensors in the blocks was strategically chosen to capture the differences in temperature between the centre of the block and the exposed surface. The ambient temperature was also monitored to determine if it would contribute to thermal cracking of the concrete. The temperature sensors at the side and bottom of the block were also placed to validate the effectiveness of the insulation, and to also assess the boundary conditions used in the proposed finite element model (Do, Lawrence, Tia, & Bergin, 2014).

3.4. Mechanical properties of concrete

In order to accurately model early-age stress development in concrete members, it was necessary to determine the mechanical properties (Atrushi, 2003; Bernard, Ulm, & Lemarchand, 2003). Many forms of equations have been developed to relate the compressive strength to the maturity development. Two commonly used equations according to Viviani (2005), are given as Equations (1) and (2):

$$f_c(t) = a + b\log(\log(M(t))), \quad f_c \geq 0 \tag{1}$$

$$f_c(t_c) = f_{\text{cult.}} \exp\left[-\left(\frac{\tau_s}{T_e}\right)^{\beta_s}\right] \tag{2}$$

where, f_c is the compressive strength value (MPa), a is a fit parameter which is usually negative (MPa), b is a fit parameter (MPa/°C/h), $f_{\text{cult.}}$ is the ultimate compressive strength parameter fit from the compressive strength tests (MPa), τ_s is a fit parameter (h), T_e is equivalent age at the reference curing temperature (h), β_s is a fit parameter, $M(t)$ is the maturity or temperature-time factor at age t, tc is average concrete temperature during the time interval, t is time

For thermal stress analysis, Equation (2) was preferred since is not discontinuous at setting and this functional form is similar to hydration models. The fit parameters f_{Cult}, τ_s and β_s were found to be 27.4, 12.35 and 1.52 respectively with an R^2 of 0.98.

In terms of concrete tensile strength, values used for the study were estimated from their respective compressive strengths (Folliard et al., 2008). Most current method for calculating the splitting tensile strength, Equation (3) assumes a power type function based on the compressive strength according to Raphael (1984):

$$f_{ct} = a(f_c)^b \tag{3}$$

where, f_{ct} is the concrete splitting tensile strength value, a and b are fit parameters, and f_c is the concrete compressive strength. Fit parameters a and b were found to be 0.06 and 1.09, respectively with an R^2 of 0.99.

Moreover, for accurately modelling of the thermal behaviour (stresses) of mass, it is important to consider the changes in the mechanical properties (elastic modulus) with time of the concrete (De Schutter, 2002; Lawrence et al., 2012). The elastic modulus is also commonly calculated from the concrete's compressive strength. Most models of this type follow a form of Equation (4):

$$E = k(f_c)^n \tag{4}$$

where, E is the elastic modulus, f_c is the compressive strength (MPa), and k and n are model parameters.

Equation (4) was used in calculating the elastic modulus from the compressive strengths developed because most engineers are familiar with it from prior experience in structural design, and readily accept its use.

4. Finite element thermal modelling

The Fourier heat transfer equation is the underline mathematical model can be used to compute the temperature for an elemental volume at a particular instance of time. The generalized governing Equation (5) expressed in the Cartesian coordinate system, was used in the three dimensional heat flow analysis.

$$\rho c_p \frac{\delta_T}{\delta_t} = k\left(\frac{d^2T}{dx^2} + \frac{d^2T}{dy^2} + \frac{d^2T}{dz^2}\right) + Q_H \tag{5}$$

where, c_p is the specific heat capacity, ρ is the density of the concrete, t is the time, k is the thermal conductivity, T is the temperature and Q_H, the rate of internal heat evolution, x, y, z are the coordinates at a particular point in the structure.

This finite element model that was used to simulate the thermal behaviour in mass concrete was verified or calibrated so that its temperature distribution for the entire volume closely match with that of the experimental block. The main modelling parameters utilized in the thermal analysis were:

- Convection coefficient.
- Ambient temperature.
- Internal heat generation rate of concrete.
- Placing temperature.

4.1. Input parameters for thermal analysis

The input parameters were either modelled as deterministic or stochastic based on the type of analysis, the type of element and the reference temperature, the heat generation function etc. A nonlinear formulation of the transient thermal analysis was adopted to account for the variations of boundary and loading conditions with time. Also, at the element level in the finite element analysis, a utilization of solid elements capable of providing reliable estimates of the thermal quantities were simulated using an eight-node isoparametric element having a single degree of freedom at each node. Under, the ANSYS platform, the PLANE 70–3D thermal solid element type was chosen from the library of constitutive material elements. The main output parameters that were of interest are the maximum in-place temperature and the thermal gradient. We defined thermal gradient as the change in temperature with respect to change in distance across a section of the concrete. Since the thermal properties for an elemental volume changed with time, coupled with convection at the surface above the formwork, an initial boundary temperature referred to as the reference temperature was chosen as the placing temperature. This parameter needed to be defined so that the time-stepping algorithm be initiated in the analysis.

4.2. Poisson's ratio

According to Mehta and Monteiro (2013), Poisson's ratio has no consistent relationship with the curing age of the concrete. A value of 0.18 was used, which was within the universally accepted range of 0.15 and 0.20 for concrete.

4.3. Thermal conductivity

A characteristic value of thermal conductivity of concrete is in the range 9 to 10.5 kJ/mh°C per the Korean standards and in the range 7.1 to 10.6 kJ/mh°C per the American Standards (ACI Committee 207, 2005b). A constant thermal conductivity value of 9 kJ/mh°C was adopted in this study per the assumption that it will not vary with location across a section, and time during the analysis.

4.4. Specific heat of concrete

A typical value of specific heat capacity for concrete ranges from 1.13 to 1.3 kJ/kg°C according to JCI, and 0.92–1.00 kJ/kg°C according to ACI Committee 207 (2005b). The specific heat values chosen in this study are 0.9 kJ/kg°C

4.5. Coefficient of thermal expansion

The coefficient of thermal expansion used in this thermal stress analysis was 2×10^{-6}/°C.

4.6. Initial boundary conditions consideration

Boundary heat transfer conditions, which are time or temperature dependent, are important for solving the Fourier differential equations. The four major boundary heat transfer mechanisms are conduction, convention, solar absorption, and irradiation. Each point of a concrete element has a different rate of heat of hydration due to the effects of the environment. We considered the overall convection heat transfer caused by air motion using Newton's law of cooling (Equation (6)).

$$Q = \bar{h}A(T_S - T_a) \tag{6}$$

where, Q is heat flow (kJ/h), $\bar{h}$ is mean convection heat transfer coefficient (kJ/m² h°C), A is surface area (m²), T_S is surface temperature (°C), T_a is air temperature (°C).

The combined heat transfer coefficient used to account for convection as well as irradiation throughout the thermal analysis was 29.3 kJ/m²h°C.

4.7. Modelling of heat of hydration from cementitious material

The discrete time-dependent temperature profile at the core of the experimental block was used to predict a continuous heat of cement hydration in the modelling. Due to the nature of the heat curve, the adiabatic hydration model that is usually defined by exponential function has been proposed by a number of researchers. The empirical adiabatic hydration model proposed by Suzuki, Tsuji, Maekawa, and Okamura (1990) was used to relate the rate of heat evolution of the cementitious material. Equation (7) was used to compute internal heat evolution to define the temperature rise curve.

$$T_a(t) = T_\infty \left(1 - e^{-\alpha t}\right) \tag{7}$$

where, T_∞ is the ultimate temperature rise, $T_a(t)$ is the Adiabatic temperature rise at t days after casting, α is the coefficient of temperature rise (reaction rate), t is time in days.

4.8. Modelling of heat generating rate
The time—dependent temperature distribution at the core of the concrete block was used to simulate adiabatic temperature rise. These temperature were then normalized by subtracting the initial placing temperature. Suzuki's model was adopted, by which a univariate nonlinear regression analysis was performed to establish the temperature—time relationship. The software package MATLAB was used to perform this analysis. Upon comparing the predicted model with experimentally determined results at a 95% confidence interval, R^2 Value of 0.988 was obtained. The proposed model quantified the volumetric for the thermal analysis using Equation (8):

$$q(t) = 20,000 \times e^{\left(-0.9398 - \frac{t}{24}\right)} \tag{8}$$

where, $q(t)$ is the heat generation per unit volume in t days, t is time in days.

4.9. Model geometry
A 1.1 m × 1.1 m × 1.1 m finite element model of the experimental block was constructed in the software package, ANSYS. A mass concrete block was then used to measure the effect of the heat evolution rate on the temperature development with time and location. A discretization scheme of 50 mm along each edge of the block was chosen while a temperature based convergence criteria was set to give reliable estimates of the temperature distribution. The element type concrete 65 was used to represent the element definition to produce 10,648 elements and nodes that were meshed in other to solve the problem. Figure 5 shows the ANSYS model block and Table 2 shows the summarised input parameters values used in the analysis.

Figure 5. Analytically modelled block of mass concrete.

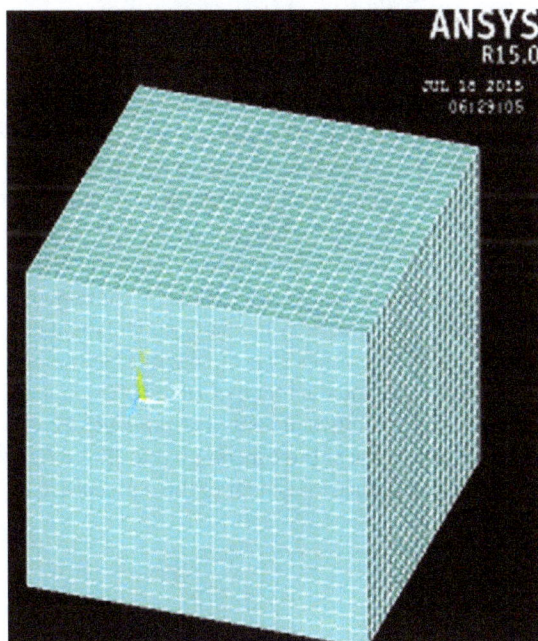

Table 2. Execution parameters used for both thermal analysis and stress analysis

Items	Unit	Value
Specific heat	kJ/kg/°C	0.96
Thermal conductivity	kJ/mh/°C	9
Thermal expansion	°C	2×10^{-6}
Poisson's ratio	Unitless	0.18
Placement temperature	°C	31.8
Analysis time	h	144
Convection (Plywood form)	kJ/m²h°C	29.3
Density	kg/m³	2,400

The transient thermal analysis model was implemented in the commercial finite element program, ANSYS, a general purpose program capable of numerical simulation of a variety of physical problems. For concrete structures with surface area to volume ratio (S_v) being less than 2, it can be classified as massive and as such the thermal effect should be considered (Flaga, 2011). Thus, we simulated analytically the specified heat generation rate, mechanical and thermal properties of the experimental block of concrete having geometric configuration of 1.1 m × 1.1 m × 1.1 m and S_v of 0.91. The parameters that were treated mainly as random variables were the ambient temperature, the hydration rate of the cementitious material as well as the heat transfer coefficient for the different surfaces of the mass concrete block. The ambient temperature and the heat evolution from the hydrating cement played major roles in the temperature rise in mass concrete (Ayotte, Massicote, Houde, & Gocevski, 1997; Truman, Petruska, Ferhi, & Fehl, 1991). Though these variables were random in nature, the temperature distribution at that particular time was assumed to emanate from a stochastic process. Also, because strength properties of concrete increased with time, coupled with the large thermal gradient between the core and surface of a mass concrete section, the damaging effect of the thermal behaviour of the hydrating cement was considered critical at early age. We defined early age of mass concrete as the first 6 days after placing of concrete. Therefore, the transient thermal analysis was estimated for 6 days. In order to monitor the temperature distribution with respect to time, the experimental concrete block was well instrumented with thermocouples at specific location (Figure 4). Measurements were then taken at discrete time steps of 2, 12, 18, 24, 48, 72, 96, 120 and 144 h.

5. Validation of analytical model

Figure 6 shows the analytical output of the temperature variation at the centre of the concrete block. Spline interpolation was adopted in order to obtain a continuous function of the time-temperature development. Figure 7 shows comparatively results of the observed temperature distribution and the simulated analytical results at the core of the concrete. A 75.4°C temperature rise at 24 h was observed at the core of the mass concrete section considered, with a gradual decline to 40.36°C at 144 h after placing of concrete. A temperature rise of 75.4°C meant that there was Delayed Ettringite Formation (DEF) which led to massive cracking of the experimental blocks (Gajda & Vangeem, 2002). Other contributing factors of the DEF development may include; temperature, alkalis in the cement, SO_3 and C_3A contents of the cement, aggregate mineralogy, and high humidity conditions (Loïc, 2003). The value of 75.4°C is comparable to temperature rise within 24 h of concrete placement value of 74.32°C obtained by Prasanna and Subhashini (2010), and the range of 79 to 42°C obtained by Bartojay (2012). A maximum temperature differential at a particular instant of time was 22.9°C throughout the analysis period (observed at the core and surface of the concrete block). This difference is attributed to the fact that hydrating concrete at the surface loses heat to the atmosphere at a higher rate than the core element of concrete.

Figure 6. Temperature variation at the Centre of the concrete block.

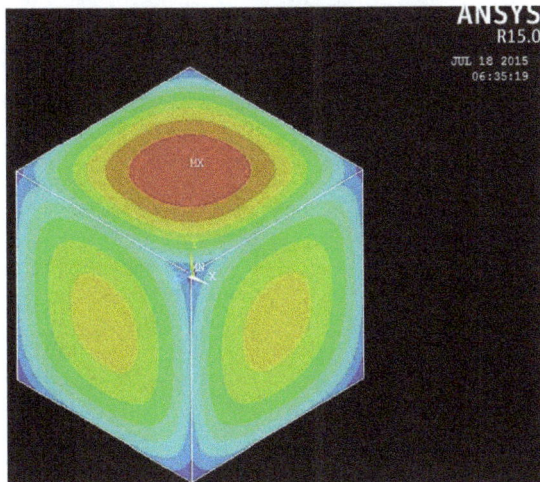

Figure 7. Temperature variation at the core of the concrete block with respect to time.

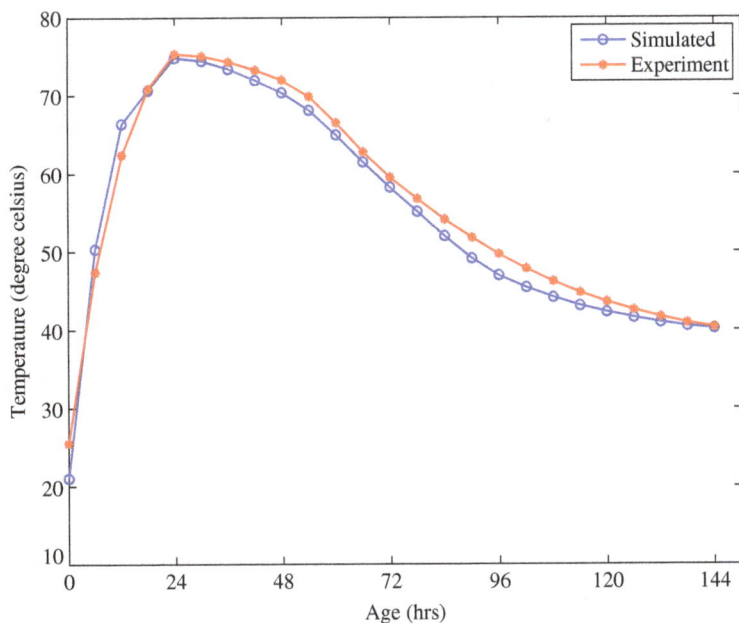

6. Thermal stress analysis

The temperature distribution and the thermal gradient are primarily the thermal quantities that were of interest in our transient thermal analysis model (Khan et al., 1998; Waller, D'Aloïa, Cussigh, & Lecrux, 2004). However, to actually ascertain the potential for cracking due to this external state of loading, a stress analysis was required to assess whether the limiting state of stress has been exceeded. The limit state condition occurred when the thermally induced stresses were greater than the tensile strength of concrete at a given age. Internal restraints caused the maximum temperature differential, the uneven expansion and the contraction of the two extremes for a given section, thereby producing cracks at the upper surfaces (Khan et al., 1998; Lachemi & Aïtcin, 1997). The procedure adopted for the stress analysis is similar to that of the thermal model. Definitions of element type, analysis type and quantification of the variations in mechanical properties and boundary conditions were made. The temperature distribution over the entire concrete block was applied as load in this model at every time step.

Figure 8. Stress analysis results.

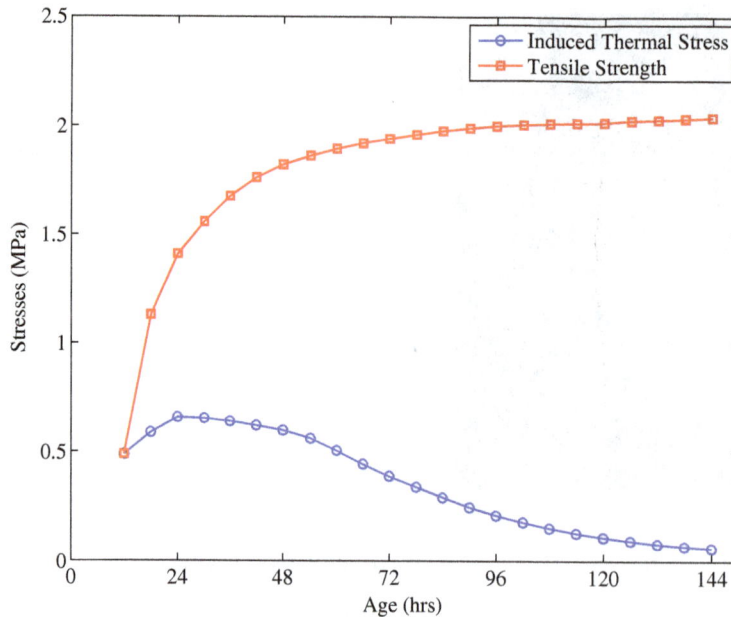

6.1. Thermal stress analysis results

Figure 8 presents the results of thermal stress analysis for the analytical model. The maximum ten-sile stress in all the orthogonal directions, due to thermal distribution was found to occur around the surface. The maximum thermal stresses coincided with the time corresponding to the peak tem-perature distribution. For the validated analytical model, it occurred at 24 h after placing concrete. Beyond this time, there was a gradual decline in the thermal stress because the corresponding tem-perature development was also decreasing. Due to this phenomenon, the temperature at early age was key in estimating the occurrence of cracks in mass concrete. The crack index, defined as the ratio of tensile strength of concrete to the induced thermal stress at early ages, was used to meas-ure the probability of cracks developing in the mass concrete structure. For crack index values less than unity, thermal crack is expected to occur. The validated analytical model yielded a crack index value of 2.1 signifying that stress induced by the temperature development were not large enough to cause thermal cracks.

7. Parametric study and probabilistic prediction of thermal crack

7.1. Effect of specimen size

The standard specimen size used in this study was a block size of 1.1 m × 1.1 m × 1.1 m. To study the effect of size on the behaviour of concrete, four additional block sizes were modelled. The sizes cho-sen were: 2 m × 2 m × 2 m, 3 m × 3 m × 3 m, 4 m × 4 m × 4 m and 5 m × 5 m × 5 m. A comparison of the temperature profiles at the centre of blocks containing concrete is shown in Figure 9. The surface area to volume ratio of each specimen was used as a parameter to evaluate the temperature distri-bution. It is evident that as the surface area to volume ratio decreased, there was a corresponding increase in temperatures with respect to time of each specimen.

Also, Figure 9 shows the progression of the peak temperatures, as the surface area to volume ratio decreased (indirectly increased in block size). It has been observed that, mass concrete structures with peak temperatures higher than 70°C are expected to experience an undesirable phenomenon known as DEF (Kishi & Maekawa, 1995; Lawrence, 2009; Tim, 2014; Wang, Ge, Grove, Ruiz, & Rasmussen, 2006). From Figure 9, surface area to volume ratios for all the specimens under study yielded a maximum observed temperature rise exceeding 70°C.

Figure 9. Comparison of temperature profiles calculated at the centre of each block.

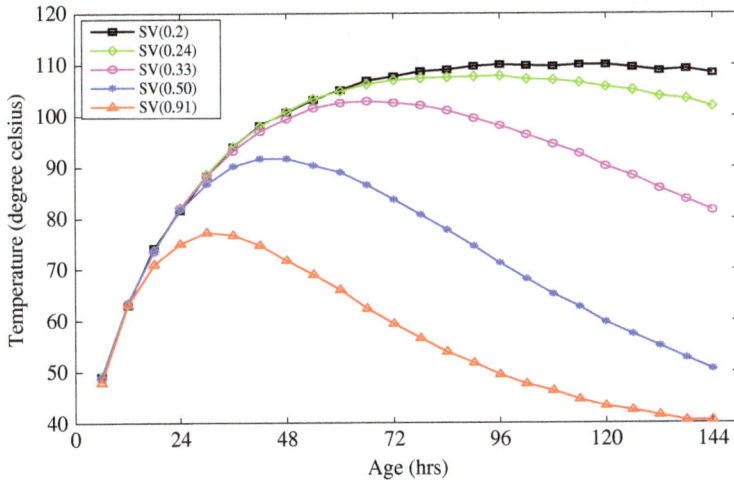

Figure 10. Calculated peak temperature values with respect to block size.

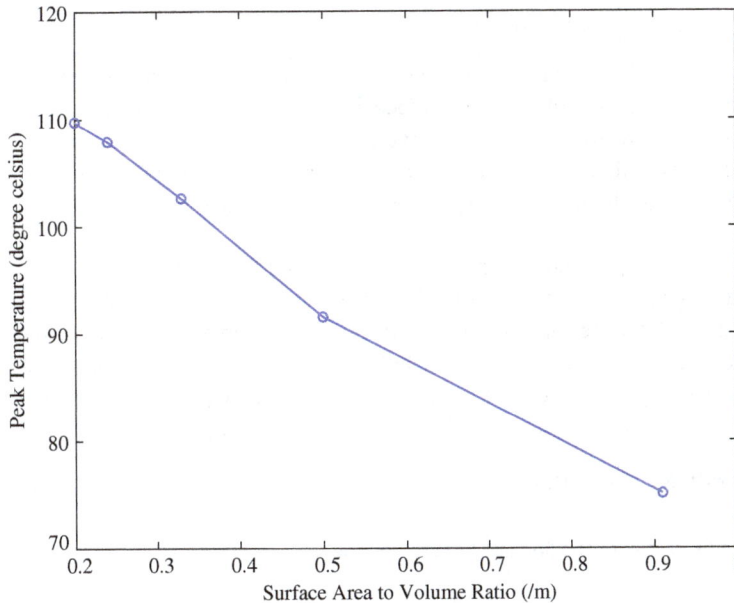

The effect of the surface area to volume ratio (indirectly the block size) on the maximum temperature differentials is presented in Figure 10. The maximum temperature differential between the centre and top surface edge increased from 22.9°C in the 1.1 m block to 70.1°C for the 5 m block. The temperature differentials and the thermal gradient (change in temperature with respect to distance) were the major parameters used to measure the degree of internal restraint caused by the temperature rise in mass concrete. The maximum temperature differential limiting value of 20°C was assumed between the surface and centre of mass concrete structure to achieved desirable strength characteristics with no thermal crack. From Figure 10, both the experimental and the analytical results produced temperatures higher than 20°C for all the specimens under study. A univariate regression analysis revealed Equation (9):

$$T_{max} = 76.144 \, S_v^{-0.24149}, \text{ with } R^2 \text{ value of } 0.957 \tag{9}$$

where, T_{max} is the maximum temperature differential, S_v is the surface area to volume ratio.

Figure 11. Maximum induced stress with respect to maximum temperature differential due to increasing block size.

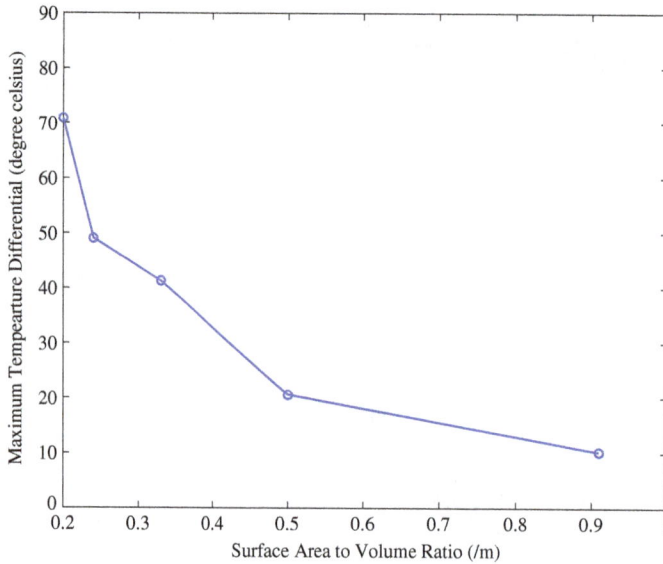

The relationship between the maximum induced stress and the increasing maximum temperature differential caused by increasing block size (decreasing surface area to volume ratio) is presented in Figure 11. For a given concrete mixture, the maximum induced stress increased with increasing maximum temperature differential (correlation coefficient of 0.99). Also, the maximum temperature difference and resulting stress in concrete elements were highly dependent on the type of concrete used.

Figure 12 shows a propagation of maximum induced stress with respect to surface area to volume ratio. When these stresses are higher than the tensile strength, thermal crack may occur. Superimposed on Figure 12 is the tensile strength for the various specimen under study at times when the thermal stresses are expected to be maximum. It was revealed that at surface area to volume ratio roughly higher than 0.36, thermal crack may not be observed. A univariate nonlinear regression was done to provide a relation between both quantities as provided in Equations (10) and (11).

$$\sigma_{tensile} = 1.4726\, S_v^{-0.2209}, \text{ with } R^2 \text{ value of } 0.867 \tag{10}$$

Figure 12. Maximum induced stresses to the Surface area to volume of each block.

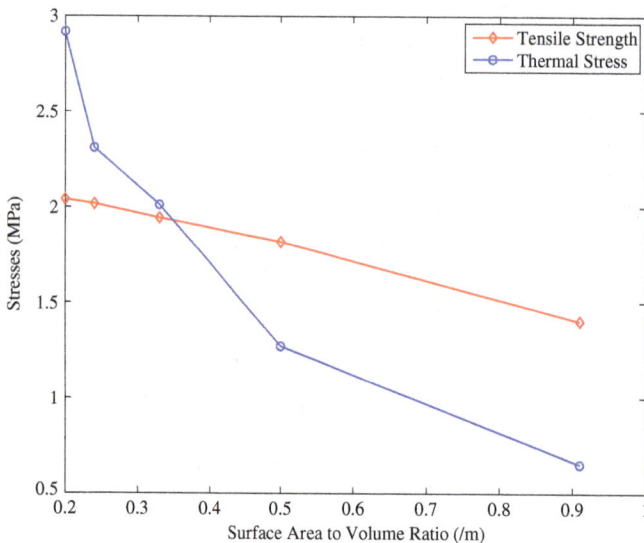

Figure 13. Crack Index to the surface area to volume ratio of each block.

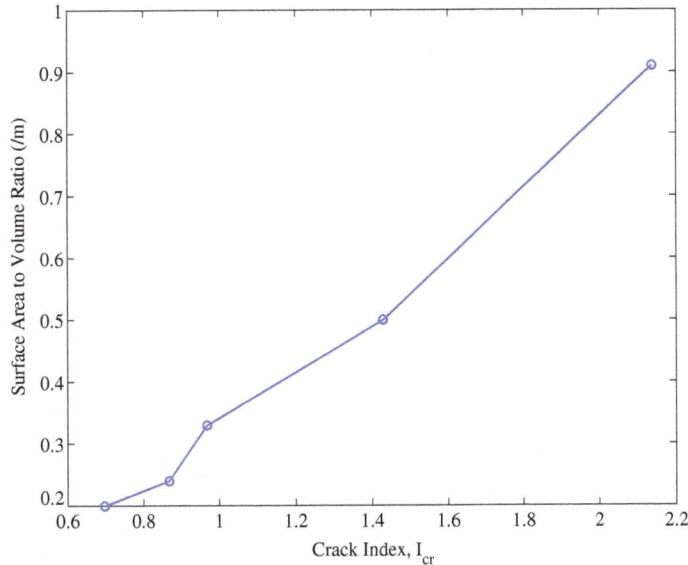

$$\sigma_{thermal} = 0.68192\ S_v^{-0.89648}, \text{ with } R^2 \text{ value of } 0.976 \tag{11}$$

where, $\sigma_{tensile}$ is the tensile strength, $\sigma_{thermal}$ is the thermal stresses, S_v is the SVR.

Figure 13 relates the crack index to the surface area to volume ratio. It can be seen that these two parameters are highly correlative (R^2 value 0.992). For surface area to volume ratio less than 0.36, the crack index was less than 1, signifying that the induced thermal stresses have exceeded the tensile strength at that concrete age, therefore the probability of thermal cracks developing become certain. The regression analysis yielded Equation (12):

$$I_{cr} = 2.2982\ S_v^{0.72216}, \text{ with } R^2 \text{ value of } 0.992 \tag{12}$$

where, I_{cr} is the crack index, S_v is the SVR.

8. Conclusions

Investigations were conducted on predicting early age thermal cracks in mass concrete structures at tropical atmospheric conditions using Ghana as the geographical scope for the research. The current state-of-the-art practice is to perform adiabatic calorimetry testing on concrete mixtures and use finite element analysis to predict temperature distribution during the time. The predicted temperature distribution is then used to quantify the induced thermal stresses. Given that the tensile strength at a particular age of concrete is less than the thermal stress, the probability of thermal cracks developing becomes certain. The research adopted both experimental and analytical modelling (Finite Element Model) scenarios to predict the early age thermal cracking in mass concrete structures in the tropics. The experimental program involved the construction of two mass concrete blocks of dimension 1.1 m × 1.1 m × 1.1 m that were intended to simulate both adiabatic and semi adiabatic conditions. The temperature distributions from these experimental blocks were used to calibrate and validate finite element models that were implemented in the commercial software ANSYS. A parametric study on the effect of the size on various concrete blocks were also investigated to help estimate the propagation of crack growth during the early ages of mass concrete structures. From the study, construction of mass concrete structures with SVR less than 0.36 is not adequate enough to prevent the occurrence of cracks. Also, the age of concrete at peak temperature was observed to increase with an increase in the size of the mass concrete block. The specific conclusions drawn were:

(1) The widely accepted limiting temperature differential of 20°C that is likely to cause thermal cracking may not be valid in the tropics for mass concrete structures.

(2) In the tropics mass concrete structures with a SVR value less than 0.36 are expected to cause thermal cracking at early ages of cement hydration.

(3) As such to provide reliable estimate of the likelihood of cracking, this assertion should be supplemented by performing a finite element stress analysis.

(4) In the tropics, we recommend a minimum of 5 days for de-shuttering of the formwork to prevent the occurrence of thermal shock.

List of Abbreviation for symbols

f_c	compressive strength value (MPa)
a	fit parameter which is usually negative (MPa)
b	fit parameter (MPa/°C/h)
$f_{cult.}$	ultimate compressive strength parameter fit from the compressive strength tests (MPa)
τ_s	fit parameter (h)
Te	equivalent age at the reference curing temperature (h)
β_s	fit parameter
$M(t)$	maturity or temperature-time factor at age t average concrete temperature during the time interval
t	time
f_{ct}	concrete splitting tensile strength value
E	elastic modulus
f_c	compressive strength (MPa)
n	model parameter
c_p	specific heat capacity
ρ	density of the concrete
k	thermal conductivity
T	temperature
Q_H	rate of internal heat evolution
x, y, z	coordinates at a particular point in the structure
Q	heat flow (kJ/h)
$\bar{h}$	mean convection heat transfer coefficient (kJ/m²h°C)
A	surface area (m²)
T_S	surface temperature (°C)
T_a	air temperature (°C)
T_∞	ultimate temperature rise
$T_a(t)$	Adiabatic temperature rise at t days after casting
α	coefficient of temperature rise (reaction rate)
$q(t)$	heat generation per unit volume in t days
T_{max}	maximum temperature differential
$\sigma_{tensile}$	tensile strength
$\sigma_{thermal}$	thermal stresses
S_v	surface area to volume ratio
I_{cr}	crack index

Acknowledgement
We will like to acknowledge the supports of Jack Osei Banahene, Noble Obeng-Ankamah and Jojo Appiah-Adinkrah as well as the Laboratory Technicians of the Civil Engineering Department, KNUST who played various roles during our Laboratory works and data analysis stage.

Funding
The authors received no direct funding for this research.

Author details
Herbert Abeka[1]
E-mail: herbertabeka@gmail.com
Stephen Agyeman[1]
E-mail: agyengo44@gmail.com
ORCID ID: http://orcid.org/0000-0002-1985-7456
Mark Adom-Asamoah[2]
E-mail: markadomasamoah@gmail.com
[1] Department of Civil Engineering, Sunyani Technical University, Box 206, Sunyani, Ghana.
[2] Department of Civil Engineering, Kwame Nkrumah University of Science & Technology, Kumasi, Ghana.

Citation information
Cite this article as: Thermal effect of mass concrete structures in the tropics: Experimental, modelling and parametric studies, Herbert Abeka, Stephen Agyeman & Mark Adom-Asamoah, *Cogent Engineering* (2017), 4: 1278297.

References
ACI Committee 207. (2005a). *Guide to mass concrete, ACI 207.1R-05*. Farmington Hills, MI: American Concrete Institute.
ACI Committee 207. (2005b). *Report on thermal and volume change effects on cracking of mass concrete, 207.2R-07*. Farmington Hills, MI: American Concrete Institute.
ACI Committee 318. (2005). *Building code requirements for structural concrete (ACI 318-05) and commentary (318R-05)*. Farmington Hills, MI: American Concrete Institute.
Atrushi, D. S. (2003). *Tensile and compressive creep of early age concrete: Testing and modelling*. Trondheim: Department of Civil Engineering, The Norwegian University of Science and Technology. URN:NBN:no-3377.
Ayotte, É., Massicote, B., Houde, J., & Gocevski, V. (1997). Modelling the thermal stresses at early ages in a concrete monolith. *ACI Materials Journal, 94*, 577–587.
Ballim, Y. (2004). A numerical model and associated calorimeter for predicting temperature profiles in mass concrete. *Cement and Concrete Composites, 26*, 695–703. doi:10.1016/S0958-9465(03)00093-3
Bartojay, K. (2012). Thermal properties of reinforced structural mass concrete. *Dam Safety Technology Development Program, Bureau of Reclamation, OMB No. 0704-0188*. Denver, CO: U.S. Department of the Interior.
Bernard, O., Ulm, F., & Lemarchand, E. (2003). A multiscale micromechanics-hydration model for the early-age elastic properties of cement-based materials. *Cement and Concrete Research*, 1293–1309. doi:10.1016/S0008-8846(03)00039-5
Bobko, P. C., Edwards, J. A., Seracino, R., & Zia, P. (2015). Thermal cracking of mass concrete bridge footings in coastal environments. *Journal of Performance of Constructed Facilities, 29*, 0401–4171. doi:10.1061/(ASCE)CF.1943-5509.0000664
Cao, Y., Zavaterri, P., Youngblood, J., Moon, R., & Weiss, J. (2014). The influence of cellulose nano crystal additions on the performance of cement paste. *Cement & Concrete Composites, 56*, 73–83. doi:10.1016/j.cemconcomp.2014.11.008

Cervera, M., Faria, R., Oliver, J., & Prato, T. (2002). Numerical modelling of concrete curing, regarding hydration and temperature phenomena. *Computers & Structures*, 1511–1521. doi:10.1016/S0045-7949(02)00104-9
Chini, A. R., & Parham, A. (2005). *Adiabatic temperature rise of mass concrete in Florida* (final report). GainesVille, FL: University of Florida.
de Borst, R., & van den Boogaard, A. H. (1994). Finite-element modeling of deformation and cracking in early-age concrete. *Journal of Engineering Mechanics, 120*, 2519–2534. http://dx.doi.org/10.1061/(ASCE)0733-9399(1994)120:12(2519)
De Freitas, T., Cuong, J., Faria, R., & Azenha, M. (2013). Modelling of cement hydration in concrete structures with hybrid finite elements. *Finite Elements in Analysis and Design, 77*, 16–30. doi:10.1016/j.finel.2013.07.008.
De Schutter, G. (2002). Finite element simulation of thermal cracking in massive hardening concrete elements using degree of hydration based material laws. *Computers & Structures, 80*, 2035–2042. http://dx.doi.org/10.1016/S0045-7949(02)00270-5
De Schutter, G., Yuan, Y., Liu, X., & Jiang, W. (2014). Degree of hydration-based creep modeling of concrete with blended binders: from concept to real applications. *Journal of Sustainable Cement-Based Materials*, 1–14, doi:10.1080/21650373.2014.928808
Do, A. T., Lawrence, M. A., Tia, M., & Bergin, M. J. (2015). Effects of thermal conductivity of soil on temperature development and cracking in mass concrete footings. *Journal of Testing and Evaluation, 43*, 1078–1090. doi:https://doi.org/10.1520/JTE20140026
Do, T., Lawrence, A., Tia, M., & Bergin, M. (2014). Determination of required insulation for preventing early-age cracking in mass concrete footings. *Transportation Research Record: Journal of the Transportation Research Board, 2441*, 91–97. doi:10.3141/2441-12
Edwards, A. J. (2013). *Early age thermal cracking of mass concrete footings on bridges in coastal environments* (Unpublished MSc Graduate Thesis). North Carolina State University, Raleigh.
Flaga, K. (2011). *Shrinkage stresses and surface reinforcement in concrete structures*. Monography 391. Krakow: Cracow Technical University.
Folliard, K. J., Juenger, M., Schindler, A., Riding, K., Poole, J., Kallivokas, L. F. ... Meadows, J. L. (2008). *Prediction model for concrete behavior - final report* (No. Report No. FHWA/TX-08/0-4563-1). Austin, TX: Texas Department of Transportation and the Federal Highway Administration.
Gajda, J. (2007). *Mass concrete for buildings and bridges*. Skokie, IL: Portland Cement Association.
Gajda, J., & Vangeem, M. (2002). Controlling temperatures in mass concrete. *Concrete International, 24*, 59–62.
Gawin, D., Pesavento, F., & Schrefler, B. (2006a). Hygro-thermo-chemo-mechanical modelling of concrete at early ages and beyond. Part II: shrinkage and creep of concrete. *International Journal for Numerical Methods in Engineering*, 332–363. doi:10.1002/nme.1636
Gawin, D., Pesavento, F., & Schrefler, B. (2006b). Hygro-thermo-chemo-mechanical modelling of concrete at early ages and beyond. Part I: hydration and hygro-thermal phenomena. *International Journal for Numerical Methods in Engineering*, 299–331. doi:10.1002/nme.1615
Ilc, A., Turk, G., Kavčič, F., & Trtnik, G. (2009). New numerical procedure for the prediction of temperature development in early age concrete structures. *Automation in Construction, 18*, 849–855. doi:10.1016/j.autcon.2009.03.009.Manuscript
Ishikawa, M. (1991). Thermal stress analysis of a concrete dam. *Computers & Structures, 40*, 347–352. http://dx.doi.org/10.1016/0045-7949(91)90360-X

Jaafar, M. S. (2007). Development of finite element computer code for thermal analysis of roller compacted concrete dams. *Advances in Engineering Software, 38*, 886–895. http://dx.doi.org/10.1016/j.advengsoft.2006.08.040

Khan, A. A., Cook, W. D., & Mitchell, D. (1998). Thermal properties and transient thermal analysis of structural members during hydration. *ACI Materials Journal, 95*, 293–300.

Kim, S. G. (2010). *Effect of heat generation from cement hydration on mass concrete placement* (No. Paper 11675) (Graduate Theses and Dissertations). Ames, IA: Graduate College Digital Repository.

Kishi, T., & Maekawa, K. (1995). Multi-component model for hydration heating of Portland cement. *Translation from Proceedings of JSCE, 29*, 97–109.

Klemczak, A. B. (2014). Modeling thermal-shrinkage stresses in early age massive concrete structures – Comparative study of basic models. *Archives of Civil and Mechanical Engineering, 83*, 1–13. doi:10.1016/j.acme.2014.01.002

Lachemi, M., & Aïtcin, P.-C. (1997). Influence of ambient and fresh concrete temperatures on the maximum temperature and thermal gradient in a high performance concrete structure. *ACI Materials Journal, 94*, 102–110.

Lawrence, A. M. (2009). *A finite element model for the prediction of thermal stresses in mass concrete* (Unpublished Ph.D. Dissertation). Gainesville, FL: University of FLorida.

Lawrence, A. M., Tia, M., Ferraro, C., & Bergin, M. (2012). Effect of early age strength on cracking in mass concrete containing different supplementary cementitious materials: Experimental and finite-element investigation. *Journal of Materials in Civil Engineering, 24*, 362–372. http://dx.doi.org/10.1061/(ASCE)MT.1943-5533.0000389

Lee, Y., & Kim, J. (2009). Numerical analysis of the early age behavior of concrete structures with a hydration based microplane model. *Computers & Structures*, 1085–1101. doi:10.1016/j.compstruc.2009.05.008

Lin, Y., & Chen, H.-L. (2015). Thermal analysis and adiabatic calorimetry for early-age concrete members: Part 2. Evaluation of thermally induced stresses. *Journal of Thermal Analysis and Calorimetry, 124*, 227–239. doi:10.1007/s10973-015-5131-x

Loïc, D. (2003). Delayed ettringite formation in massive concrete structures: Summary of studies conducted on deteriorated bridges. *Bulletin Des Laboratories Des Ponts et Chaussées, 4473*, 91–111.

Mehta, P. K., & Monteiro, P. J. M. (2013). *Concrete: Microstructure, properties and materials* (4th ed., pp. 259–264). New York, NY: The McGraw-Hill Education.

Milestone, N. B., & Rogers, D. E. (1981). Use of an isothermal calorimeter for determining heats of hydration at early ages. *World Cement Technology, 12*, 374–380.

Muhammad, A. N. (2009). Simulation of the thermal stress in mass concrete using a thermal stress measuring device. *Cement and Concrete Research, 139*, 154–164.

Noorzaei, J., Bayagoob, K. H., Thanoon, W. A., & Jaafar, M. S. (2006). Thermal and stress analysis of Kinta RCC dam. *Engineering Structures, 28*, 1795–1802. http://dx.doi.org/10.1016/j.engstruct.2006.03.027

Pofale, A. D., Tayade, K. C., & Deshpande, N. V. (2013). Calorimetric studies on heat evolution and temperature rise due to hydration of cementitious materials in concrete using semi-adiabatic calorimeter. *Special Issue for National Conference on Recent Advances in Technology and Management for Integrated Growth 2013 (RATMIG 2013)* (Vol. 2013, pp. 1–7). Retrieved from www.ijaiem.org

Prasanna, W. G., & Subhashini, P. A. (2010). Cracking due to temperature gradient in concrete. *International Conference on Sustainable Built Environment (ICSBE-2010) on 13th to 14th December 2010* (pp. 496–504). Kandy.

Radovanovic, S. (1998). *Thermal and structural finite element analysis of early age mass concrete structures*. Winnipeg, Manitoba: University of Manitoba.

Raphael, J. M. (1984). Tensile strength of concrete. *ACI Journal, 81*, 158–165.

Riding, K. A., Poole, J. L., Folliard, K. J., Juenger, M. C., & Schinder, A. K. (2012). Modelling hydration of cementitious systems. *ACI Materials Journal, 109*, 225–234. Retrieved from http://hdl.handle.net/2009/15459

Suzuki, Y., Tsuji, Y., Maekawa, K., & Okamura, H. (1990). Quantification of hydration-heat generation process of cement in concrete. *Japan Society of Civil Engineers (JSCE), 16*, 111–124.

Tanabe, T., Kawasumi, M., & Yamashita, Y. (1986). Thermal stress analysis of massive concrete. *Seminar Proceedings For Finite Element Analysis of Reinforced Concrete Structures. Tokyo, Japan on 21-24 May 1986*. New York, NY: ASCE.

Tang, K., Millard, S., & Beattie, G. (2015). Early-age heat development in GGBS concrete structures. *Structures and Buildings, 168*, 541–553. doi:10.1680/stbu.14.00089

Tatro, S., & Schrader, E. (1992). Thermal analysis for RCC-a practical approach. In K. D. Hansen & F. G. McLean (Eds.), *Roller compacted concrete III* (pp. 389–406). New York, NY: American Society of Civil Engineers.

Tia, M., Lawrence, A., Ferraro, C., Do, T. A., & Chen, Y. (2013). *Pilot project for maximum heat of mass concrete* (Report number: 00093793). Gainesville, FL: The Florida Department of Transportation.

Tim, C. T. (2014). Challenges and opportunities in tropical concreting. *2nd International Conference on Sustainable Civil Engineering Structures and Construction Materials 2014 (SCESCM 2014)* (pp. 348–355). doi:10.1016/j.proeng.2014.12.193

Truman, K. Z., Petruska, D., Ferhi, A., & Fehl, B. (1991). Nonlinear, incremental analysis of mass-concrete lock monolith. *Journal of Structural Engineering, 117*, 1834–1851. http://dx.doi.org/10.1061/(ASCE)0733-9445(1991)117:6(1834)

Ulm, F., & Coussy, O. (1995). Modeling of Thermochemomechanical Couplings of Concrete at Early Ages. *Journal of Engineering Mechanics, 121*, 785–794. doi:10.1061/(ASCE)0733-9399(1995) 121:7(785))

van Breugel, K. (1991). Simulation model for development of properties of early-age concrete. In L. Taerwe & H. Lambotte (Eds.), *Quality Control of Concrete Structures – Proceedings of the Second International RILEM/CEB Symposium on June 1991* (pp. 139–151). Ghent, RILEM Proceedings 14: E & FN Spon: London.

Viviani, M. (2005). *Monitoring and modeling of construction materials during hardening* (Doctoral Thesis). Lausanne: Swiss Federal Institute of Technology.

Waller, V. D'Aloia, L., Cussigh, F., & Lecrux, S. (2004). Using the maturity method in concrete cracking control at early ages. *Cement and Concrete Composites*, 589–599. doi:10.1016/S0958-9465(03)00080-5

Wang, K., Ge, Z., Grove, J., Ruiz, J. M., & Rasmussen, R. (2006). Developing a simple and rapid test for monitoring the heat evolution of concrete mixtures for both laboratory and field applications. *Center for Transportation Research and Education, Iowa State University* (Report No. FHWA DTF61-01-00042). Washington, DC: National Concrete Pavement Technology Center.

Yuan, Y., & Wan, Z. L. (2002). Prediction of cracking within early-age concrete due to thermal, drying and creep behavior. *Cement and Concrete Research, 32*, 1053–1059. doi:10.1016/S0008-8846(02)00743-3

Zhai, X., Wang, Y., & Wang, H. (2015). Thermal stress analysis of concrete wall of LNG tank during construction period. *Materials and Structures*. doi:10.1617/s11527-015-0656-9

Permissions

All chapters in this book were first published in CE, by Cogent OA; hereby published with permission under the Creative Commons Attribution License or equivalent. Every chapter published in this book has been scrutinized by our experts. Their significance has been extensively debated. The topics covered herein carry significant findings which will fuel the growth of the discipline. They may even be implemented as practical applications or may be referred to as a beginning point for another development.

The contributors of this book come from diverse backgrounds, making this book a truly international effort. This book will bring forth new frontiers with its revolutionizing research information and detailed analysis of the nascent developments around the world.

We would like to thank all the contributing authors for lending their expertise to make the book truly unique. They have played a crucial role in the development of this book. Without their invaluable contributions this book wouldn't have been possible. They have made vital efforts to compile up to date information on the varied aspects of this subject to make this book a valuable addition to the collection of many professionals and students.

This book was conceptualized with the vision of imparting up-to-date information and advanced data in this field. To ensure the same, a matchless editorial board was set up. Every individual on the board went through rigorous rounds of assessment to prove their worth. After which they invested a large part of their time researching and compiling the most relevant data for our readers.

The editorial board has been involved in producing this book since its inception. They have spent rigorous hours researching and exploring the diverse topics which have resulted in the successful publishing of this book. They have passed on their knowledge of decades through this book. To expedite this challenging task, the publisher supported the team at every step. A small team of assistant editors was also appointed to further simplify the editing procedure and attain best results for the readers.

Apart from the editorial board, the designing team has also invested a significant amount of their time in understanding the subject and creating the most relevant covers. They scrutinized every image to scout for the most suitable representation of the subject and create an appropriate cover for the book.

The publishing team has been an ardent support to the editorial, designing and production team. Their endless efforts to recruit the best for this project, has resulted in the accomplishment of this book. They are a veteran in the field of academics and their pool of knowledge is as vast as their experience in printing. Their expertise and guidance has proved useful at every step. Their uncompromising quality standards have made this book an exceptional effort. Their encouragement from time to time has been an inspiration for everyone.

The publisher and the editorial board hope that this book will prove to be a valuable piece of knowledge for researchers, students, practitioners and scholars across the globe.

List of Contributors

Quoc-Bao Bui
University Savoie Mont Blanc, LOCIE, CNRS, POLYTECH Annecy-Chambery, Chambery 73000, France

Tan-Trung Bui and Ali Limam
University of Lyon, LGCIE, INSA Lyon, 69621 Villeurbanne Cedex, France

Sakdirat Kaewunruen and Ratthaphong Meesit
Birmingham Centre for Railway Research and Education, School of Civil Engineering, University of Birmingham, Birmingham, UK

Kasun Nandapala and Rangika Halwatura
Department of Civil Engineering, University of Moratuwa, Moratuwa, Sri Lanka

O. G. Okeola
Department of Water Resources & Environmental Engineering, University of Ilorin, Ilorin, Nigeria

S. O. Balogun
National Centre for Hydropower Research & Development, Ilorin, Nigeria

Francis Atta Kuranchie, Sanjay Kumar Shukla, Daryoush Habibi and Alireza Mohyeddin
School of Engineering, Edith Cowan University, 270 Joondalup Drive, Joondalup, Western Australia 6027, Australia

Naveen Kumar ChikkaKrishna, Manoranjan Parida and Sukhvir Singh Jain
Department of Civil Engineering, Transportation Engineering Group, Indian Institute of Technology Roorkee, Roorkee 247667, Uttarakhand, India

Mostafa Ahmed Moawad Abdeen
Department of Engineering Mathematics and Physics, Faculty of Engineering, Cairo University, Giza 12211, Egypt

Alaa El-Din Abdin
National Water Research Center, Ministry of Water Resources and Irrigation, Cairo, Egypt

W. Abbas
Basic and Applied Science Department, College of Engineering and Technology, Arab Academy for Science, Technology, and Maritime Transport (Cairo Branch), Cairo, Egypt

Chanchal Verma and Sangeeta Madan
Department of Environmental Sciences, Gurukul Kangri University, Haridwar 249407, Uttarakhand, India

Athar Hussain
Civil Engineering Department, School of Engineering, Gautam Buddha University, Greater Noida 201310, Uttar Pradesh, India

Sergio M. R. Lopes
FCTUC, University of Coimbra, Portugal

Luis F. A. Bernardo
Department of Civil Engineering and Architecture, University of Beira Interior, Edifício II das Engenharias, Calçada Fonte do Lameiro, 6201-001 Covilhã, Portugal

Devanjan Bhattacharya and Hakan Kutoglu
Faculty of Engineering, Geomatics Engineering Department, Bulent Ecevit University, Zonguldak 67100, Turkey

Jayanta Kumar Ghosh
Department of Civil Engineering, Indian Institute of Technology Roorkee, Roorkee, Uttarakhand 247667, India

Jitka Komarkova
Faculty of Economic & Administrative Science, Institute of System Engineering & Informatics, University of Pardubice, Pardubice, Czech Republic

Santo Banerjee
Institute for Mathematical Research, University Putra Malaysia, Serdang, Malaysia

Jianping Luo, Anfeng Li, Dan Huang and Jianju Ma
National Engineering Research Center for Urban Environmental Pollution Control, Beijing Municipal Research Institute of Environmental Protection, Beijing 100037, China

Jiqiang Tang
Beijing Cement Plant Co. Ltd, Beijing 102202, China

Paratibha Aggarwal, Rahul Pratap Singh and Yogesh Aggarwal
Civil Engineering Department, National Institute of Technology, Kurukshetra, Kurukshetra, India

Fariborz M. Tehrani
Department of Civil and Geomatics Engineering, California State University, Fresno, 2320 E. San Ramon Avenue, M/S EE94, Fresno, CA 93740, USA

Petri Juntunen, Mika Liukkonen, Markku Lehtola and Yrjo Hiltunen
Department of Environmental Science, University of Eastern Finland, Kuopio FI-70210, Finland

Shuo Yang, Halil Ceylan, Sunghwan Kim and Kasthurirangan Gopalakrishnan
Department of Civil, Construction & Environmental Engineering (CCEE), Iowa State University of Science and Technology, Ames, IA 50011-3232, USA

Keyan Shen and Daji Qiao
Department of Electrical and Computer Engineering (ECpE), Iowa State University of Science and Technology, Ames, IA 50011-3232, USA

Muzaffar Ahmad Mir and Athar Hussain
Environmental Engineering Division, School of Engineering, Gautam Buddha University, Greater Noida, India

Chanchal Verma
Environmental Sciences Department, Gurukul Kangri University, Hardwar, India

Index